Power Systems

For further volumes:
http://www.springer.com/series/4622

Jahangir Hossain · Hemanshu Roy Pota

Robust Control for Grid Voltage Stability: High Penetration of Renewable Energy

Interfacing Conventional and Renewable Power Generation Resources

Springer

Jahangir Hossain
Griffith School of Engineering
Griffith University
Gold Coast, QLD
Australia

Hemanshu Roy Pota
School of Engineering and Information Technology
The University of New South Wales at Australian Defence Force Academy
Canberra, ACT
Australia

ISSN 1612-1287 ISSN 1860-4676 (electronic)
ISBN 978-981-10-1367-6 ISBN 978-981-287-116-9 (eBook)
DOI 10.1007/978-981-287-116-9
Springer Singapore Heidelberg New York Dordrecht London

Printed on acid-free paper

Springer is part of Springer Science+Business Media (www.springer.com)

To my parents

Preface

Voltage stability is a major concern in many stressed power systems as shown by a number of recent blackouts caused by voltage instability. In transmission systems the voltage instability problem is likely to become more prominent because of the integration of renewable generators, mainly induction generators with wind and photovoltaic (PV) solar units. The induction generators are operated in constant reactive power mode and the PV units are operated at unity power factor. Unless the operating mode of the renewable generators is changed, additional devices are required to provide reactive power support to maintain a good voltage stability margin. In distribution systems, with high penetration of distributed generation, the proximity of the generation and dynamic induction motor loads gives rise to new dynamic phenomena that includes voltage oscillation modes. Installing devices for reactive power support in distribution networks and proper controller design can mitigate most adverse dynamic effects of distributed generation. This book comprehensively covers all the topics necessary to increase the voltage stability margin of transmission and distribution systems.

In most cases the integration of higher levels of renewable energy into an existing transmission system does not require a major redesign, provided a thorough analysis before the integration is conducted to check the viability of renewable integration. In some cases the analysis indicates the necessity of additional high-performance control and compensating equipment to enable the system to recover from severe system disturbances. In this book, first dynamic voltage instability problems which are likely to occur in future power systems are presented and then novel robust controllers are synthesised for the enhancement of the stability.

The first part of this book contains case studies for capturing the development of different types of dynamic voltage instability, in both the short- and long-term, caused by the dynamics of wind generators and PV units. In addition, an investigation is presented on the changing nature of systems and their dynamic behaviours giving rise to critical issues that limit the large-scale integration of wind generators and flexible AC transmission system (FACTS) devices.

It is common in the power industry to tune fixed structure controllers, like the proportional-integral controllers, based on single device infinite-bus dynamics. This approach is practical but these controllers do not permit the participation of reactive power support devices and renewable generators to contribute to the fault recovery at their full potential. Novel controllers are required to enable a full participation of all devices in fault recovery so that the ratings of the support devices can be kept low to increase the affordability of the support devices. This book presents detailed design and implementation of robust controllers for stability enhancement and also to reduce the required ratings of support devices for fault recovery.

In robust control design it is important to choose how to capture the knowledge of the unmodelled dynamics in the design process. Partial details such as the 'size' of the unmodelled dynamics are used in the robust control design. It is well-known that smaller 'size' unmodelled dynamics lead to higher performing controllers. The second part of the book presents a method to bound unmodeled nonlinear dynamics and to design excitation control for the enhancement of large-disturbance voltage stability in power systems with significant induction motor loads. A new technique is presented which captures the full nonlinearity of systems in the region of interest. The nonlinear power system model is reformulated with a linear and a nonlinear term. The nonlinear term is the Cauchy remainder in the Taylor series expansion and in this book its bound is used in robust control design.

The third part of the book contains the design of robust controllers which augment the low-voltage ride-through capability of FSIGs during severe disturbances. Control algorithms, using both structured and unstructured uncertainty representations, are developed for the stabilisation of faulted systems under different operating conditions. A method is presented which can be used to design a linear controller for doubly-fed induction generators (DFIGs) which is sufficiently robust to accommodate post-fault low-voltage conditions. An analysis of the possible negative interaction among PV controllers and design a robust controller to mitigate unwanted interactions is also presented. In the proposed robust control design, parts of nonlinear dynamics and control interaction are modeled as disturbances and this ensures a non-interacting robustness of control design.

The performance of the proposed control schemes fulfils the criteria for robust stability and performance, and produces adequate stability margins for a range of test cases. The effectiveness of the suggested control strategies is validated by detailed simulations with complete nonlinear model of the devices. The performances of the designed controllers are also compared with those of conventional controllers. Simulation results show that both the dynamic voltage stability and the transient stability of a power system can be improved by the use of the robust control methods presented in this book.

Contents

H_G	Inertia constant of the wind generator
K_s	Torsion stiffness
D_m	Torsion damping of the turbine
D_G	Torsion damping of the wind generator
f	Normal grid frequency
γ	Torsion angle
T_m	Mechanical torque
X'	Transient reactance of the wind generator
X_m	Magnetising reactance of the wind generator
R_s	Stator resistance of the wind generator
R_r	Rotor resistance of the wind generator
X	Rotor open-circuit reactance of the wind generator
X_r	Rotor reactance of the wind generator
X_s	Stator reactance of the wind generator
i_{ds}	*d*-Axis component stator current
i_{qs}	*q*-Axis component stator current
ω_G	Rotor speed of the wind generator
T'_o	Transient open-circuit rotor time constant
v_{tm}	Voltage sensor output
E'_{dr}	*d*-Axis transient rotor voltages
s	Slip of the induction machine
E'_{qr}	*q*-Axis transient rotor voltages
V_{ds}	*d*-Axis stator voltage of the induction machine
V_{qs}	*q*-Axis stator voltage of the induction machine
R_C	Internal resistance of the capacitor
v_{dc}	Capacitor voltage
α	Bus angle of the STATCOM in the reduced network
T_e	Electrical torque
P_s	Power supplied by the system to the STATCOM
C	Capacitor
T_m	Time constant of the voltage transducer
K_m	Sensor constant
m	Modulation index of the inverter
β	Scaling parameter
ϕ, ψ	Uncertainty gain matrix
τ	Free parameter
k	Inverter constant
C_s	Supercapacitor
v_{sc}	Supercapacitor voltage
D	Deadband of an OLTC
ε	Hysteresis band of an OLTC
h	Step size of an OLTC
n_{k+1} and n_k	Turns-ratio before and after a tap change

ATC	Available Transfer Capability
AVR	Automatic Voltage Regulator
BESS	Battery Energy Storage System
DFIG	Doubly-Fed Induction Generator
FACTS	Flexible AC Transmission System
FRT	Fault Ride-Through
FSIG	Fixed-Speed Induction Generator
GARE	Generalised Algebraic Riccati Equation
HV	High Voltage
HVDC	High Voltage Direct Current
IG	Induction Generator
IM	Induction Motor
IQC	Integral Quadratic Constraints
LV	Low Voltage
LMI	Linear Matrix Inequality
LQG	Linear Quadratic Gaussian
LQR	Linear Quadratic Regulator
LVRT	Low-Voltage Ride-Through
MSC	Mechanically-Switched Capacitor
OLTC	On-Load Tap Changer
OXL	Over-excitation Limiter
PSS	Power System Stabiliser
PWM	Pulse-Width Modulation
SMES	Superconducting Magnetic Energy Storage
SMIB	Single-Machine Infinite Bus
STATCOM	Static Synchronous Compensator
SVC	Static VAr (volt-ampere reactive) Compensator
TCSC	Thyristor Controlled Series Capacitor
VSWT	Variable-Speed Wind Turbine
WT	Wind Turbine
WTGS	Wind Turbine Generator System

Chapter 1
Introduction

Abstract The purpose of this introductory chapter is intended to provide the background to this book including contribution. An outline of which is given at the end of this chapter. The motivation behind this book on renewable energy integration is also discussed.

1.1 General

Voltage instability is a major power system concern which if not given due attention may result in severe detrimental effects of economical, technical and social dimensions. A number of recent contingencies and voltage collapses around the world have prompted a significant effort to be made towards the study and prevention of voltage instabilities [1]. The electrical power industry is undergoing major changes, both technically and politically [2]. In recent years power systems are experiencing dramatic changes in electrical power generation, transmission, distribution and end-user facilities [3].

Renewable energy is increasing in importance and has become an integral part of the energy in many countries of the world. The growing utilisation of wind power, especially in remote areas with favourable wind conditions but relatively weak transmission systems, brings new challenges for voltage control and reactive power compensation. To achieve high penetration levels, such as twenty percent intermittent renewable resources by 2020, fast responsive energy storage and flexible AC transmission system (FACTS) devices with suitable controllers are required. Although the integration of higher levels of wind power into an existing transmission system does not require a major redesign, it requires a thorough analysis before the integration can take place. This necessitates additional high-performance control and compensating equipment to enable a system to recover from severe system disturbances. This book will focus on the effects of large-scale penetration of wind power on dynamic voltage stability and the enhancement of stability margins using

J. Hossain and H. R. Pota, *Robust Control for Grid Voltage Stability: High Penetration of Renewable Energy*, Power Systems,
DOI: 10.1007/978-981-287-116-9_1, © Springer Science+Business Media Singapore 2014

robust control techniques. As recently significant photovoltaic (PV) is being integrating in the low voltage distribution systems, this book will also discuss PV integration issues and robust control design to minimise the negative impacts on system performance.

1.2 Background

The management of power systems has become increasingly complex due to several contributing factors [4]: power systems are now being operated closer to their maximum operating limits; environmental constraints restrict the expansion of transmission networks; the number of long distance power transfers has increased; and lower inertial wind turbines (WTs) and inertia-less PV units have been integrated into the existing grids. This changing nature of a power system significantly affects its dynamic behaviour: the dynamic interactions between different partly synchronised couplings of connected devices cause oscillations; lower inertial and intermittent units absorb reactive power especially during transient periods; and power systems have been restructured in many parts of the world to create competition amongst different power producers [5] which has resulted in increased complexity and the emergence of several new threats to the stable operations of power systems.

Power system instability has been an important challenge for researchers and utility companies since 1920 [6]. Historically, transient instability has been the dominant stability problem in most systems and the focus of much of the power industry's attention concerning system stability. Continuing electrical load growth and higher power transfer demands in a largely interconnected network lead to complex and less secure power system operations. As the complexities of power systems have evolved, different forms of instability have emerged. As one of the dominant instability problem of power systems, voltage stability is becoming an increasing source of concern for the secure operation of present-day power systems.

Voltage instability has been considered to be a new type of instability for the last three decades. In recent years, the voltage instability of large electrical power systems has caused costly blackouts [7]. Several incidences of voltage collapse resulting from voltage instability have been observed in different parts of the world; e. g., France (1978), Belgium (1982), Sweden (1983 and 2003), Western France (1987), Tokyo (1987), São Paulo (1997) and Greece (2004) [8]. Voltage control and stability problems were prominent in the August 14, 2003 blackouts in Ohio, Michigan, New York and Ontario. The total costs of the 2003 blackout were estimated to be between $4 billion and $10 billion [9].

Voltage stability will remain a challenge for the foreseeable future and indeed it is likely to increase in importance. One reason is the need for more intensive use of available transmission facilities as load demand will continue to increase. Almost all the developed countries have set a target of producing 20 % electricity from renewable energy by 2020 [10]. Wind generation is the dominant source of renewable energy. The restructuring of power systems will create new problems, such as reduced available transfer capability (ATC) and congestion of some transmission lines. This may further reduce the voltage stability margin. Due to the free-energy trading market, power flows on transmission lines could be increased to their thermal limits. The thermal limit is normally higher than the surge impedance loading of a line and leads to the degradation of system voltage characteristics.

Voltage stability has been studied for a long time as it is an important subject for secure power system operations [11]. However, the influencing factors and possible mechanisms of instability is changing in power systems under a deregulated environment in which new types of instabilities, such as immediate instabilities, tend to increase and, therefore, changing requirements for analyses [11]. This has highlighted the need to analyse the preconditions, influencing factors and developing process mechanisms of voltage instability.

Wind generation is a growing energy resource which boasts the benefits of clean energy and a reduction in the dependence of energy on fossil fuel. There has been a rapid increase in WT connections to distribution and transmission networks in recent years which makes the power network more dependent on, and susceptible to, wind energy production. The introduction of significant intermittent generation will affect the way an electricity system operates [12]. One of the major challenges faced by the electricity industry is how to effectively integrate a significant amount of wind power into an electricity system [13]. The dynamics of power systems will likely be dominated by the dynamics of WTs and FACTS devices in the near future. Therefore, the impacts of wind power and FACTS devices on a power system's stability and reliability need to be thoroughly analysed before they are integrated into existing power system networks. Although a changing power system does not need to be redesigned, it requires controllers to ensure stability under a wide range of operating conditions.

The large-scale integration of PV (photovoltaic) generators into sub-transmission and distribution grids can have a significant impact on a power system's operation and stability [14]. It is well known that a sudden change in sunlight can initiate a rapid disconnection or reduction in a PV generating capacity. As the penetration of PV schemes increases, this can lead to a problem of voltage variation and transient voltage instability in the case of a weak coupling with the grid [15]. The large-scale penetration of PV units also has an impact on the short-term voltage and transient stability of a system, which is not only restricted to the distribution network but begins to influence the whole system [16]. Therefore, it is significantly important to control the output power to ensure stability and security. The implementation of appropriate controls for PV units can improve the system's performance without violating network constraints and can facilitate the effective participation of PV penetration [17, 18].

Control is vital for maintaining the stability of modern interconnected power systems. Their control and operation must be evaluated in order to meet the changing characteristics of plants and the increasing dependence of present society on a cheap and reliable supply of electrical power. The control of a system has to be viewed as a single complex integrated operation involving all types of plants, with sufficient flexibility to accommodate new plants and the changing roles of earlier plants, as well nonlinear interactions and ensure stability for large disturbances, thus keeping the wind farm connected to the main grid under both fault and post-fault conditions.

The main objective of installing controllers in power systems is to achieve the desired stability and security at a reasonable cost. Robust control theories have been significantly developed in the past few years [36]. The key idea of a robust control paradigm is to check whether the design specifications are satisfied for even the worst-case uncertainty. Many efforts have been made to investigate the application of robust control techniques to power systems. Although there have been developments in the area of robust control design in order to enhance power system stability, so far methods for quantifying the magnitude of uncertainties have not been addressed.

Robust control in power systems deals with the application of new techniques in linear system theory to enhance the voltage and transient stabilities of power systems [36]. With robust control, we can capture not only the typical, or nominal, behaviour of a plant, but also the amount of its uncertainty and variability. Plant model uncertainty can result from: (i) model parameters with approximately known or varying values; (ii) neglected or poorly known dynamics, such as high-frequency dynamics; (iii) changes in operating conditions; (iv) linear approximations of nonlinear behaviours; and (v) estimation errors in a model identified from measured data. By quantifying the level of uncertainty in each element, the overall fidelity and variability of a plant model can be captured. The ways in which each uncertain element affects performance can be analysed and the worst-case combinations of uncertain elemental values identified. Robust control designed for the worst-case scenarios can ensure stability during a large disturbance.

Although only recently has voltage instability come to be considered a dominant problem, in many systems, power oscillations of small magnitudes and low frequencies have been persisting for a long time. In some cases, this also presented a limitation on the transmitted power over tie lines. Traditionally, such oscillations are damped by conventional power plants with synchronous generators which are equipped with power system stabilisers. Because of the characteristics of a WT, it is possible that wind power, under some circumstances, may contribute to a reduction in system damping. In fact, wind power resources are generally located far from large centres of consumption, while generators that are not to be committed, typically thermal units, are located closer to major loads [20]. This may involve transmitting energy over long distances through congested lines and, perhaps, a significant change in the generation profile and typical power flows which may have a significant impact on damping. The replacement of conventional generators by WTs also reduces the damping of overall power systems. In such circumstances, it is essential to control both the reactive and active power.

However, as a STATCOM has the capability for only reactive power control, its application is limited to reactive power support in a power system. To overcome this limitation, coordinated STATCOM and pitch angle control proposed in this book can be used effectively to enhance the LVRT capability of an induction generator by controlling both the active and reactive power. However, the response of pitch control is slow. A STATCOM with a energy storage system (STATCOM/ESS) has emerged as a more promising device for power system applications, as it has both real and reactive power control abilities [37]. A STATCOM enhanced with energy storage can provide additional effects, such as an increased capability to damp electromechanical oscillations [38], improved power quality and reliability of supply [39].

Power systems are large-scale systems consisting of several interconnected subsystems. This interconnection has been constructed in order to maximise their economic benefits. For a large-scale system, it is often desirable to have some form of decentralisation in its control structure [40]. For such a system, it is not realistic to assume that all output measurements can be transmitted to every local control station. In practice, large-scale dimensional and severe model uncertainties lead to significant difficulties in control strategy design. With the emphasis being on robustness and system performance, there is a need to analyse and design controllers in an integrated manner, taking into consideration the interactions between the various subsystems and controllers in a system. Any successful strategy for the control of a large-scale power system must satisfy two fundamental requirements. Firstly, the control must be decentralised since only local measurements are normally available to any given machine. Secondly, the control needs to be robust, in the sense that it must guarantee satisfactory performance over a wide range of operating conditions and disturbances. Next, an overview of renewable energy integration into power systems and the control techniques of STATCOMs for wind generators are discussed.

1.3 Renewable Energy Integration in Different Countries

The growing level of penetration of non-traditional renewable generation—especially wind and solar—has led to the need for renewable generation to contribute more significantly to power system voltage control and reactive power capacity. Modern wind-turbine generators, and increasingly PV inverters as well, have considerable dynamic reactive power capability, which can be further enhanced with other reactive support equipment at the plant level to meet interconnection requirements. The integration of renewable energy has requirements for intermittent renewable energy integration such as active and reactive power capability, voltage and frequency tolerance, and fault current contribution. If needed to meet interconnection requirements, the reactive power capability of solar and wind plants can be further enhanced by adding of a static var compensator (SVC), static synchronous compensators (STATCOMs), and other reactive support equipment at the plant level.

It is evident that renewable based generation will be comparable to conventional power generation in the coming decades. Therefore, many transmission system operators (TSOs) and regulators around the world have come up with interconnection rules or codes to request these volatile renewable resource-based power plants to have more or less the same operating competence as conventional power plants. Power factor requirements and reactive power capabilities are more or less similar in the grid codes of different countries. Table 1.1 illustrates those requirements [41].

Table 1.1 Power factor/reactive power requirement

Countries	PF requirements	Voltage range	Frequency range
USA	0.95 leading to 0.95 lagging	±10%	Not specified
China	0.97 lagging to 0.97 leading	−3 to +7%	49.5–50.5 Hz
Germany	For a rated active power capacity <100 MW:		
	0.95 lagging to 0.95 leading	220 kV −13 to +12 %	47.5–51.5Hz
	For a rated active power capacity >100 MW:	400 kV −8 to +10 %	
	Power factor is voltage dependent		
Spain	0.98 leading to 0.98 lagging without any penalty	±10 %	49.5–50.5 Hz
India	0.95 lagging to 0.95 leading	− 10 to +5 %	47.5–51.5 Hz
Denmark	At full production:	−3 to +13 %	50.5–52.0Hz
	Q/Prated = 0 to Q/Prated = 0.1		
	and at zero production		and 49.5–47.5Hz
	From Q/Prated = −0.1 through a		
	straight line to Q/Prated= 0		
Ireland	At full production:	−9 to +12 %	50.5–52.0Hz
	0.95 lagging to 0.95 leading,		and 49.5-47.5Hz
	For active capacity from 100 %		
	production to 50 % production:		
	32.6MVAr per 100MW installed,		
	From 50 % production to idle:		
	0.95 lagging to 0.95 leading		
UK	0.95 lagging to 0.95 leading	−10 to +5 %	47.5–52 Hz
Canada	0.90 lagging to 0.95 leading	provided by TSO	WECC requirement
Australia	0.93 lagging to 0.93 leading	±10 %	47.0–52.0 Hz

1.4 Overview of Integration of Large-Scale Wind Generators and PV Units

The integration of renewable energy sources (RESs) has increased in recent years due to several technical, environmental, and social benefits which include the use of these sources as supplement and alternative of conventional fossil fuel generation to meet the increased energy demand, keep the environment free from fuel exhaustion. There are different types of RESs such as solar, wind, biogas, tidal, etc., and among these RESs, wind energy and PV unit are the fastest growing sources.

Wind power is the most promising and mature technology among the non-hydro renewable energies. The present progressive scale of integration has brought to a head serious concerns about the impact of such a scale of wind penetration on the future safety, stability, reliability and security of an electricity system. One of the major concerns related to the high-level penetration of integrated WTs is their impact on power system stability and is concerned with transient angular and voltage stability.

In the years to come, the focus will be on maintaining power system stability and voltage stability, for example, at a short-circuit fault, ensuring that power supply safety and other important tasks [42] are maintained as the amount of wind power increases. This situation makes it necessary to find solutions with respect to maintaining the dynamic stability and reliable operation of a power system with large amounts of wind power. However, detailed analyses are increasingly important for determining the correct solutions. In recent years, a lot of research work has been done to determine the effects of replacing the output of a conventional generator by wind power [43–45]. A modeling and control for off-shore wind farms in order to improve short-term voltage is discussed in [46]. The results of a systematic study of the effects of several design and operational parameters on the transient stability of a FSWT are presented in [47]. The effects of wind power on oscillations is investigated by gradually replacing the power generated by the synchronous generators in the system with power from either constant or variable-speed WTs, while observing the movement of the eigenvalues through the complex plane [45]. Variable speed wind turbines equipped with DFIGs are becoming more widely used due to their advanced reactive power and voltage control capability. The converter model of a DFIG is a multi-input multi-output nonlinear model, and the difficulty of controlling it is mainly due to its nonlinear behaviour

Residential PV installations are increasing significantly due to their small relative size, noiseless operation, and feed-in tariff [48]. The major concerns of integrating PV into the grid is the stochastic behavior of solar irradiations and interfacing of inverters with the grid. The intermittent PV generation varies with changes in atmospheric conditions. Due to the high initial investment and reduced life time of PV system as compared to traditional energy sources, it is essential to extract maximum power from PV systems [49]. Controllers on grid-connected PV systems are applied to achieve the desired performance under disturbances like changes in atmospheric conditions, changes in load demands, or external faults within the system.

Several system operators are publishing severe requirements for the connection of new wind farms and PV units in order to ensure their proper behaviour after network faults. Their dynamic behaviours must be adapted to the new requirements. New studies must be performed in order to evaluate the behaviour of wind farms and PV units after severe faults and to improve their design in efficient and economical ways.

1.5 Overview of Control for Wind Generators and PV Units

The rapid growth of renewable energy and its large-scale integration fostered by renewable portfolio standards coming into effect are accompanied by opportunities and challenges alike. Electricity providers are increasingly faced with the challenge to integrate variable renewable generation with the existing generation portfolio and the electricity grid. To operate reliably, grids need to continuously balance energy supply and demand a task complicated by the variability of renewable energy resources, such as solar and wind The output of a wind power plant and its total load vary continuously throughout the day. Reactive power compensation is required to maintain normal voltage levels in a power system as reactive power imbalances can seriously affect it. In the past, there was no requirement under the grid-code for a wind farm to remain connected to the grid during a fault or voltage disturbance. The protection of wind farms has been focussed mainly on turbine protection without considering the impact this might have on a power system. This implies that a WT is disconnected from the grid as soon as a violation of voltage or frequency operating limits is exceeded. Worldwide, the new grid connection requirements have identified three areas to be considered in the operation of wind farms: voltage and reactive power control; FRT capability; and frequency range of operation [50]. These requirements can be fulfilled by reactive power compensation devices such as FACTS devices.

Recently, various voltage-source or current-source inverter-based FACTS devices have been used for flexible power-flow control, secure loading, damping of power system oscillation and even for stabilising a wind generator. In this book, we propose a novel STATCOM controller to stabilise a grid-connected wind turbine generator system (WTGS). A STATCOM can also contribute to the LVRT requirement because it can operate at full capacity even at lower voltages.

The problems of voltage stability and power smoothing in wind generation systems have been extensively studied. The application of STATCOMs to WTGSs with induction generators have been described in various studies, demonstrating improved steady-state reactive power and voltage control [31, 51]. A comprehensive study has been undertaken to investigate how STATCOMs can be used with FSWTs, which use IGs, to improve both the steady-state and dynamic impacts of a wind farm on a network [52]. A study of the control of a self-excited IG is discussed in [53] in which the authors use a state-space linear model. In [53], the nonlinear model of the wind generator is linearised around an operating point. The voltage and frequency control are investigated by varying the effective rotor resistance of a self-excited slip-ring IG in [54]. However, the controller in [53] cannot stabilise the system under severe large disturbances. This motivates the use of advanced control techniques that consider nonlinear interactions and ensure stability for large disturbances.

Different control systems for accommodating PV units in a network are currently being investigated. The optimum power flow technique has been used for distribution systems with a few PV units [55]. A new voltage control procedure that includes a series reactor is proposed in [56]. Another droop control approach for frequency control using a capacitor is discussed in [57]. Grid-connected PV systems suffer from nonlinear behaviors where most of the nonlinearities occur due to the variation of solar irradiance and nonlinear switching functions of inverters. for nonlinear PV systems as presented in [55–57], provide satisfactory operation over a fixed set of operating points as the system is linearised at an equilibrium point.

1.6 Robust Control of Wind Turbines and PV Units

Adaptive Control is the control method used by a controller which must adapt to a controlled system with parameters which vary, or are initially uncertain [58]. It provides a systematic approach for automatic adjustment of controllers in real time, in order to achieve or to maintain a desired level of control system performance when the parameters of the plant dynamic model are unknown and/or change in time. For example, in power system, its load varies continuously; a control law is needed that adapts itself to such changing conditions. The reasons for using adaptive controller are: (i) variations in process dynamics, (ii) variations in the character of disturbances and (iii) engineering efficiency. Generally, it is not easy to control a system subjected to variations in dynamics. There are two methods to handle that, robust control and adaptive control. The tuning of a robust design for the true nominal model using an adaptive control technique will improve the achieved performance of the robust controller design. Therefore, robust control design will benefit from the use of adaptive control in terms of performance improvements and extension of the range of operation. On the other hand, using an underlying robust controller design for building an adaptive control system may drastically improve the performance of the adaptive controller.

Conventional proportional-integral (PI) controllers with fixed parameters are mainly used in power systems as they are easy to design and implement. As this controller requires less feedback information, this can be employed for most plants for which a dynamic model can be achieved. Useful details of the PI control of wind turbines are given in [59]. However, these controllers cannot ensure stability for changing operating conditions, for example, in different loading conditions. The adaptive control has been used to tackle the parameter variation of wind turbines. An adaptive controller for wind turbines is proposed in [60, 61] to compensate for unknown and time varying parameters. In the context of gain-scheduling, adaptive control structures which allow different control goals to be formulated; depending on the considered operating points were also introduced. To decide which controller to apply to the plant varies from simply switching between the controllers associated to the various operating points to quite sophisticated interpolation strategies [62]. In addition to its complexity and time-consuming determination of the adequate loading schedule for the complete operating range, the adaptive structure has a major drawback in that the rotor characteristics should be known quite accurately. The dynamic behavior of a wind turbine or a PV unit is highly dependent on variable wind speed and solar irradiance. In this book, a systematic reformulation technique is introduced to capture the nonlinearity and an advanced technique is used to design the robust controller for the extended operating range.

1.7 Contributions of This Book

In this book, firstly dynamic voltage instability problems which are likely to occur in future power systems are presented and then the novel robust controllers are synthesised for the enhancement of stability. This research work is aimed at providing deeper insights into the mechanisms of different types of dynamic voltage instability caused by the dynamics of power system devices and large disturbances. This book also intends to improve the present power system control design methodology. The proposed controller design is focused on improving the voltage stability, transient stability and LVRT capability of induction generators and the robustness provided by the power system controller. The major contributions of the book in this direction are:

Analysis:

- developing comprehensive explanations of dynamic voltage instability problems and the mechanisms of voltage collapse problems in multi-machine power systems, with induction machines and PV units, to provide more insight into voltage instability problems;
- analysing and mitigating the impacts of renewable energy sources on power systems;
- investigating the changing nature of systems and their dynamic behaviours to identify critical issues that limit the large-scale integration of wind generators and FACTS devices into existing power systems;
- studying the effects of STATCOM ratings and wind farm integration on the FRT capability of FSIGs analytically using the power-voltage and torque-slip relationships as well as through simulations.
- investigating and quantifying the degree of interaction among physically close photovoltaic (PV) units.

Control Design:

- modeling and excitation control design to enhance large disturbance voltage stability in power systems with significant IM loads;
- designing decentralised robust output-feedback controllers for STATCOMs in the presence of uncertainty and interconnection effects resulting in the increase of dynamic ATCs for power systems with FSWGs;
- presenting a novel robust controller for a STATCOM to enhance the FRT capability of FSIGs;
- presenting a procedure for designing simultaneous robust multivariable controllers for a STATCOM and for the pitch angle of a FSIG with the objective of enhancing the LVRT capability of wind farms;
- designing a new voltage controller to augment the LVRT capability of FSWTs which produces a less conservative result and reduces the calculation burden and controller dimensions;
- proposing a systematic procedure for designing decentralised multi-variable controllers for large interconnected power systems using a minimax output-feedback control design method and formulating the controller design procedure as an optimisation problem involving rank-constrained LMIs;
- focusing on cost-effective integration of emerging technologies such as a STATCOM/ESS system for challenging system applications, and developing reliable, stable and effective models and controllers to utilise STATCOM/ESS concept for the specific system enhancement objectives.
- designing of non-interacting controllers for PV Systems in distribution networks

The proposed power system control methodologies are validated through simulations. Controllers are designed for test-case power systems and simulations are carried out under large disturbances, such as symmetrical and unsymmetrical three-phase faults, outages of key transmission lines, large variations of load, etc.

1.8 Overview of the Book

Based on the above objectives, the outline of this book is as follows:

This chapter is intended to provide the background to this book including contribution and motivation.

Chapter 2 introduces the concepts of voltage instability and the distinctions between voltage and angle instability. The driving force and main causes of voltage instability are analysed. Different methods and devices used to enhance voltage stability are also explained. The steady-state and dynamic modeling of the power system devices under consideration have been discussed.

Chapter 3 briefly reviews the conventional linearisation and modal analysis technique. The analytical tools that are commonly used in small-signal stability analyses are presented. The proposed linearisation technique using the mean-value theorem with Cauchy's remainder is introduced and compared with the conventional linearisation technique.

Chapter 4 concentrates on the possible mechanisms of voltage instability due to large disturbances. The detailed case studies introduced in this chapter are used to illustrate some of the key instability mechanisms caused by the dynamics of wind generators. In addition the impact of wind power and FACTS devices on power system dynamics are presented. The strengths and limitations of previous works, the advantages of this research and different case studies and discussions are followed by a summary.

Chapter 5 presents a robust excitation control design algorithm for a synchronous generator with dynamic induction motor loads for enhancing dynamic voltage stability. The power system model, with dynamic induction motors, test cases, control tasks, uncertainty modeling and bounding, is followed by the minimax control approach and control design algorithm. Controller performances are evaluated through simulations and discussions, and conclusions are provided. Finally, the closed loop performances of the designed controller and a conventional exciter control are compared.

Chapter 6 provides the design of a decentralised robust STATCOM controller using the minimax LQ control synthesis technique which results in the augmentation of the ATC of a transmission line. Firstly, a method for determining the ATC is presented. Then, a short description of the output-feedback LQ decentralised control approach is given. Different case studies on the impacts of wind generators on the ATC of transmission are analysed and, finally, the performance of the designed controller is validated and compared with that of a conventional STATCOM controller.

Chapter 7 deals with different robust control techniques for the augmentation of the LVRT capability of WTs in different operating conditions. This chapter contains four distinct control methods for enhancing LVRT capabilities of systems with wind turbines. This chapter is divided into four main parts which present: firstly, a robust controller for a STATCOM to enhance the FRT capability of FSIGs; secondly, a designed simultaneous pitch angle and STATCOM control with its performance compared with only a STATCOM controller; thirdly, STATCOM controller design, including the structured uncertainty representation; and, fourthly, a systematic procedure for the design of decentralised multi-variable controllers for large interconnected power systems.

Also, previously developed control techniques and algorithms used for the voltage regulation and reactive power control of wind farms are reviewed. The basic ideas behind the critical clearing time (CCT), critical speed (CS) and critical voltage are given, followed by linearisation and uncertainty modeling, a control design algorithm and simulation cases and their results.

Chapter 8 provides the design of a decentralised robust controller using the minimax LQ control synthesis technique which results in the augmentation of the LVRT capability of DFIGs. Firstly, the complete nonlinear model of DFIG is presented. Then, a short description of the output-feedback LQ decentralised control approach is given. A systematic control algorithm is presented and, finally, the performance of the designed controller is validated and compared with that of a conventional PI controller.

Chapter 9 presents a design of non-interacting controller for PV units. In the first part of this chapter, it is shown that there may exist oscillations in distribution networks with physically close photovoltaic (PV) units due to their control interactions. The next part comprises the uncertainty modeling, bounding and controller design approach. Simulation studies and discussions to validate the proposed controllers are included in the final part of this chapter.

Chapter 10 provides the book summary, conclusions and recommendations for areas of future research.

Chapter 11 contains the appendices.

References

1. Kundur P (1994) Power system stability and control. Mcgraw-Hill, New York
2. Crow ML (1997) The new-but is it improved? Power system [deregulation issues]. IEEE Potentials 16(5):9–10
3. Kilani KB, Schlueter RA (March 2000) Trends in model development for stability studies in power systems. Electric Power Syst Res 53(3):207–215
4. Leung JSK, Hill DJ, Ni Y (2005) Global power system control using generator excitation, PSS, FACTS devices and capacitor switching. Int J Electr Power Energy Syst 27(5–6):448–464
5. Hossain MJ, Pota HR, Mahmud MA, Ramos RA (2011) Impacts of large-scale wind generators penetration on the voltage stability of power systems. In: IEEE Power and Energy Society general meeting, pp 1–8
6. Ganjefar S, Rezaei M (May 2009) A new method to control dynamic stability of power system through wave variables and signal prediction via internet. International J Recent Trends Eng 11(1):62–66
7. IEEE, Special Publication 90TH0358-2-PWR (1990) Voltage stability of power systems: concepts, analytical tools, and industry experience
8. Singh B, Scholar R, Sharma NK, Tiwari AN (2010) Prevention of voltage instability by using FACTS controllers in power systems: a literature survey. Int J Eng Sci Technol 2(5):980–992
9. U.S.-Canada Power System Outage Task Force (2004) Final report on the August 14, 2003 blackout in the United States and Canada: causes and recommendations
10. Greenpeace and Wind Industry Unveil Global Energy Blueprint: (1999) The European wind energy association (EWEA)). http://www.ewea.org/src/press.htm
11. Kundur P, Paserba J, Ajjarapu V, Andersson G, Bose A et al (2004) Definition and classification of power system stability ieee/cigre joint task force on stability terms and definitions. IEEE Trans Power Syst 19(3):1387–1401
12. The IEA WIND Task 25 (2007) Design and operation of power systems with large amounts of wind power: state-of-the-art report. VIT Working Papers 82
13. Georgilakis PS (2008) Technical challenges associated with the integration of wind power into power systems. Renew Sustain Energy Rev 12(3):852–863
14. Zhou W, Lou C, Li Z, Lu L, Yang H (2010) Current status of research on optimum sizing of stand-alone hybrid solar-wind power generation systems. Appl Energy 87(2):380–389
15. Mahmud MA, Pota HR, Hossain MJ (2013) Partial feedback linearizing current control scheme for a single-phase grid-connected photovoltaic system. IEEE Trans Sustain Energy. doi:10.1109/TSTE.2013.2279884
16. Mahmud MA, Pota HR, Hossain MJ (2012) Dynamic stability of three-phase grid-connected photovoltaic system using zero dynamic design approach. IEEE J Photovoltaics 12(4):564–571
17. Roy NK, Pota HR, Mahmud MA, Hossain MJ (2013) D-STATCOM control in distribution networks with composite loads to ensure grid code compatible performance of photovoltaic generators. In: The 8th IEEE conference on industrial electronics and applications, Melbourne, Australia, pp 55–60

18. Hossain MJ, Saha TK, Mithulananthan N (2012) Robust control strategy for PV system integrations in distribution systems. Appl Energy 99:355–362
19. Ni H, Heydt GT, Mili L (2002) Power system stability agents using robust wide area control. IEEE Trans Power Syst 17(4):1123–1131
20. Ackermann T (2005) Wind power in power systems. Wiley, England
21. Hansen LH, Helle L, Blaabjerg F, Ritchie E, Munk-Nielsen S, Bindner H, Sorensen P, Bak-Jensen B (2001) Conceptual survey of generators and power electronics for wind turbines. Technical Report, Riso National Laboratory, Roskilde, Denmark
22. Hossain MJ, Pota HR, Ugrinovskii V, Ramos RA (2009) Excitation control for large disturbances in power systems with dynamic loads. In: IEEE Power and Energy Society general meeting, Calgary, Canada, pp 1–8
23. IEEE Task Force on Load Representation for Dynamic Performance (1995) Standard load models for power flow and dynamic performance simulation. IEEE Trans Power Syst 10(3):1302–1313
24. Overbye TJ (1994) Effects of load modelling on analysis of power system voltage stability. IEEE Trans Power Syst 16(5):329–338
25. Taylor CW (1994) Power system voltagestability. McGraw-Hill, New York
26. Gardner P, Tremblay M, Price D (2009) Technical requirements for high-penetration wind: What system operators need, and what wind technology can deliver. Integration of wide-scale renewable resources into the power delivery system. CIGRE/IEEE PES Joint SymposiumCalgary, Canada, pp 1–1
27. de Alegria IM, Andreu J, Martin JL, Ibañez P, Villate JL, Camblong H (2007) Connection requirements for wind farms: a survey on technical requirements and regulation. Renew Sustain Energy Rev 11(8):1858–1872
28. Johnson GL (2004) Wind energy systems. Prentice-Hall, Manhattan
29. Rathi MR, Mohan N (2005) A novel robust low voltage and fault ride through for wind turbine application operating in weak grids. In: 31st annual conference of IEEE Industrial Electronics Society, pp 2481–2486
30. Qi L, Langston J, Steurer M (2008) Applying a STATCOM for stability improvement to an existing wind farm with fixed-speed induction generators. In: IEEE Power and Energy Society general meeting- Conversion and delivery of electrical energy in the 21st Century, Pittsburgh, PA, pp 1–6
31. Chompoo-inwai C, Yingvivatanapong C, Methaprayoon K, Lee WJ (2005) Reactive compensation techniques to improve the ride-through capability of wind turbine during disturbance. IEEE Trans Ind Appl 41(3):666–672
32. Rao P, Crow ML, Yang Z (2000) STATCOM control for power system voltage control applications. IEEE Trans Power Delivery 15(4):1311–1317
33. Salle SADL, Reardon D, Leithead WE, Grimble MJ (1990) Review of wind turbine control. Int J Control 52(6):1295–1310
34. Hossain MJ, Pota HR, Ugrinovskii V, Ramos RA (2009)A novel STATCOM control to augment LVRT capability of fixed-speed induction generators. In: 48th IEEE conference on decision and control, Shanghai, China, pp 7843–7848
35. Sun J, Czarkowski D, Zabar Z (2002) Voltage flicker mitigation using PWM-based distribution STATCOM. In: IEEE Power Engineering Society summer meeting, vol 1, pp 616–621
36. Pal B, Chaudhuri B (2005) Robust control in power systems. Springer, New York
37. Muyeen SM, Takahashi R, Murata T, Tamura J, Ali MH (2009) Application of STATCOM/BESS for wind power smoothening and hydrogen generation. Electric Power Syst Res 79(2):365–373
38. Reed G, Greaf J, Matsumoto T, Sidell AP, Chervus RE, Nebecker CK (2000) Application of a 5 MVA, 4.16 kV D-STATCOM system for voltage flicker compensation at seattle iron and metals. In: IEEE Power Engineering Society summer meeting, Seattle, WA, pp 1605–1611
39. Ribeiro PF, Johnson BK, Crow ML, Arsoy A, Liu Y 2001) Energy storage systems for advanced power applications. Proc IEEE 89(12):1744–1756

40. Siljak DD, Zecevic AI, Neskovic G (2004) Robust decentralized exciter control with linear feedback. IEEE Trans Power Syst 19(2):1096–1103
41. Ellis A, Nelson R, Engeln EV et al (2012) Reactive power interconnection requirements for pv and wind plants recommendations to nerc. Sandia National Laboratories, Albuquerque, New Mexico 87185 and Livermore, California 94550
42. Akhmatov V, Knudsen H, Nielsen AH (2000) Advanced simulation of windmills in the electrical power supply. Int J Electr Power Energy Syst 22(6):421–434
43. Akhmatov V, Knudsen H (2002) An aggregate model of a grid-connected, large-scale, offshore wind farm for power stability investigations-importance of windmill mechanical system. International J Electr Power Energy Syst 24(9):709–717
44. Chi Y, Liu Y, Wang W, Dai H (2006) Voltage stability analysis of wind farm integration into transmission network. In: International conference on power system technology, vol 1, Chongqing, pp 1–7
45. Slootweg JG, Kling WL (2003) The impact of large scale wind power generation on power system oscillations. Electric Power Syst Res 67(1):9–20
46. Akhmatov V, Knudsena H, Nielsenb AH, Pedersenb JK, Poulsenc NK (2003) Modelling and transient stability of large wind farms. Int J Electr Power Energy Syst 25(2):123–144
47. Ledesma P, Usaola J, Rodriguez JL (2003) Transient stability of a fixed speed wind farm. Renew Energy 28(9):1341–1355
48. Zahedi A (2009) Development of an economical model to determine an appropriate feed-in tariff for grid-connected solar pv electricity in all states of australia. Renew Sustain Energy Rev 13(4):871–878
49. Mahmud MA, Pota HR, Hossain MJ (2012) Nonlinear controller design for dingle-phase grid-connected photovoltaic systems using partial feedback linearization. In: Australian control conference, 15–16 Nov 2012, Sydney, Australia, pp 30–35
50. de Alegria IM, Andreu J Martn JL, Ibañez P, Villate JL, Camblong H (2007) Connection requirements for wind farms: a survey on technical requirements and regulation. Renew Sustain Energy Rev 11(8):1858–1872
51. Singh B, Murthy SS, Gupta S (2004) Analysis and design of STATCOM-based voltage regulator for self-excited induction generators. IEEE Trans Energy Conv 19(4):783–790
52. Saad-Saoud Z, Lisboa ML, Ekanayake JB, Jenkins N, Strbac G (1998) Application of STATCOMs to wind farms. IEE Proc Gener Transm Distrib 45(15):511–516
53. Abdin ES, Xu W (2000) Control design and dynamic performance analysis of a wind turbine-induction generator unit. IEEE Trans Energy Conv 15(1):91–96
54. Chan TF, Nigim KA, Lai LL (2004) Voltage and frequency control of self-excited slip-ring induction generators. IEEE Trans Energy Conv 19(1):81–87
55. Delfino F, Procopio R, Rossi M, Ronda G (2010) Integration of large-scale photovoltaic systems into the distributed grids: a pq chart approach to access reactive support capability. IET Renew Power Gener 4:329–340
56. Kakimoto N, Piao QY, Ito H (2011) Voltage control of photovoltaic generator in combination with series reactor. IEEE Trans Sustaina Energy 2:374–382
57. Kakimoto N, Takayama S, Satoh H, Nakamura K (2009) Power modulation of photovoltaic generators for frequency control of power systems. IEEE Trans Energy Conv 24:943–949
58. Landau ID, Lozano R, M'Saad M, Karimi A (2011) Adaptive control: algorithm, analysis and applications. Springer, New York
59. Bossanyi EA (2000) The design of closed loop controllers for wind turbines. Wind Energy 3:149–163
60. Johnoson KE, Pao LY, Balas MJ, Fingersh LJ (2006) Control of variable speed wind turbines, standard and adaptive techniques for maximizing energy capture. IEEE Control Syst Mag 26(3):70–81
61. Song Y, Dhinakaran B, Bao X (2000) Variable speed control of wind turbines using nonlinear and adaptive algorithms. J Wind Eng Ind Aerodyn 85:293–308
62. Shamma J, Athans M (1991) Guaranteed properties of gain scheduled control for linear parameter-varying plants. Automatica 27(3):559–564

Chapter 2
Power System Voltage Stability and Models of Devices

Abstract This chapter introduces the concepts of voltage instability and the distinctions between voltage and angle instability. The driving force and main causes of voltage instability are analysed. Different methods and devices used to enhance voltage stability are also explained. The steady-state and dynamic modelling of the power system devices including wind generators and photovoltaic units have been discussed.

2.1 Introduction

Power system stability has been recognised as an important problem for secure system operation since the beginning of last century. Many major blackouts caused due to power system instability have illustrated the importance of this phenomenon [1, 2]. Angle stability had been the primary concern of the utilities for many decades. However, in the last two decades power systems have operated under much more stressed conditions than they usually had in the past. There are number of factors responsible for this: continuing growth in interconnections; the use of new technologies; bulk power transmissions over long transmission lines; environmental pressures on transmission expansion; increased electricity consumption in heavy load areas (where it is not feasible or economical to install new generating plants); new system loading patterns due to the opening up of the electricity market; growing use of induction machines; and large penetration of wind generators and local uncoordinated controls in systems. Under these stressed conditions a power system can exhibit a new type of unstable behaviour, namely, voltage instability.

In recent years, voltage instability has become a major research area in the field of power systems after a number of voltage instability incidents were experienced around the world [3, 4]. In Japan, a large-scale power failure occurred in the Tokyo metropolitan area in 1987 (about an 8-GW loss) because of voltage instability [5]. In Tokyo, the capacitance of 275-kV underground cables created adverse effects on

J. Hossain and H. R. Pota, *Robust Control for Grid Voltage Stability: High Penetration of Renewable Energy*, Power Systems,
DOI: 10.1007/978-981-287-116-9_2, © Springer Science+Business Media Singapore 2014

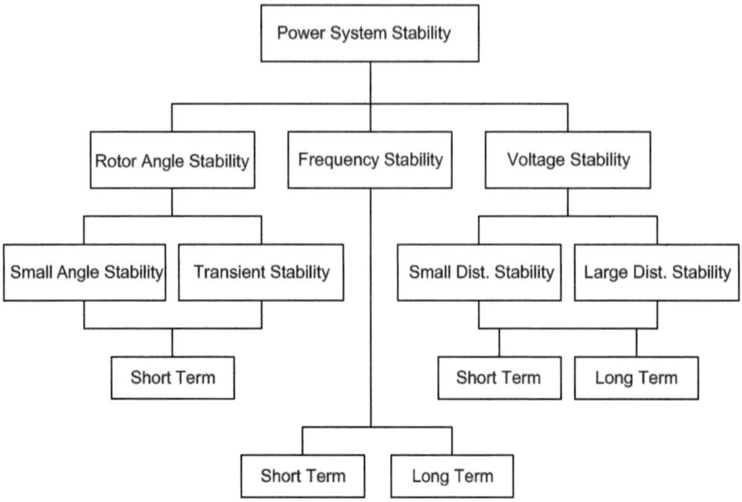

Fig. 2.1 Classification of power system stability

voltage-stability characteristics, making voltage stability one of the most important issues regarding system security. It has even been suggested that part of the problems that led to the North American blackout of August 2003 might be linked to short-term voltage instability [6]. In recent years, voltage instability has been responsible for several network collapses and blackouts [7] and is now receiving special attention in many systems.

This chapter will provide an overview of voltage stability problems and methods of effectively addressing them in the design and operation of electrical power systems. This includes the basic concepts, physical aspects of the phenomenon, methods of analysis, examples of major power grid blackouts due to voltage instability and methods of preventing voltage instability. This chapter addresses issues of power system voltage stability and identifies different categories of voltage stability behaviour that are important in power system stability analyses. In addition, the modeling of power system devices under consideration will be discussed.

2.2 Power System Stability and Voltage Stability

Power system stability is the ability of an electrical power system, for given initial operating conditions, to regain a state of operating equilibrium after being subjected to a physical disturbance, with most system variables bounded so that practically the entire system remains intact. Figure 2.1 gives the overall picture of the power system stability problem, identifying its categories and subcategories.

The concept of voltage stability addresses a large variety of different phenomena depending on which part of the power system is being analysed; for instance, it can be a fast phenomenon if induction motors, air conditioning loads or high-voltage DC transmission (HVDC) links are involved or a slow phenomenon if, for example, a mechanical tap changer is involved. Today, it is well accepted that voltage instability is a dynamic process since it is related to dynamic loads [8, 9].

Voltage stability refers to the ability of a power system to maintain steady voltages at all buses in the system and maintain or restore equilibrium between load demand and load supply from its given initial operating conditions after it has been subjected to a disturbance. Instability may result in progressive voltage falls or rises at some buses. A possible outcome of voltage instability is the loss of load in an area, and possible tripping of transmission lines and other elements by their protective systems which can lead to cascading outages.

Voltage collapse is more complex than voltage instability and is the process by which the sequence of events accompanying voltage instability lead to a blackout or abnormally low voltages in a significant part of a power system. The main symptoms of voltage collapse are: low voltage profiles; heavy reactive power flows; inadequate reactive support; and heavily loaded systems. The collapse is often precipitated by low-probability single or multiple contingencies. When a power system is subjected to a sudden increase of reactive power demand following a system contingency, the additional demand is met by the reactive power reserves of generators and compensators. Generally, there are sufficient reserves and the system settles to a stable voltage level. However, it is possible, due to a combination of events and system conditions, that the lack of additional reactive power may lead to voltage collapse, thereby causing a total or partial breakdown of the system.

2.3 Voltage and Angle Instability

Power system instability is essentially a single problem; however, the various forms of instability that a power system may undergo cannot be properly understood and effectively dealt with by treating it as such. Because of the high dimensionality and complexity of stability problems, it helps to simplify models in order to analyse specific types of problems using an appropriate degree of detail of the system representation and appropriate analytical techniques.

There is no clear distinction between voltage and angle instability problems but, in some circumstances, one form of instability predominates over the other. Distinguishing between the two types is important for understanding their underlying causes in order to develop appropriate design and operating procedures but, although this is effective, the overall stability of the system should be kept in mind. Solutions for one problem should not be at the expense of another. It is essential to look at all aspects of the stability phenomena and at each aspect from more than one viewpoint.

However, there are many cases, in which, one form of instability predominates. An IEEE report [10] points out the extreme situations of: (1) a remote synchronous

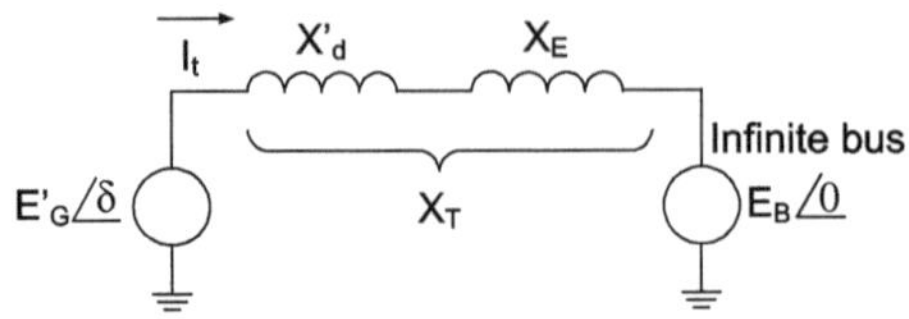

Fig. 2.2 Pure angle stability

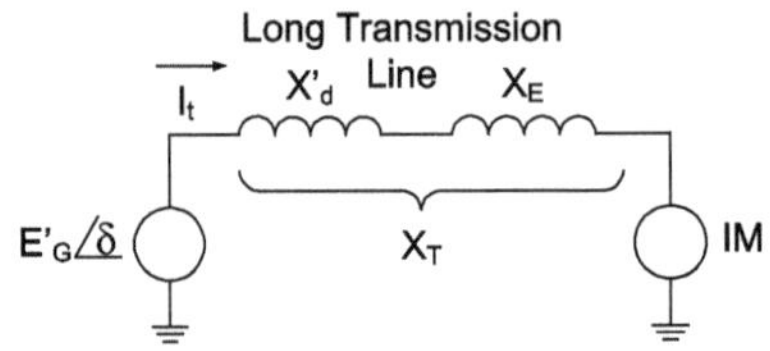

Fig. 2.3 Pure voltage stability

generator connected by transmission lines to a large system–angle stability dominates (one machine to an infinite-bus problem); and (2) a synchronous generator or large system connected by long transmission lines to an asynchronous load–voltage stability dominates. Figures 2.2 and 2.3 show these extremes. Details of the relationship between voltage and angle stability are given in [11].

Voltage stability is concerned with load areas and load characteristics. For rotor angle stability, we are often concerned with integrating remote power plants to a large system over long transmission lines. Basically, voltage stability is load stability and rotor angle stability is generator stability. In a large interconnected system, voltage collapse of a load area is possible without the loss of synchronism of any generator. Transient voltage stability is usually closely associated with transient rotor angle stability but longer-term voltage stability is less linked with rotor angle stability. It can be said that if voltage collapses at a point in a transmission system remote from the load, it is an angle instability problem. If it collapses in a load area, it is mainly a voltage instability problem.

2.4 Wind Power Generation and Power System Stability

In most countries, the amount of wind power generation integrated into large-scale electrical power systems is only a small part of the total power system load. However, the amount of electricity generated by wind turbines (WTs) is continuously increasing. Therefore, wind power penetration in electrical power systems will increase in future and will start to replace the output of conventional synchronous generators. As a result, it may also begin to influence overall power system behaviour. WTs use generators, such as squirrel-cage induction generators (IGs) or generators that are grid-coupled via power electronic converters. The interactions of these generator types with the power system are different from that of a conventional synchronous generator. As a consequence, WTs affect the dynamic behaviour of a power system in a way that might be different from that of synchronous generators. Therefore, the impact of wind power on the dynamics of power systems should be studied thoroughly in order to identify potential problems and to develop measures to mitigate those problems.

In grid impact studies of wind power integration, voltage stability is the main problem that will affect the operation and security of wind farms and power grids [12]. Voltage stability deterioration is mainly due to the large amount of reactive power absorbed by the WTs during their continuous operation and system contingencies. The various WT types presently in use behave differently during grid disturbances. Induction generators consume reactive power and behave similarly to induction motors for the duration of system contingency and will deteriorate the local grid voltage stability. Also, variable-speed wind turbines (VSWTs) equipped with doubly-fed induction generators (DFIGs) are becoming more widely used for their advanced reactive power and voltage control capability. DFIGs make use of power electronic converters and are, thus, able to regulate their own reactive power so as to operate at a given power factor or to control grid voltage. But, because of the limited capacity of a pulse-width modulation (PWM) converter [13], the voltage control capability of a DFIG cannot match with that of a synchronous generator. When the voltage control requirement is beyond the capability of a DFIG, the voltage stability of the grid is also affected.

When dealing with power system stability and wind power generation these questions may be raised, How does wind power generation contribute to power system stability? What are the factors that limit the integration of WTs into existing power systems? How many additional wind generators can be integrated by using static and dynamic compensations? Some cases of system stability problems related to wind power generation are presented in this book.

2.5 Voltage Instability and Time Frame of Interest

The time-frame of interest for voltage stability problems may vary from a few seconds to tens of minutes. Therefore, voltage stability may be either a short- or long-term phenomenon. Short-term voltage stability involves the dynamics of fast-acting load components such as induction motors, electronically controlled loads and HVDC converters. The study period of interest here is in the order of several seconds, and any analysis requires solutions of the appropriate system differential equations; this is similar to the analysis of rotor angle stability. Dynamic modeling of loads is often essential. In contrast to angle stability, short-circuits near loads are important.

Long-term voltage stability involves slower-acting equipment, such as tap-changing transformers, thermostatically controlled loads and generator current limiters. Here, the study period of interest may extend to several or many minutes, and long-term simulations are required for the analysis of a system's dynamic performance [14]. Stability is usually determined by the resulting outage of equipment, rather than the severity of the initial disturbance. Instability is due to the loss of long-term equilibrium (e.g., when loads try to restore their power beyond the capability of the transmission network and connected generation), the post-disturbance steady-state operating point being small-disturbance unstable, and/or a lack of attraction toward the stable post-disturbance equilibrium (e.g., when a remedial action is applied too late) [15, 16]. The disturbance could also be a sustained load build-up (e.g., motoring load increase).

Large-disturbance voltage stability refers to a system's ability to maintain steady voltages following large disturbances, such as system faults, loss of generation or circuit contingencies. This ability is determined by the system and load characteristics, and the interactions of both continuous and discrete controls and protections. Determination of large-disturbance voltage stability requires the examination of the nonlinear response of the power system over a period of time sufficient to capture the performance and interactions of such devices as motors, underload transformer tap changers and generator field-current limiters. The study period of interest may extend from a few seconds to tens of minutes.

Small-disturbance voltage stability refers to a system's ability to maintain steady voltages when subjected to small perturbations, such as incremental changes in system load. This form of stability is influenced by the characteristics of loads, continuous controls and discrete controls at a given instant of time. This concept is useful for determining, at any instant, how the system voltages will respond to small system changes. With appropriate assumptions, system equations can be linearised for analysis, thereby allowing the computation of valuable sensitivity information which is useful for identifying the factors influencing stability. However, this linearisation cannot account for nonlinear effects, such as tap-changer controls (dead-bands, discrete tap steps, and time delays). Therefore, a combination of linear and nonlinear analyses can be used in a complementary manner [8] to study voltage stability.

2.6 Voltage Stability

The practical importance of voltage stability analysis is that it helps in designing and selecting countermeasures which will avoid voltage collapse and enhance stability. Voltage stability analysis has gained increasingly importance in recent years due to:

- generation being centralised in fewer, larger power plants which means fewer voltage-controlled buses, and longer electrical distances between generation and load;
- the integration of large-scale induction generators;
- the extensive use of shunt capacitor compensation;
- voltage instability caused by line and generator outages;
- many incidents having occurred throughout the world (France, Belgium, Sweden, Japan, USA, etc) [3, 4]; and
- the operation of a system being closer to its limits.

2.7 Voltage Stability and Nonlinearity

Historically, power systems were designed and operated conservatively. It was comparatively easy to match load growth with new generation and transmission equipment. So, systems were operated in regions where behaviour was fairly linear. Only occasionally would systems be forced to extremes where nonlinearities could begin to have significant effects. However, the recent trend is for power systems to be operated closer to their limits. Also, as the electricity industry moves towards an open-access market, operating strategies will become much less predictable. Hence, the reliance on fairly linear behaviour which was adequate in the past, must give way to an acceptance that nonlinearities are going to play an increasingly important role in power system operation.

One important aspect of the voltage stability problem, making its understanding and solution more difficult, is that the phenomena involved are truly nonlinear. As the stress on a system increases, this nonlinearity becomes more and more pronounced. The nonlinearity of loads and generator dynamics are important factors when determining voltage instability. Therefore, it is essential that the nonlinear behaviour of power system devices should be taken into account when designing controllers and analysing dynamic behaviours.

2.8 Main Causes of Voltage Instability

The driving force for voltage instability is usually the loads; in response to a disturbance, power consumed by the loads tends to be restored by the action of motor slip adjustment, distribution voltage regulators, tap-changing transformers and thermostats. Restored loads increase the stress on a high-voltage network by increasing the reactive power consumption and causing further voltage reduction. A run-down situation causing voltage instability occurs when load dynamics attempt to restore power consumption beyond the capability of the transmission network and the connected generation [15–18].

A major factor contributing to voltage instability is the voltage drop that occurs when both active and reactive power flow through the inductive reactances of a transmission network; this limits the capabilities of the transmission network, in terms of power transfer and voltage support, which are further limited when some of the generators hit their field, or armature current, time-overload capability limits. It is worth noting that, in almost all voltage instability incidents, one or several crucial generators were operating with a limited reactive capability [16]. Voltage stability is threatened when a disturbance increases the reactive power demand beyond the sustainable capacity of the available reactive power resources.

While the most common form of voltage instability is progressive drops in bus voltages, the risk of over-voltage instability also exists and has been experienced in at least one system [19]. This is caused by the capacitive behaviour of a network

(EHV transmission lines operating below surge impedance loading) as well as by underexcitation limiters preventing generators and/or synchronous compensators from absorbing the excess reactive power. In this case, instability is associated with the inability of the combined generation and transmission systems to operate below some load level. In their attempt to restore this load power, transformer tap changers may cause long-term voltage instability.

Voltage stability problems may also be experienced at the terminals of HVDC links used for either long-distance or back-to-back applications [20]. They are usually associated with HVDC links connected to weak AC systems and may occur at rectifier or inverter stations, and are associated with the unfavourable reactive power load characteristics of converters. A HVDC link's control strategies have a very significant influence on such problems, since the active and reactive power at the AC/DC junction is determined by the controls. If the resulting loading on an AC transmission stresses it beyond its capability, voltage instability occurs. Such a phenomenon is relatively fast with the time frame of interest being in the order of one second or less. Voltage instability may also be associated with converter transformer tap-changer controls which is a considerably slower phenomenon. Recent developments in HVDC technology (voltage-source converters and capacitor-commutated converters) have significantly increased the limits for the stable operation of HVDC links in weak systems compared with the limits for line-commutated converters.

One form of the voltage stability problem, that results in uncontrolled over-voltages, is the self-excitation of synchronous machines. This can arise if the capacitive load of a synchronous machine is too large. Examples of excessive capacitive loads that can initiate self-excitation are open-ended high-voltage lines, and shunt capacitors and filter banks from HVDC stations. The over-voltages that result when a generator load changes to a capacitive load are characterised by an instantaneous rise at the instant of change followed by a more gradual rise. This latter rise depends on the relationship between the capacitive load component and the machine reactance, together with the excitation system of the synchronous machine. The negative field current capability of an exciter is a feature that has a positive influence on its limits for self-excitation. A voltage collapse may be aggravated by the excessive use of shunt capacitor compensation, due to the inability of the system to meet its reactive demands, or large sudden disturbances, such as the loss of either a generating unit or a heavily loaded line, or cascading events or poor coordination between various control and protective systems.

2.9 Methods for Improving Voltage Stability

The control of voltage levels is accomplished by controlling the production, absorption and flow of reactive power at all levels in a system. In order to function properly, it is essential that the voltage is kept close to the nominal value throughout the entire power system. Traditionally, this has been achieved differently for transmission networks and distribution grids. In transmission networks, a large-scale centralised power plant keeps the node voltages within an allowed deviation from their nominal values and the number of dedicated voltage control devices is limited.

In contrast, distribution grids incorporate dedicated equipment for voltage control and the generators connected to the distribution grid are hardly, if at all, involved in controlling the node voltages. The most frequently used voltage control devices in distribution grids are tap-changer transformers that change their turns ratio but switched capacitors and reactors are also applied. However, a number of recent developments challenge this traditional approach. One of these is the increased use of WTs for generating electricity. When large-scale wind farms are connected to the grids, it will be difficult to maintain node voltages using the traditional reactive power control devices. In these cases, some dedicated equipment, such as flexible AC transmission system (FACTS) devices will have to be used as well. FACTS devices offer fast and reliable control over the three AC transmission system parameters, i.e., voltage, line impedance and phase angle, and make it possible to control voltage stability dynamically.

2.9.1 Voltage Stability and Exciter Control

Automatic voltage regulators (AVRs) with synchronous machines are the most important means of voltage control in a power system. A synchronous machine is capable of generating and supplying reactive power within its capability limits to regulate system voltage. For this reason, it is an extremely valuable part of the solution to the collapse-mitigation problem.

The performance requirements of excitation systems are determined by considerations of the synchronous generator as well as the power system. The basic requirement is that the excitation system supplies and automatically adjusts the field current of the synchronous generator in order to maintain the scheduled terminal voltage as the output varies within the continuous capability of the generator. An excitation system must be able to respond to transient disturbances by field forcing consistent with the generator's instantaneous and short-term capabilities. The generator capability is limited by several factors: rotor insulation failure due to high field voltage; rotor heating due to high field current; stator heating due to high armature current loading; core end heating during underexcited operation; and heating due to excess flux (volts/Hz).

The role of an excitation system for enhancing power system performance has been continually growing. Early excitation systems were controlled manually to maintain the desired generator terminal voltage and reactive power loading. When the voltage control was first automated, it was very slow, basically filling the role of an alert operator [18]. Many research works have been undertaken in the area of voltage control using efficient excitation control. Modern excitation systems are capable of providing practically instantaneous responses with high ceiling voltages.

The combination of a high field-forcing capability and the use of auxiliary stabilising signals contributes to the substantial enhancement of overall system dynamic performance.

2.9.2 *Voltage Stability and FACTS Devices*

During the past two decades, the increase in electrical energy demand has presented higher requirements for the power industry. In recent years, the increases in peak load demands and power transfers between utilities have elevated concerns about system voltage security. Voltage instability is mainly associated with a reactive power imbalance. Improving a system's reactive power-handling capacity via FACTS devices is a remedy for the prevention of voltage instability and, hence, voltage collapse.

With the rapid development of power electronics, FACTS devices have been proposed and installed in power systems. They can be utilised to control power flow and enhance system stability. Particularly with the deregulation of the electricity market, there is an increasing interest in using FACTS devices for the operation and control of power systems with new loading and power flow conditions. For a better utilization of existing power systems, i.e., to increase their capacities and controllability, installing FACTS devices becomes imperative.

In the present situation, there are two main aspects that should be considered when using FACTS devices: the flexible power system operation according to their power flow control capability; and improvements in the transient and steady-state stability of power systems. FACTS devices are the right equipment to meet these challenges and different types are used in different power systems.

The most commonly used devices in present power grids are shunt capacitors and mechanically-controlled circuit breakers (MCCBs). Within limits, static reactive sources, such as shunt capacitors, can assist in voltage support. However, unless they are converted to pseudo-dynamic sources by being mechanically switched, they are not able to help support voltages during emergencies, when more reactive power support is required. In fact, shunt capacitors suffer from a serious drawback of providing less reactive support at the very time that more support is needed, i. e., during a voltage depression volt-ampere-reactive (VAr) output being proportional to the square of the applied voltage.

Long switching periods and discrete operation make it difficult for MCCBs to handle the frequently changing loads smoothly and damp out the transient oscillations quickly. In order to compensate for these drawbacks, large operational margins and redundancies are maintained in order to protect the system from dynamic variation and recover from faults. However, this not only increases the cost and lowers the efficiency, but also increases the complexity of a system and augments the difficulty of its operation and control. Severe black-outs in power grids which have happened recently worldwide have revealed that conventional transmission systems are unable to manage the control requirements of complicated interconnections and variable power flows.

More smoothly controlled, and faster, reactive support than mechanically switched capacitors can be provided by true dynamic sources of reactive power such as static VAr compensators (SVCs), static synchronous compensators (STATCOMs), synchronous condensers and generators. The application of SVCs and STATCOMs, in the context of voltage stability, has been discussed in recent literature [21]. The main differences between these two devices are that the SVC becomes a shunt capacitor when it reaches the limit of its control and all capacitance is fully switched in, and its reactive power output decreases as the square of the voltage when the maximum range of control is reached. The main advantage of the STATCOM over the thyristor type SVC is that the compensating current does not depend on the voltage level of the connecting point and thus the compensating current is not lowered as the voltage drops [22]. STATCOMs help to meet the wind farm interconnection standards and also provide dynamic voltage regulation, power factor correction and a low-voltage ride-through capability for an entire wind farm.

2.10 Modeling of Power System Devices

Power systems are large interconnected systems consisting of generation units, transmission grids, distribution systems and consumption units. The stability of a power system is dependent on several components, such as conventional generators and their exciters, wind generators, PV units, dynamic loads and FACTS devices. Therefore, an understanding of the characteristics of these devices and the modeling of their performances are of fundamental importance for stability studies and control design. There are numerous dynamics associated with a power system which may affect its large-signal stability and cause other kinds of stability problems. The large-signal stability technique analyses a system's stability by studying detailed simulations of its dynamics.

Modern power systems are characterised by complex dynamic behaviours which are due to their size and complexity. As the size of a power system increases, its dynamic processes become more challenging for analysis as well as for an understanding of its underlying physical phenomena. Power systems, even in their simplest form, exhibit nonlinear and time-varying behaviours. Moreover, there is a wide variety of equipment in today's power systems, namely: (1) synchronous generators, PV units and wind generators; (2) loads; (3) reactive-power control devices, such as capacitor banks and shunt reactors; (4) power-electronically switched devices, such as SVCs, and currently developed FACTS devices, such as STATCOMs; (5) series capacitors, thyristor-controlled series capacitors (TCSCs), among others. Though the kinds of equipment found in today's power systems are well-established and quite uniform in design, their precise modeling plays an important role in analysis and simulation studies of a whole system.

Different approaches to system modeling lead to different analytical results and accuracy. Improper models may result in over-estimated stability margins which can be disastrous for system operation and control. On the contrary, redundant models will greatly increase computation costs and could be impractical for industrial application. To study the problem of modeling, all the components of a power system should be considered for their performance. Based on the requirements of stability study, different modeling schemes can be used for the same device; for example, three kinds of models of a system or device are necessary in order to study a power system's long term, midterm and transient stabilities.

Traditional system modeling has been based on generators and their controls as well as the transmission system components. Only recently load modeling has received more and more attention for stability analysis purposes. Test systems considered in this dissertation consist of conventional generators, wind generators, PV units, generator control systems including excitation control, automatic voltage regulators (AVRs), power system stabilisers (PSSs), transmission lines, transformers, reactive power compensation devices, newly developed FACTS devices and loads of different kinds. Each piece of equipment has its own dynamic properties that may need to be modelled for a stability study.

The dynamic behaviours of these devices are described through a set of nonlinear differential equations while the power flow in the network is represented by a set of algebraic equations. This gives rise to a set of differential-algebraic equations (DAEs) describing the behaviour of a power system. After suitable representations of these elements, one can arrive at a network model of a system in terms of its admittance matrix. Generally because of a large number of nodes in the system, this matrix will be large but can be reduced by making suitable assumptions. Different types of models have been reported in the literature for each type of power system component depending upon its specific applications [18]. In this chapter, the relevant equations governing the dynamic behaviours of the specific types of models used in this dissertation are described.

2.10.1 Modeling of Synchronous Generators

A synchronous machine is one of the most important power system components. It can generate active and reactive power independently and has an important role in voltage control. The synchronising torques between generators act to keep large power systems together and make all generator rotors rotate synchronously. This rotational speed is what determines the mains frequency which is kept very close to the nominal value of 50 or 60 Hz.

Generally, the well-established Park's model for a synchronous machine is used in system analysis. However, some modifications can be employed to simplify it for stability analysis. Depending on the nature of the study, several models of a synchronous generator, having different levels of complexity, can be utilised [18]. In the simplest case, a synchronous generator is represented by a second-order differential equation, while studying fast transients in a generator's windings requires the use of a more detailed model, e.g., a sub-transient $6th$-order model. Throughout this book, sub-transient and third-order transient generator models are used.

The IEEE recommended practice regarding the d–q axis orientation of a synchronous generator is followed here [18]. This results in a negative d-axis component of stator current for an overexcited synchronous generator delivering power to the system. The differential equations, governing the sub-transient dynamic behaviour of generators in a multi-machine interconnected system, are given by [23]:

$$\dot{\delta}_k = \omega_k \omega_s - \omega_s, \tag{2.1}$$

$$\begin{aligned}\dot{\omega}_k = \frac{1}{2H_k}\Bigg[T_{m_k} - \frac{X''_{d_k} - X_{ls_k}}{X'_{d_k} - X_{ls_k}} E'_{q_k} I_{q_k} - \frac{X'_{d_k} - X''_{d_k}}{X'_{d_k} - X_{ls_k}} \psi_{1d_k} I_{q_k} + \frac{X'_{q_k} - X''_{q_k}}{X'_{q_k} - X_{ls_k}} \psi_{2q_k} I_{d_k} \\ - \frac{X''_{q_k} - X_{ls_k}}{X'_{q_k} - X_{ls_k}} E'_{d_k} I_{d_k} + (X''_{q_k} - X''_{d_k}) I_{q_k} I_{d_k} - D_k \omega_k\Bigg],\end{aligned} \tag{2.2}$$

$$\begin{aligned}\dot{E}'_{q_k} = \frac{1}{T'_{do_k}}\Bigg[-E'_{q_k} - (X_{d_k} - X'_{d_k})\{-I_{d_k} - \frac{X'_{d_k} - X''_{d_k}}{(X'_{d_k} - X_{ls_k})^2}(\psi_{1d_k} \\ -(X'_{d_k} - X_{ls_k}) I_{d_k} - E'_{q_k})\} + K_{a_k}(V_{ref_k} - V_{t_k} + V_{s_k})\Bigg],\end{aligned} \tag{2.3}$$

$$\begin{aligned}\dot{E}'_{d_k} = -\frac{1}{T'_{qo_k}}\Bigg[E'_{d_k} + (X_{q_k} - X'_{q_k})\{I_{q_k} - \frac{X'_{q_k} - X''_{q_k}}{(X'_{q_k} - X_{ls_k})^2}(-\psi_{2q_k} \\ +(X'_{q_k} - X_{ls_k}) I_{q_k} - E'_{d_k})\}\Bigg],\end{aligned} \tag{2.4}$$

$$\dot{\psi}_{1d_k} = \frac{1}{T''_{do_k}}\left[-\psi_{1d_k} + E'_{q_k} + (X_{d_k} - X_{ls_k}) I_{d_k}\right], \tag{2.5}$$

$$\dot{\psi}_{2q_k} = -\frac{1}{T''_{qo_k}}\left[\psi_{2q_k} + E'_{d_k} - (X_{q_k} - X_{ls_k}) I_{q_k}\right], \tag{2.6}$$

for $k = 1, 2, \ldots, m$, where m is the total number of generators, K_{a_k} the AVR gain, V_{ti_k} the terminal voltage, V_{s_k} the auxiliary input signal to the exciter, δ_k the power angle of the generator, ω_k the rotor speed with respect to a synchronous reference, E'_{q_k} the transient emf due to field flux linkage, E'_{d_k} the transient emf due to flux linkage in the d-axis damper coil, ψ_{1d_k} the sub-transient emf due to flux linkage in the d-axis damper, ψ_{2q_k} the sub-transient emf due to flux linkage in the q-axis damper, ω_s the absolute value of the synchronous speed in radians per second, H_k the inertia constant of the generator, D_k the damping constant of the generator, T'_{do_k} and T''_{do_k} the direct-axis open-circuit transient and sub-transient time constants, T'_{qo_k} and T''_{qo_k} the q-axes open-circuit transient and sub-transient time constants, I_{d_k} and I_{q_k} the d- and q-axes components of the stator current, X_{ls_k} the armature leakage reactance, X_{d_k}, X'_{d_k} and X''_{d_k} the synchronous, transient and sub-transient reactances along the d-axis, X_{q_k}, X'_{q_k} and X''_{q_k} the synchronous, transient and sub-transient reactances along the q-axis, respectively.

For stability analysis, the stator transients are assumed to be much faster compared to the swing dynamics [23]. Hence, the stator quantities are assumed to be related to the terminal bus quantities through algebraic equations rather than differential equations. The stator algebraic equation is given by:

$$V_i \cos(\delta_i - \theta_i) - \frac{X''_{d_i} - X_{ls_i}}{X'_{d_i} - X_{ls_i}} E'_{q_i} - \frac{X'_{d_i} - X''_{d_i}}{X'_{d_i} - X_{ls_i}} \psi_{1d_i} + R_{s_i} I_{q_i} - X''_{d_i} I_{d_i} = 0, \quad (2.7)$$

$$V_i \sin(\delta_i - \theta_i) + \frac{X''_{q_i} - X_{ls_i}}{X'_{q_i} - X_{ls_i}} E'_{d_i} - \frac{X'_{q_i} - X''_{q_i}}{X'_{q_i} - X_{ls_i}} \psi_{2q_i} - R_{s_i} I_{d_i} - X''_{q_i} I_{d_i} = 0, \quad (2.8)$$

where V_i is the generator terminal voltage. Under typical assumptions, the single-axis synchronous generator can be modeled by the following set of nonlinear differential equations [24]:

$$\dot{\delta}_k = \omega_k \omega_s - \omega_s, \quad (2.9)$$

$$\dot{\omega}_k = \frac{1}{2H_k} \left[P_{m_k} - E'_{q_k} I_{q_i} - D_k \omega \right], \quad (2.10)$$

$$\dot{E}'_{q_k} = \frac{1}{T'_{d0_k}} \left[E_{fd_k} - E'_{q_k} - (X_{d_k} - X'_{d_k}) I_{d_k} \right], \quad (2.11)$$

where E_{fd_i} is the equivalent emf in the exciter coil. The mechanical input power, P_{m_i}, to the generator is assumed to be constant.

2.10.2 *Modeling of Excitation Systems*

Control of the excitation system of a synchronous machine has a very strong influence on its performance, voltage regulation and stability [25]. Not only is the operation of a single machine affected by its excitation but, also, the behaviour of the whole system is dependent on the excitation system of the generators; for example, inter-area oscillations are directly connected to the excitations of the generators [26]. In general, the whole excitation control system includes:

- a PSS;
- an excitation system stabilizer;
- an AVR; and
- a terminal voltage transducer and load compensator.

There are different types of excitation systems commercially available in the power industry. However, one of the most commonly encountered models is the so-called IEEE Type ST1A excitation system. Other excitation system models for large-scale power system stability studies can be found in [27]. The main equations describing IEEE Type ST1A excitation are listed below:

$$\dot{V}_{tr_k} = \frac{1}{T_{r_k}} \left[-V_{tr_k} + V_{t_k} \right], \quad (2.12)$$

$$E_{fd_k} = K_{a_k} (V_{ref_k} - V_{tr_k}), \quad (2.13)$$

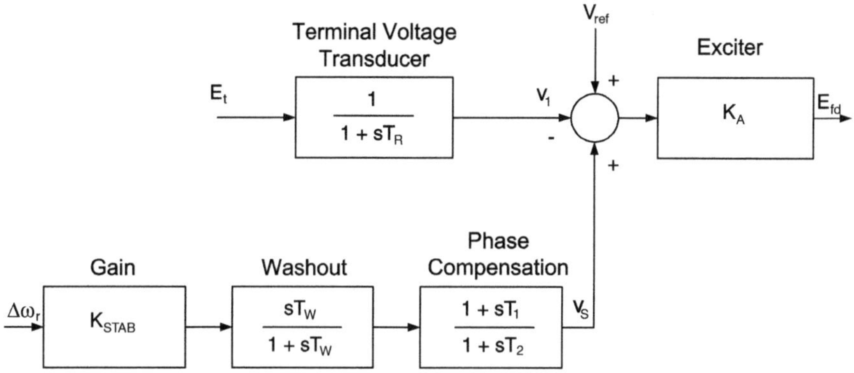

Fig. 2.4 PSS with AVR block diagram

where V_{tr_k} is the measured voltage state variable after the sensor lag block, V_{t_k} the measured terminal voltage, K_{a_k} the AVR gain and T_{r_k} the sensor time constant. In this dissertation, a robust excitation system is designed later and its performance is compared with that of the above excitation system.

2.10.3 Power System Stabilisers

The AVR plays an important role in keeping a generator synchronised with other generators in the grid. To achieve this, it should be fast-acting. Using high AVR gain to increase the action time often leads to unstable and oscillatory responses. To increase the damping of a lightly damped mode, the AVR uses a signal proportional to the rotor speed, although generator power and frequency may also be used [28]. The dynamic compensator used to modify the input signal to an AVR is commonly known as a PSS. Most generators have a PSS to improve stability and damp out oscillations.

Synchronous machines connected to a grid employ PSSs to enhance the damping of rotor oscillations. A typical PSS uses the change in speed, $\Delta\omega$, as the feedback variable and its output, V_s, is mixed with the reference voltage, V_{ref}, to produce the excitation signal. The block diagram in Fig. 2.4 shows the excitation system with an AVR and a PSS [18]. The amount of damping provided by a PSS depends on the value of the gain block, K_{STAB}. The phase compensation block introduces the phase lead necessary to compensate for the phase lag that is introduced between the exciter input and the generator electrical torque. The wash-out block serves as a high-pass filter, with the time constant, T_W, being high enough to allow signals associated with oscillations in ω_r to pass unchanged and block slowly varying speed changes. It allows the PSS to respond only to fast changes in speed.

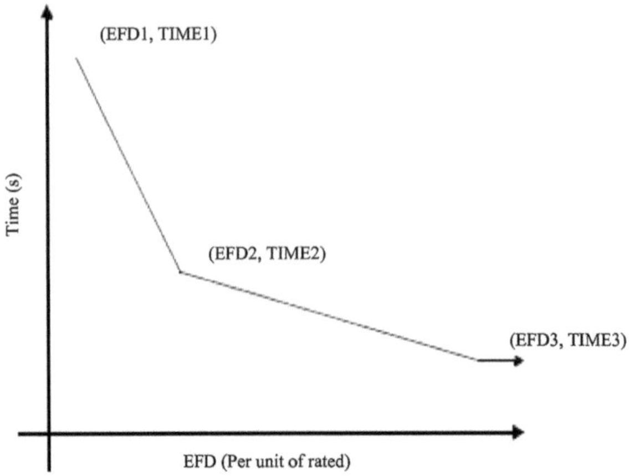

Fig. 2.5 Over-excitation limiter operating principle

2.10.4 Over-Excitation Limiters

An over-excitation limiter (OXL) can take two forms: (1) a device that limits the thermal duty of the rotor field circuit on a continuous current basis; and (2) a device that limits the effects of stator or transformer core iron saturation due to excessive generator terminal voltage, under-frequency, or the combination of both. An OXL to protect the rotor from thermal overload, is an important controller in system voltage stability. It is usually disabled in the transient time-frame to allow the excitation system to force several times the rated voltage across the rotor winding and more than the rated continuous current to help retain transient stability.

After a few seconds, the limiter is activated in an inverse time function–the higher the rotor current, the sooner the limiter is activated. This brings the continuous rotor current down to, or just below, the rated level to ensure the rotor is not overheated by excessive current. The limiter acts without regard to the actual rotor temperature. Even if the rotor is very cool before the over-excitation event, the time characteristic of the limiter is not changed. The over-excitation operating principle is shown in Fig. 2.5.

Note:

- below EFD1, the device is inactive;
- above EFD3, the time to operate is constant and equal to TIME3; and
- if EFD goes below EFD1 at any time before the device has timed-out, the timer resets.

2.10.5 Load Modeling

Several studies, [15, 29], have shown the critical effect of load representation in voltage stability studies and, therefore, the need to find more accurate load models than those traditionally used. Given a power system topology, the behaviour of a system following a disturbance, or the possibility of voltage collapse occurring, depends to a great extent on how the loads are represented.

Loads can be classified into different groups that are generally represented as an aggregated model. The main classifications are as static and dynamic models. As a static load model is not dependent on time, it describes the relationship of the active and reactive power at any time to the voltage and/or frequency at the same instant of time. The characteristics of load with respect to frequency are not critical for the phenomena of voltage stability but those with respect to voltage are. On the other hand, a dynamic load model expresses this active/reactive power relationship at any instant of time as a function of the voltage and/or frequency at a past instant of time, usually including the present moment. Static load models have been used for a long time for both purposes, i.e., to represent static load components, such as resistive and lighting loads, and also to approximate dynamic components. This approximation may be sufficient in some of the cases but for the fact that load representation has critical effects in voltage stability studies. This situation may become worse due to the traditional static load models being replaced with dynamic ones.

The modeling of load is complicated because a typical load bus represented in a stability analysis is composed of a large number of devices, such as fluorescent and incandescent lamps, refrigerators, heaters, compressors, motors and furnaces, etc. The exact composition of load is difficult to estimate. Also, its composition changes depending on many factors, including time, weather conditions and the state of the economy. An example of the composite load model representation used in this dissertation is shown in Fig. 2.6.

Common static load models for active and reactive power are expressed in polynomial or exponential forms and can include, a frequency dependence term. In this book, we use the exponential form to represent static loads as:

$$P(V) = P_0 \left(\frac{V}{V_0}\right)^a \tag{2.14}$$

$$Q(V) = Q_0 \left(\frac{V}{V_0}\right)^b \tag{2.15}$$

where P and Q are active and reactive components of load, respectively, when the bus voltage magnitude is V. The subscript 0 identifies the values of the respective variables at the initial operating condition. The parameters of this model are the exponents a and b. With these exponents equal to 0, 1 or 2, the model represents the constant power, constant current or constant impedance characteristics of load components, respectively.

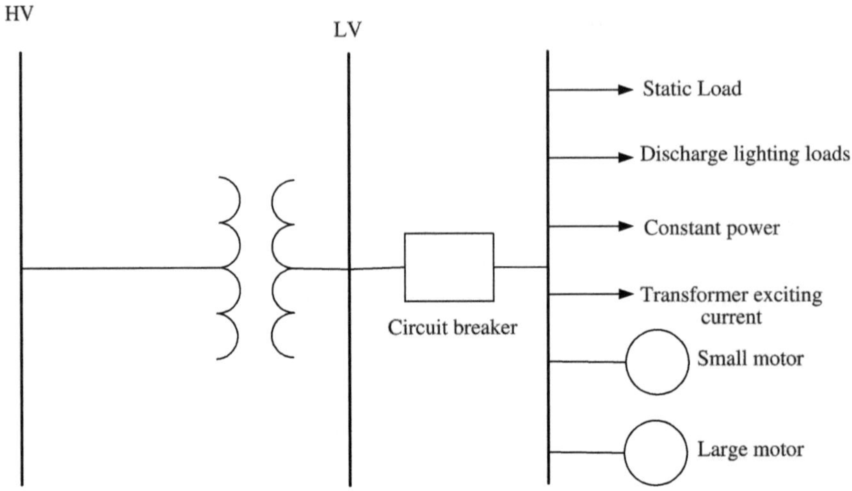

Fig. 2.6 Example of mixed load

2.10.6 Modeling of Induction Motors

A large amount of power consumption is by induction motors (IMs) in residential, commercial and industrial areas, commonly for the compressor loads of air conditioning and refrigeration in residential and commercial areas [18]. These loads require nearly constant torque at all speeds and are the most demanding from a stability viewpoint. On the other hand, pumps, fans and compressors account for more than half of industrial motor use. Typically, motors consume 60 to 70 % of the total power system energy and their dynamics are important for voltage stability and long-term stability studies. Therefore, the dynamics attributed to motors are usually the most significant aspects of the dynamic characteristics of system loads.

For modeling of induction motor in power system stability studies, the transients in stator voltage relations can be neglected [15], which corresponds to ignoring the DC components in stator transient currents, thereby permitting representation of only the fundamental frequency components. The transient model of a squirrel-cage induction motor is described by the following DAEs written in a synchronously-rotating reference frame [15]:

$$(v_{ds_i} + jv_{qs_i}) = (R_{s_i} + jX_i')(i_{ds_i} + ji_{qs_i}) + j(e'_{qr_i} - je'_{dr_i}), \tag{2.16}$$

$$\dot{s} = \frac{1}{2H_{m_i}}\left[T_e - T_L\right], \tag{2.17}$$

$$T'_{do_i}\dot{e}'_{qr_i} = -e'_{qr_i} + (X_i - X_i')i_{ds_i} - T'_{do}s\omega_s e'_{dr}, \tag{2.18}$$

$$T'_{do}\dot{e}'_{dr_i} = -e'_{dr_i} - (X_i - X_i')i_{qr_i} + T'_{do_i}s\omega_s e'_{qr_i}, \tag{2.19}$$

where for $i = 1, \ldots, p$, p is the number of induction motor, $X'_i = X_{s_i} + X_{m_i} X_{r_i} / (X_{m_i} + X_{r_i})$ the transient reactance, $X_i = X_{s_i} + X_{m_i}$ the rotor open-circuit reactance, $T'_{do_i} = (L_{r_i} + L_{m_i}) / R_{r_i}$ the transient open-circuit time constant, $T_{e_i} = e'_{qr_i} i_{qs_i} + e'_{dr_i} i_{ds_i}$ the electrical torque, s_i the slip, e'_{dr_i} the direct-axis transient emf, e'_{qr_i} the quadrature-axis transient emf, T_{L_i} the load torque, X_{s_i} the stator reactance, X_{m_i} the magnetising reactance, H_{m_i} the inertia constant of the motor, i_{ds_i} and i_{qs_i} the d- and q-axis components of the stator current, respectively.

There are two ways to obtain aggregation in load models. One is to survey the customer loads in a detailed load model, including the relevant parts of the network, and carry out system reduction. Then, a simple load model can be chosen so that it has similar load characteristics to the detailed load model. Another approach is to choose a load model structure and identify its parameters from measurements.

2.10.7 Modeling of On-Load Tap Changers

Load tap-changing transformers do not correspond to a load component but, seen from a transmission system viewpoint, they may be considered as part of the load. After a disturbance, they restore the sub-transmission and distribution voltages to their pre-disturbance values, but they also affect the status of the voltage-sensitive loads. The restoration of the voltage and, consequently, the increase in these loads may lead the system to voltage instability and collapse. The restoration process takes several minutes. A tap changer is governed by its step size, time constant, reference voltage and deadband. In this model, a tap changing takes place (after some built-in time delay) if the load voltage, V_{rms}, falls outside of a voltage range of $[Vref - D - \varepsilon, Vref + D + \varepsilon]$. The dynamic model of an OLTC is given by:

$$n_{k+1} = n_{k+d}(V_{ref} - V), \tag{2.20}$$

where n_{k+1} and n_k are the turns-ratios before and after a tap change, respectively, and ε, D and d are the hysteresis band, dead-band and step size of the tap, respectively.

2.10.8 Modeling of Wind Generators

The generation of electricity using wind power has received considerable attention world-wide in recent years. With the increasing penetration of wind-derived power in interconnected power systems, it has become necessary to model complete wind energy systems in order to study their impact, and also wind plant controls. Wind energy conversion systems comprise mechanical and electrical equipment and their controls. Modeling these systems for power system stability studies requires careful

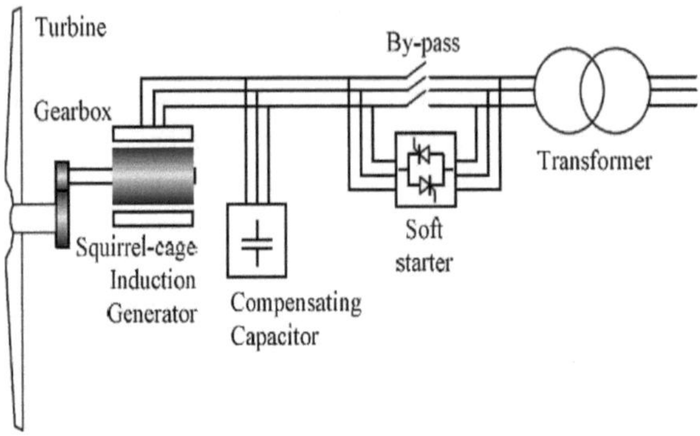

Fig. 2.7 System structure of wind turbine with directly connected squirrel-cage induction generator (source [31])

analysis of the equipment and controls to determine the characteristics that are important in the time frame and bandwidth of such studies.

The response of a wind farm or, alternatively, a model of a wind farm, is very dependent on the type of equipment used. The four concepts of operation of currently used grid-connected wind turbines (WTs) are: constant speed; limited variable speed; variable-speed with partial-scale frequency converter; and variable-speed with full-scale frequency converter [30]. At the moment, the majority of installed WTs are of the fixed-speed types, with SCIGs, known as the 'Danish concept' while, from a market perspective, the dominating technology WTs with doubly-fed induction generators (DFIGs). This book, however, focusses on the fixed-speed wind turbine (FSWT) technology.

FSWTs dominated the first ten years of WT development during the 1990. Operation at constant speed means that, regardless of the wind speed, the WT's rotor speed is fixed and is determined by the frequency of the grid, the gear ratio and the generator design. Usually, a FSWT is equipped with a SCIG connected to the grid, and a soft starter and capacitor bank for reducing the reactive power consumption. It is designed to achieve maximum efficiency at a particular wind speed. Although wound rotor synchronous generators have also been applied, at present, the most common generator is the induction generator (IG).

The schematic structure of a FSWT with a SCIG is depicted in Fig. 2.7. It is the simplest type of WT technology and has a turbine that converts the kinetic energy of wind into mechanical energy. The generator then transforms the mechanical energy into electrical energy and then delivers the energy directly to the grid. It needs to be noted that the rotational speed of the generator, depending on the number of poles, is relatively high (in the order of 1,000–1,500 rpm for a 50 Hz system frequency). Such a rotational speed is too high for the turbine in terms of turbine efficiency and mechanical stress. For this reason, a gear box is used to transform the rotational speed. The fixed-speed induction generator (FSIG) technology operates by drawing reactive power from the external grid via the stator to flux the rotor circuits. This results

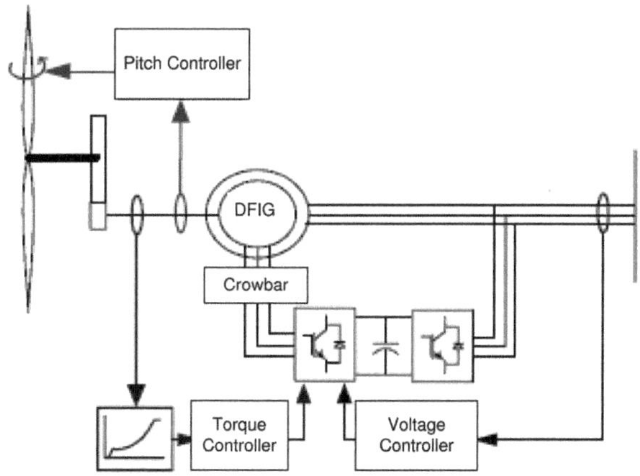

Fig. 2.8 Schematic diagram of variable-speed doubly-fed induction generator (source [32])

in the unit demonstrating a low full-load power factor. Switched capacitor banks or power electronically-controlled reactive power compensation devices (SVCs or STATCOMs) are installed to compensate for the reactive power consumed in order to reduce the intake of reactive power from the grid, hence reducing transmission losses and, in some instances, improving grid stability. The main concern for utilising a FSIG in wind generation is its absorption of excessive reactive power from the power system to magnetise the generator rotor circuit during voltage sag conditions arising from switching-in or system short-circuit fault events. These effects are more pronounced in a weak power system where reactive power reserves are scarce.

The schematic diagram of a variable-speed wind turbine (VSWT) is shown in Fig. 2.8. In this concept, a gear-box is also used. These types of WTs have back-to-back voltage-source converters (VSCs) for feeding the rotor windings and a pitch angle control to limit the power extracted in high wind speed conditions. No compensation capacitors are used.

A power collection and transmission system is required in a wind farm to connect the WTs arrays with the other components of the farm and to transmit the generated power to either distribution or transmission networks depending on the farm's capacity and voltage level [33]. The most common configuration is one in which each turbine unit has a transformer connected to it. However, in some configurations, two or three turbine units are connected together to one transformer. The output power of the transforms is carried by medium-voltage underground cables to overhead or underground collection lines that transmit the power to the wind farm sub-station. Here, the primary transformer steps up the voltage to the required voltage level of the grid.

Wind power has evolved rapidly over the last two decades with regard to the WT power ratings and, consequently, the rotor diameters of WTs. In the past few years, a different type of development has taken place: instead of a continuous increase in WT rated power, the WT manufacturers have focussed on developing WTs that are more reliable, grid code-compliant and suitable for different installation environments–onshore and offshore. Recently, the commercial offer from the wind industry with the majority of WTs has been rated at around 2–3 MW.

As wind farms become a larger part of the total generation of power systems worldwide, issues related to integration, stability effects and voltage impacts become increasingly important. Adequate load flow and dynamic simulation models (encompassing all significant air-dynamical, mechanical and electrical factors) are necessary to evaluate the impact of wind farms on power systems.

2.10.9 Load Flow Representation

Usually, a wind farm comprises a large number of individual turbine units that are interconnected in a radial or parallel arrangement. When studying the impact of a wind farm on a system, it is reasonable to construct an equivalent of the wind farm with a reduced number of aggregated units connected to the network. Such an aggregated representation is advantageous since it saves the user time and effort in modeling the wind farm. The program available from Siemens PTI allows the user to model a wind farm in PSS/E by merging groups of individual identical units into one or more equivalent machines. These equivalent machines are placed, along with their step-up transformers, at collector buses designated by the user.

The real power output of a WT unit is a function of the wind speed felt by the turbine blades and the site-dependent air density which is related by a so-called power curve. The program mentioned above has the capability to either calculate the MW output based on a given wind speed or, as is more reasonable for system studies, to allow the user to directly dispatch the individual or equivalent units. The reactive power injection or consumption of a WT unit is determined by its dispatch and the AC voltage or power factor control. Based on its control strategy, the program calculates the reactive output and determines the amount of additional shunt capacitors required to be added to provide the desired power factor. In general, the wind farm is represented as a PQ bus in a load-flow study.

The following equations are used to estimate the reactive power output from induction generator [34]:

$$K_1 = X_r + X_m, \; Ax^2 + xB + C = 0, \tag{2.21}$$

where $x = \frac{r_r}{s}$, $A = P(r_s^2 + K_3^2) - V^2 r_s$, $B = 2P(r_s K_2 + K_3 K_4) - V^2(K_2 + K_1 + K_3)$, $C = P(K_2^2 + K_4^2) - V^2 K_1 K_4$, $K_2 = -X_s K_1 - X_r X_m$, $K_3 = X_s + X_m$, $K_4 = r_s K_1$, and $P = \frac{V^2(xT_1 + K_1 T_2)}{T_3}$. Then, the reactive power of the IG is given by

$$Q_g = -\frac{V^2(K_1 T_1 - x_1 T_2)}{T_3}, \tag{2.22}$$

where $T_1 = xR_s - X_s K_1 - X_r X_m$, $T_2 = x(X_m + X_s) + r_s K_1$, $T_3 = T_1^2 + T_2^2$.

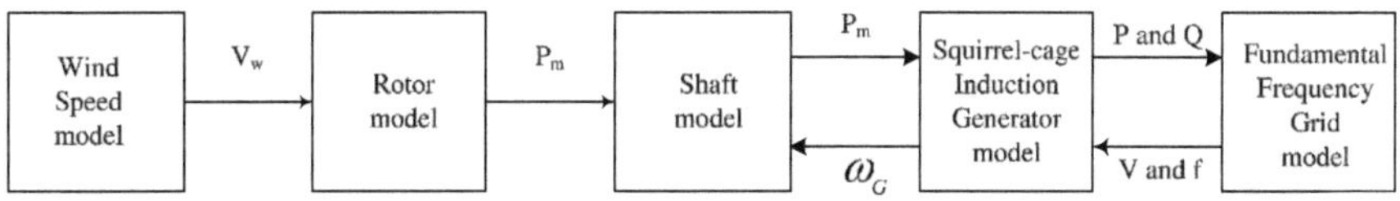

Fig. 2.9 General structure of constant-speed wind turbine model

2.10.10 Dynamic Model of Wind Generators

This book uses a model of the induction generator written in appropriate d-q reference frame to facilitate investigation of control strategies. Figure 2.9 depicts the general structure of a model of a constant-speed wind turbine. The most important components of a constant speed wind turbine are rotor, drive train and the generator, combined with a wind speed model.

2.10.11 Rotor Model

WTs are the main components of wind farms. They are usually mounted on towers to capture the most kinetic energy. Because the wind speed increases with height, taller towers enable turbines to capture more energy and generate more electricity. The three bladed rotor, consisting of the blades and a hub, is the most important and most visible part of a WT. It is through the rotor that the energy of the wind is transformed into mechanical energy that turns the main shaft of a WT.

The rotor of a WT, with radius R_i, converts energy from the wind to the rotor shaft, rotating at the speed of ω_{m_i}. The power from the wind depends on the wind speed, V_{w_i}, the air density, ρ_i, and the swept area, A_{wt_i}. From the available power in the swept area, the power on the rotor is given based on the power coefficient, $c_{p_i}(\lambda_i, \theta_i)$, which depends on the pitch angle of the blade, θ_i, and the ratio between the speed of the blade tip and the wind speed, denoted as the tip-speed ratio, $\lambda_l = \frac{\omega_{m_i} R_i}{V_{w_i}}$. The aerodynamic torque applied to the rotor for the *ith* turbine by the effective wind speed passing through the rotor is given as [30]:

$$T_{ae_i} = \frac{\rho_i}{2\omega_{m_i}} A_{wt_i} c_{p_i}(\lambda_i, \theta_i) V_{w_i}^3, \tag{2.23}$$

where c_{p_i} is approximated by the following relationship [35]:

$$c_{p_i} = (0.44 - 0.0167\theta_i) \sin\left[\frac{\pi(\lambda_i - 3)}{15 - 0.3\theta_i}\right] - 0.00184(\lambda_i - 3)\theta_i,$$

where $i = 1, \cdots, n$ and n is the number of WTs.

A controller equipped with a WT starts up the machine at wind speeds of about 8–16 miles per hour (mph) and shuts it off at about 55 mph. Turbines do not operate at wind speeds above about 55 mph because they might be damaged. The radius of a 2 MW wind turbine is about 80m, the typical value of air density is 1.225 kg/m^3, c_p is in the range of 0.52–0.55, towers range from 60 to 90 m (200 to 300 feet) tall and the blades rotate at 10–22 revolutions per minute.

Equation 2.23 shows that aerodynamic efficiency is influenced by variation in the blade's pitch angle. Regulating the rotor blades provides an effective means of regulating or limiting the turbine power during high wind speeds or abnormal conditions. A pitch controlled turbine performs power reduction by rotating each blade about its axis in the direction of the angle of attack. In comparison with the passive stall, the pitch control provides greater energy capture at the rated wind speed and above. The aerodynamic braking facility of the pitch control can reduce extreme loads on a turbine and also limit its power input so as to control possible over-speed of the machine if the loading of the turbine-generator system is lost, for instance, because of a power system fault. On a pitch-controlled WT, electronic controllers check the power output of the turbine several times per second. When the power output becomes too high, a message is sent to the blade-pitch mechanism which immediately turns the rotor blades slightly in an attempt to restore this output to an acceptable value. In this work, the pitch-rate limit is set to the typical value of 12 deg s^{-1}.

2.10.12 Shaft Model

A two-mass drive train model of a WT generator system (WTGS) is commonly used as drive train modeling can satisfactorily reproduce the dynamic characteristics of a WTGS because the low-speed shaft of a WT is relatively soft [36]. Therefore, although, it is essential to incorporate a shaft representation into the constant-speed wind turbine model, only a low-speed shaft is included. The gearbox and high-speed shaft are assumed to be infinitely stiff. The resonance frequencies associated with gearboxes and high-speed shafts usually lie outside the frequency bandwidth of interest [37]. Therefore, we use a two-mass representation of the drive train.

The drive train attached to the WT converts the aerodynamic torque, T_{ae_i}, on the rotor into the torque on the low-speed shaft, which is scaled down through the gear-box to the torque on the high-speed shaft. The first mass term stands for the blades, hub and low-speed shaft and the second for the high-speed shaft with inertia constants, H_{m_i} and H_{G_i}, respectively. The shafts are interconnected by a gear ratio, N_{g_i}, combined with torsion stiffness, K_{s_i}, and torsion damping, D_{m_i} and D_{G_i}, resulting in the torsion angle, γ_i. The normal grid frequency is f. The dynamics of the shaft are represented as in [30]:

$$\dot{\omega}_{m_i} = \frac{1}{2H_{m_i}}\left[T_{ae_i} - K_{s_i}\gamma_i - D_{m_i}\omega_{m_i}\right], \tag{2.24}$$

$$\dot{\omega}_{G_i} = \frac{1}{2H_{G_i}} \left[K_{s_i}\gamma_i - T_{e_i} - D_{G_i}\omega_{G_i} \right], \tag{2.25}$$

$$\dot{\gamma}_i = 2\pi f(\omega_{m_i} - \frac{1}{N_{g_i}}\omega_{G_i}). \tag{2.26}$$

The generator receives the mechanical power from the gear-box through the stiff shaft. The relationship between the mechanical torque and the torsional angle is given by:

$$T_{mi} = K_{s_i}\gamma_i. \tag{2.27}$$

The gear-box connects the low-speed shaft to the high-speed shaft and increases the rotational speeds from about 30 to 60 rotations per minute (rpm) to about 1,000–1,800 rpm, which is the rotational speed required by most generators to produce electricity.

2.10.13 Induction Generator Model

An IG can be represented in different ways, depending on the level of detail characterised mainly by the number of phenomena, such as stator and rotor flux dynamics, magnetic saturation, skin effects and mechanical dynamics included. Although, a very detailed model which includes all these dynamics is a possibility, it may not be beneficial for stability studies because it increases the complexity of the model and requires time-consuming simulations. More importantly, not all of these dynamics are shown to have significant influence in stability studies.

A comparison of different induction generator models can be found in [30]. Accordingly, as the inclusion of iron losses in a model is a complicated task, its influence for stability studies is neglected. The main flux saturation is only of importance when the flux level is higher than the nominal level. Hence, this effect can be neglected for most operating conditions. The skin effect should only be taken into account for a large slip operating condition which is not the case for a FSWT.

Another constraint of including dynamics in a model is the availability of relevant data. Typically, saturation and skin effect data are not provided by manufacturers. Therefore, in general, it is impractical to use them in WT applications. For the representation of FSIG models in power system stability studies [38], the stator flux transients can be neglected in the voltage relations.

All of these arguments lead to the conclusion that rotor dynamics are only the major factors required to be considered in an IG model for a voltage stability analysis. Representation of the third-order model of an IG offers a compatibility with the network model and provides more efficient simulation time. The main drawbacks of the third-order model is its inability to predict peak transient current and, to some extent, its less accurate estimation of speed. However, at a relatively high inertia, the third-order model is sufficiently accurate.

The transient model of a SCIG is described by the following DAEs [30, 34]:

$$\dot{s}_i = \frac{1}{2H_{G_i}}\left[T_{m_i} - T_{e_i}\right], \tag{2.28}$$

$$\dot{E}'_{qr_i} = -\frac{1}{T'_{o_i}}\left[E'_{qr_i} - (X_i - X'_i)i_{ds_i}\right] - s_i\omega_s E'_{dr_i}, \tag{2.29}$$

$$\dot{E}'_{dr_i} = -\frac{1}{T'_{o_i}}\left[E'_{dr_i} + (X_i - X'_i)i_{qs_i}\right] + s_i\omega_s E'_{qr_i}, \tag{2.30}$$

$$V_{ds_i} = R_{s_i}i_{ds_i} - X'_i i_{qs_i} + E'_{dr_i}, \tag{2.31}$$

$$V_{qs_i} = R_{s_i}i_{ds_i} + X'_i i_{qs_i} + E'_{qr_i}, \tag{2.32}$$

$$v_{t_i} = \sqrt{V_{ds_i}^2 + V_{qs_i}^2}, \tag{2.33}$$

where $X'_i = X_{s_i} + X_{m_i}X_{r_i}/(X_{m_i} + X_{r_i})$ is the transient reactance, $X_i = X_{s_i} + X_{m_i}$ the rotor open-circuit reactance, $T'_{o_i} = (L_{r_i} + L_{m_i})/R_{r_i}$ the transient open-circuit time constant, v_{t_i} the terminal voltage of the IG, s_i the slip, E'_{dr_i} the direct-axis transient voltages, E'_{qr_i} the quadrature-axis transient voltages, V_{ds_i} the d-axis stator voltage, V_{qs_i} the q-axis stator voltage, T_{m_i} the mechanical torque, $T_{e_i} = E_{dr_i}i_{ds_i} + E_{qr_i}i_{qs_i}$, the electrical torque, X_{s_i} the stator reactance, X_{r_i} is the rotor reactance, X_{m_i} the magnetising reactance, R_{s_i} the stator resistance, R_{r_i} the rotor resistance, H_{G_i} the inertia constant of the IG, and i_{ds_i} and i_{qs_i} the d- and q-axis components of the stator current, given by:

$$I_{di} = \sum_{j=1}^{n}\left[E'_{drj}(G_{ij}\cos\delta_{ji} - B_{ij}\sin\delta_{ji}) + E'_{qrj}(G_{ij}\sin\delta_{ji} + B_{ij}\cos\delta_{ji})\right], \tag{2.34}$$

$$I_{qi} = \sum_{j=1}^{n}\left[E'_{drj}(G_{ij}sin\delta_{ji} + B_{ij}cos\delta_{ji}) + E'_{qrj}(G_{ij}cos\delta_{ji} - B_{ij}sin\delta_{ji})\right]. \tag{2.35}$$

2.10.14 Modeling of DFIG

The equations that describe a SCIG are identical to those of the DFIG except that the rotor is short-circuited. The converter for DFIGs [30] used in this book consists of two VSCs connected back-to-back. This enables variable-speed operation of the WTs by using a decoupling control scheme which controls the active and reactive components of the current separately. The modeling of IGs for power-flow and dynamic analyses is discussed in [30, 34].

The DC-link dynamic of a DFIG is given by:

$$C_i v_{dc_i} \dot{v}_{dc_i} = -\frac{v_{dc_i}^2}{R_{loss_i}} - P_{r_i}(t) - P_{g_i}(t) \tag{2.36}$$

where resistor R_{loss_i} represents the total conducting and switching losses of the converter. Also, $P_{r_i}(t)$ is the instantaneous input rotor power, and $P_{g_i}(t)$ is the instantaneous output power of the GSC which are given by:

$$P_{r_i} = v_{rd_i} i_{rd_i} + v_{rq_i} i_{rq_i}, \tag{2.37}$$

$$P_{g_i} = v_{gd_i} i_{gd_i} + v_{gq_i} i_{gq_i}. \tag{2.38}$$

2.10.15 Aggregated Model of Wind Turbine

The development of aggregated models of wind farms is also an important issue because, as the sizes and numbers of turbines on wind farms increase, representing wind farms as individual turbines increases complexity and leads to a time-consuming simulation which is not beneficial for stability studies of large power systems.

For the aggregation of WTs, the models of several identical WTs (even in the incoming wind) are combined in a single turbine model with a higher rating. The parameters are obtained by preserving the electrical and mechanical parameters of each unit, and by increasing the nominal power to the equivalent of the involved turbines in the aggregation process [39].

This aggregated model reduces computation and simulation times in comparison with those of a detailed model with different representations of tens or hundreds of turbines and their interconnections. However, the aggregated model requires specific care in choosing what to aggregate in order to be as close to reality as possible. In addition, this type of modeling is very difficult for WTs without a parallel distribution (i.e., in the form of an array which is the most common distribution for offshore, but not for onshore, wind farms).

2.10.16 Modeling of PV Unit

As shown in Fig. 2.10, PV plants have mainly two parts (a) solar conversion and (b) electrical interface with the electrical network (a power electronic converter). A PV array is connected to the grid through a DC–DC converter and a DC-AC inverter. A DC–DC converter enables the transfer of maximum power from the solar module to the inverter. The PV array as shown in Fig. 2.11 is described by its current-voltage characteristics function [40, 41]:

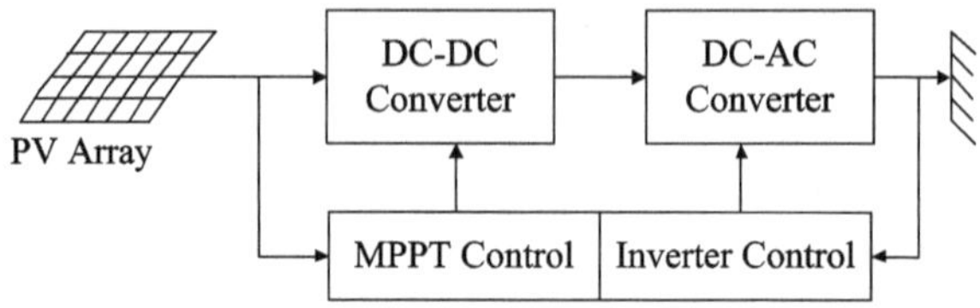

Fig. 2.10 Block Diagram of a PV System

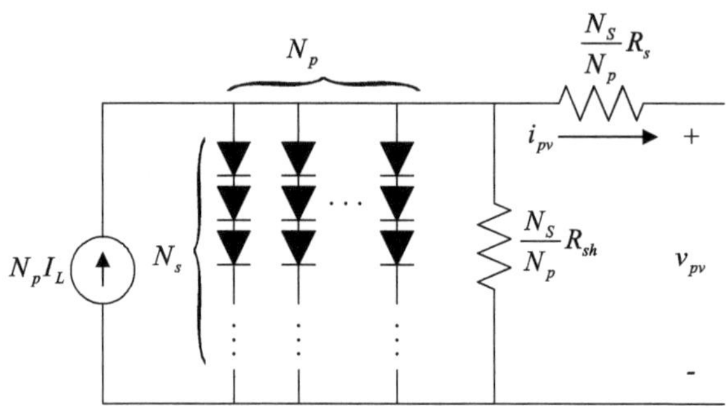

Fig. 2.11 Equivalent circuit of a PV array

$$
\begin{aligned}
i_{\mathrm{pv}_i} = N_{\mathrm{p}_i} I_{\mathrm{L}_i} - N_{\mathrm{p}_i} I_{\mathrm{s}_i} \left[\exp \left[\alpha_{p_i} \left(\frac{v_{\mathrm{pv}_i}}{N_{\mathrm{s}_i}} + \frac{R_{\mathrm{s}_i} i_{\mathrm{pv}_i}}{N_{\mathrm{p}_i}} \right) \right] - 1 \right] \\
- \frac{N_{\mathrm{p}_i}}{R_{\mathrm{sh}_i}} \left(\frac{v_{\mathrm{pv}_i}}{N_{\mathrm{s}_i}} + \frac{R_{\mathrm{s}_i} i_{\mathrm{pv}_i}}{N_{\mathrm{p}_i}} \right),
\end{aligned}
\tag{2.39}
$$

where I_{L_i} is the light-generated current, I_{s_i} is the reverse saturation current, chosen as 9×10^{-11} A, N_{s_i} is the number of cells in series and N_{p_i} is the number of modules in parallel, R_{s_i} and R_{sh_i} are the series and shunt resistances of the array respectively, i_{pv_i} is the current flowing through the array, and v_{pv_i} is the output voltage of the array. The constant α_{p_i} in Eq. (9.1) is given by

$$
\alpha_{p_i} = \frac{q_i}{A_i k_i T_{\mathrm{r}_i}}
\tag{2.40}
$$

where $k_i = 1.3807 \times 10^{-23} JK^{-1}$ is the Boltzmann constant, $q_i = 1.6022 \times 10^{-19}$ C is the charge of the electron, A_i is the p–n junction ideality factor with a value between 1 and 5, and T_{r_i} is the cell reference temperature. The schematic of a grid-connected PV system consisting of switching elements is shown in Fig. 2.12 [42, 43]. A nonlinear model of the three-phase grid connected PV system shown in Fig. 2.12 can be written as [42, 43]:

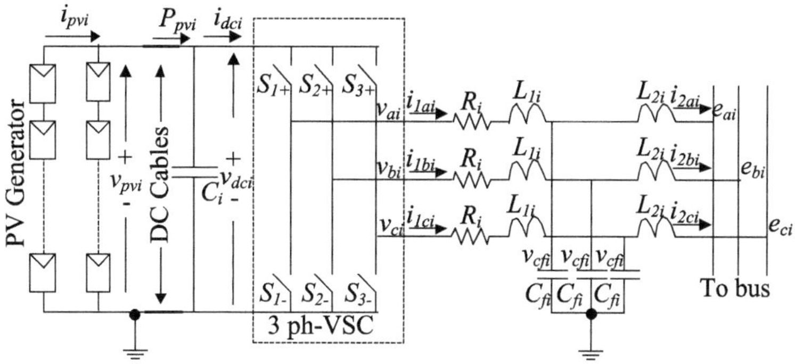

Fig. 2.12 PV system connected to the grid

$$\begin{aligned}
\dot{i}_{1a_i} &= -\frac{R_i}{L_{1_i}} i_{1a_i} - \frac{1}{L_{1_i}} e_{a_i} + \frac{v_{pv_i}}{3L_{1_i}} (2K_{a_i} - K_{b_i} - K_{c_i}) \\
\dot{i}_{1b_i} &= -\frac{R_i}{L_{1_i}} i_{1b_i} - \frac{1}{L_{1_i}} e_{b_i} + \frac{v_{pv_i}}{3L_{1_i}} (-K_{a_i} + 2K_{b_i} - K_{c_i}) \\
\dot{i}_{1c_i} &= -\frac{R_i}{L_{1_i}} i_{1c_i} - \frac{1}{L_{1_i}} e_{c_i} + \frac{v_{pv_i}}{3L_{1_i}} (-K_{a_i} - K_{b_i} + 2K_{c_i})
\end{aligned} \tag{2.41}$$

$$\begin{aligned}
\dot{v}_{cfa_i} &= \frac{1}{C_{f_i}} \left(i_{1a_i} - i_{2a_i}\right), \ \dot{v}_{cfb_i} = \frac{1}{C_{f_i}} \left(i_{1b_i} - i_{2b_i}\right) \\
\dot{v}_{cfc_i} &= \frac{1}{C_{f_i}} \left(i_{1c_i} - i_{2c_i}\right), \ \dot{i}_{2a_i} = \frac{1}{L_{2_i}} \left(v_{cfa_i} - e_{a_i}\right) \\
\dot{i}_{2b_i} &= \frac{1}{L_{2_i}} \left(v_{cfb_i} - e_{b_i}\right), \ \dot{i}_{2c_i} = \frac{1}{L_{2_i}} \left(v_{cfc_i} - e_{c_i}\right)
\end{aligned} \tag{2.42}$$

where K_{a_i}, K_{b_i}, and K_{c_i} are the binary input switching signals. By applying KCL at the node where the DC link is connected, we get

$$\dot{v}_{pv_i} = \frac{1}{C_i} \left(i_{pv_i} - i_{dc_i}\right). \tag{2.43}$$

The input current of the inverter i_{dc_i} can be written as [43]

$$i_{dc_i} = i_{a_i} K_{a_i} + i_{b_i} K_{b_i} + i_{c_i} K_{c_i}. \tag{2.44}$$

Now Eq. (2.43) can be rewritten as:

$$\dot{v}_{pv_i} = \frac{1}{C_i} i_{pv_i} - \frac{1}{C_i} \left(i_{a_i} K_{a_i} + i_{b_i} K_{b_i} + i_{c_i} K_{c_i}\right). \tag{2.45}$$

Equations (2.41) and (2.45) can be transformed into dq frame using the angular frequency ω_i of the grid as:

$$\begin{aligned}
L_{1_i}\dot{i}_{1\mathrm{d}_i} &= -R_i i_{1\mathrm{d}_i} + \omega_i L_{1_i} i_{1\mathrm{q}_i} - v_{\mathrm{cfd}_i} + K_{\mathrm{d}_i} v_{\mathrm{pv}_i} \\
L_{1_i}\dot{i}_{1\mathrm{q}_i} &= -R_i i_{1\mathrm{q}_i} - \omega_i L_{1_i} i_{1\mathrm{d}_i} - v_{\mathrm{cfq}_i} + K_{\mathrm{q}_i} v_{\mathrm{pv}_i} \\
L_{2_i}\dot{i}_{2\mathrm{d}_i} &= +\omega_i L_{2_i} i_{2\mathrm{q}_i} + v_{\mathrm{cfd}_i} - E_{\mathrm{d}_i} \\
L_{2_i}\dot{i}_{2\mathrm{q}_i} &= -\omega L_2 i_{2\mathrm{d}} + v_{\mathrm{cfq}} - E_{\mathrm{q}} \\
C_{\mathrm{f}_i}\dot{v}_{\mathrm{cfd}_i} &= \omega_i C_{\mathrm{f}_i} v_{\mathrm{cfq}_i} + C_{\mathrm{f}_i}\left(i_{1\mathrm{d}_i} - i_{2\mathrm{d}_i}\right) \\
C_{\mathrm{f}_i}\dot{v}_{\mathrm{cfq}_i} &= -\omega_i C_{\mathrm{f}_i} v_{\mathrm{cfd}_i} + C_{\mathrm{f}_i}\left(i_{1\mathrm{q}_i} - i_{2\mathrm{q}_i}\right) \\
C_i\dot{v}_{\mathrm{pv}_i} &= i_{\mathrm{pv}_i} - i_{1\mathrm{d}_i} K_{\mathrm{d}_i} - i_{1\mathrm{q}_i} K_{\mathrm{q}_i}
\end{aligned} \tag{2.46}$$

The synchronization scheme for $abc \to dq$ transformation is chosen such that the q-axis of the dq frame is aligned with the grid voltage vector, $E_{\mathrm{q}_i} = 0$, and the real and reactive power delivered to the grid can be written as $P_i = \frac{3}{2} E_{\mathrm{d}_i} I_{\mathrm{d}_i}$ and $Q_i = -\frac{3}{2} E_{\mathrm{d}_i} I_{\mathrm{q}_i}$.

2.10.17 Modeling of FACTS Devices

In general, FACTS devices can be utilised to increase the transmission capacity, the stability margin and dynamic behaviour and serve to ensure improved power quality. Their main capabilities are reactive power compensation, voltage control and power-flow control. Due to their controllable power electronics, FACTS devices always provide fast controllability in comparison with that of conventional devices, such as switched compensation or phase shifting transformers. Different control options provide high flexibility and lead to multi-functional devices.

Several kinds of FACTS devices have been developed and there are several years of documented evidence of their use in practice and research. Some of them, such as the thyristor based SVC, are widely applied technology; others, like the VSC-based STATCOMs or the VSC high voltage DC (HVDC) are being used in a growing number of installations worldwide. The most versatile FACTS devices, such as the unified power-flow controller (UFPC) are still confined primarily to research and development applications. In this book, we mainly use a STATCOM and, in few cases, SVC and thyristor-controlled switched capacitors (TCSCs).

2.10.18 STATCOM Model

The concept of the STATCOM was proposed by Gyugyi in 1976 [44]. A STATCOM is a shunt FACTS device which is mostly employed for controlling the voltage at

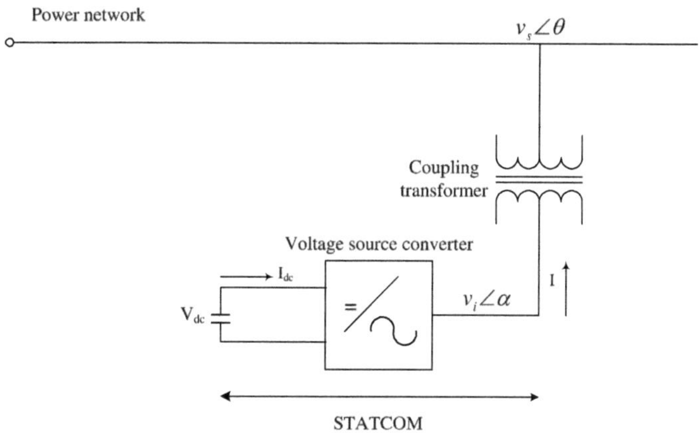

Fig. 2.13 Schematic diagram of VSC-based STATCOM

the point of connection to the network, as shown in Fig 2.13. In general, a STATCOM system consists of three main parts: a VSC; a coupling reactor or a step-up transformer; and a controller. The magnitude and phase of VSC's output voltage, V_i, can be regulated through the turn-on/turn-off of the VSC switches so that the VSC output current, I, can be controlled. Here, I is equal to the sum of V_i minus the voltage at an AC point of common coupling (PCC), V_s, divided by the impedance of the coupling reactor, X_s. In other words, the capacitive or inductive output currents of a STATCOM can be achieved through regulating the magnitude of V_i to be larger or smaller than the magnitude of V_s. Meanwhile, the phase of V_i is almost in phase with V_s but has a small phase-shift angle to compensate for the converter's internal loss, thereby keeping the system stable. Therefore, I can be controlled inherently and independently of V_s.

By controlling the magnitude and angle of its output voltage, i.e., $V_i\angle\alpha$, the STATCOM is able to control its active and reactive exchanges with the power system and, therefore, control the voltage at the PCC. The real and reactive power expressions are given by:

$$P = \frac{V_i V_s}{X}\sin(\alpha - \theta),$$

$$Q = \frac{V_i(V_i - V_s\cos(\alpha - \theta))}{X}.$$

The direction of the reactive power flow can now be determined by the magnitude of the inverter output voltage. For values of V_i larger than V_s, a STATCOM is in the capacitive mode and injects reactive power into the network while, for V_i values smaller than V_s, it is in the inductive mode and absorbs reactive power from the network. In a typical STATCOM with a capacitor as the DC link, the values of the active and reactive power depend on one another. However, a STATCOM connected

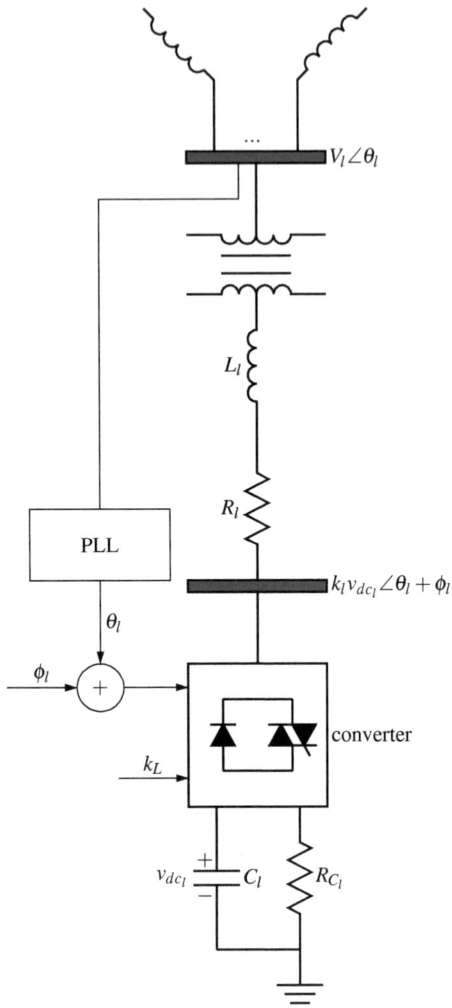

Fig. 2.14 Schematic diagram of STATCOM

to a battery energy storage system (STATCOM/BESS) is capable of controlling the values of both the active and reactive power independently [45].

For the purposes of stability studies, the STATCOM shown in Fig. 2.14 can be modelled as an AC voltage source with controllable magnitude and phase [46]. The dynamics of this voltage source are governed by the charging and discharging of a large (nonideal) capacitor. The capacitor, C_l, and its resistance, R_{C_l}, are shown in Fig. 2.14. The DC voltage across the capacitor is inverted and connected to an external bus via a short transmission line and a transformer bus. Details of various inverter schemes and how a controllable phase and magnitude are achieved are described in [44]. The phase-locked loop (PLL) block in Fig. 2.14 indicates that the phase shift of the inverter wave is adjusted with reference to the external bus voltage.

For the stability analysis we include the transformer and the transmission line (represented by R_l and L_l in Fig. 2.14) in the reduced impedance matrix. This directly interconnects the controllable inverter output with the rest of the system. The capacitor voltage can be adjusted by controlling the phase-angle difference between the line voltage, V_l, and the VSC voltage, E_l, ($E_l = k_l v_{dc_l} \angle \alpha_l$). If the phase angle of the line voltage is taken as a reference, the phase angle of the VSC voltage is the same as the firing angle, α_l, of the VSC. Thus, if the firing angles are slightly advanced, the DC voltage, v_{dc_l}, decreases and the reactive power flows into the STATCOM. Conversely, if the firing angles are slightly delayed, the DC voltage increases and the STATCOM supplies reactive power to the bus. By controlling the firing angle of the VSC, the reactive power can be generated from, or absorbed, by the STATCOM and, thus, voltage regulation can be achieved.The dynamics for l^{th} STATCOM can be described by the following equation:

$$\dot{v}_{dc_l}(t) = -\frac{P_{s_l}}{C_l v_{dc_l}} - \frac{v_{dc_l}}{R_{C_l} C_l}, \tag{2.47}$$

for $l = 1, \ldots, m$ where m is the number of STATCOMs, v_{dc_l} the capacitor voltage, C_l the DC capacitor, R_{C_l} the internal resistance of the capacitor, α_l the bus angle of the STATCOM in the reduced network, and P_{s_l} the power supplied by the system to the STATCOM to charge the capacitor which is given by:

$$\begin{aligned} P_{s_l} &= |E_l|^2 G_{ll} + \sum_{\substack{p=1 \\ p \neq l}}^{m} |E_l||E_p| \left[B_{lp} \sin \alpha_{pl} + G_{pl} \cos \alpha_{lp} \right], \\ &+ \sum_{\substack{j=1 \\ j \neq l}}^{n} |E_l||E_j'| \left[B_{lj} \sin(\delta_j - \alpha_l) + G_{lj} \cos(\delta_j - \alpha_l) \right], \end{aligned} \tag{2.48}$$

where G_{lp} and B_{lp} are the real and imaginary parts of the equivalent transfer impedances between the terminal buses of STATCOMs l and p and G_{lj} and B_{lj} are between the terminal buses of STATCOM l and IG j. The term E_j' denotes both E_{drj}' and E_{qrj}' and $\sin \alpha_{pl} = \sin(\alpha_p - \alpha_l)$.

The terminal voltage of STATCOMs is measured using a transducer with firs-order dynamic:

$$\dot{v}_{tm_l} = -\frac{v_{tm_l}}{T_{m_l}} + K_{m_l} v_{t_l}, \tag{2.49}$$

where v_{tm_l} is the sensor output, v_{t_l} the voltage at the connection point of STATCOM, K_{m_l} the constant and T_{m_l} the time constant of the voltage transducer. For the linear analysis we can assume that we know the equilibrium condition α_0 and control only $\Delta\alpha$ as shown in Fig. 2.15.

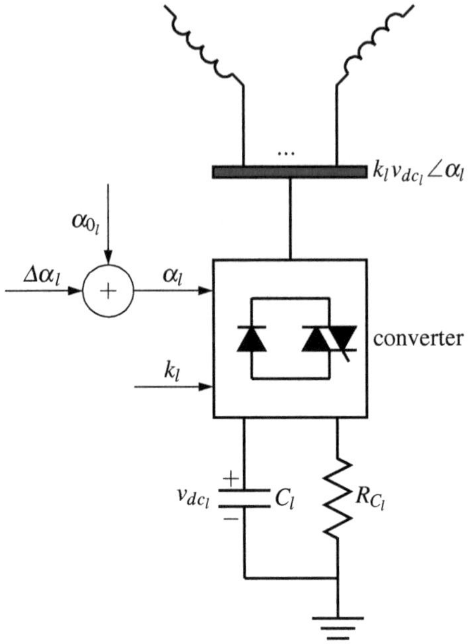

Fig. 2.15 STATCOM (equivalent)

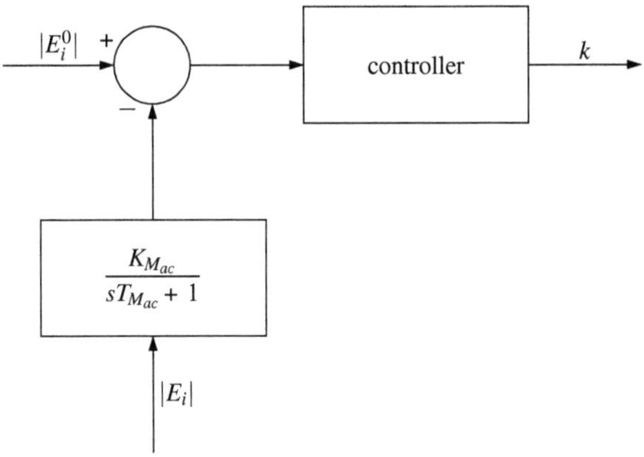

Fig. 2.16 Modulation index (k) control

The indirect conventional controllers for k and α are shown in Figs. 2.16 and 2.17. This gives an idea of the controller structure used by the industry. In this research, we will control Δk and $\Delta\alpha$ for the STATCOM in Fig. 2.15 directly instead of using conventional controllers.

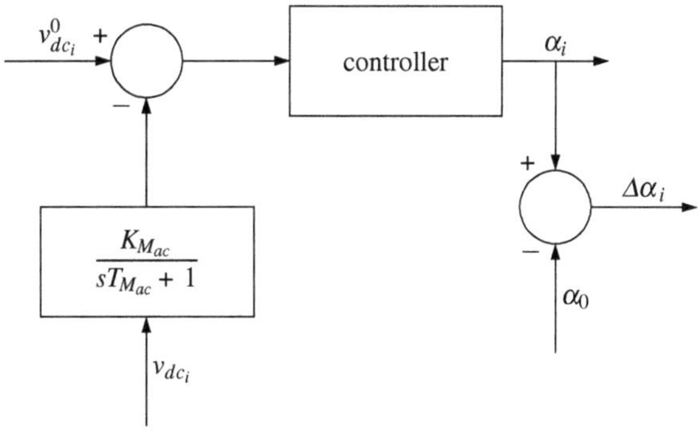

Fig. 2.17 Firing angle (α) control

2.10.19 SVC Modeling

By providing dynamic reactive power, SVC can be used for the purpose of regulating the system voltage, compensating the voltage at a reasonable level, improving the power flow capacity of a transmission line, enhancing the damping of low frequency oscillations as well as inhibiting the sub-synchronous oscillations. SVC is also capable of inhibiting the variation of busbar voltage caused by the fluctuating load, which is favourable for the recovery of transient voltage and the improvement of stabilisation of the system voltage.

For industrial users, it can effectively control the reactive power, improve the power factor, reduce the voltage influence and harmonic interference caused by the nonlinear load, balance the three-phase load, improve the power quality, productive efficiency and the product quality, and reduce the energy consumption. It is widely used in the machine, electric power, metallurgy, electrified railway, mine and wind power generation industries. The overall performance indicators are given below:

- SVC dynamic capacity: 0–400 MVAr;
- control-target bus rated voltage: 6–500 kV;
- total dynamic response time: (reactive power output): <15 ms; and
- SVC biggest loss: <0.8 %.

The SVC circuit contains the voltage measuring and voltage regulator circuits, outputs of which are fed into the thyristor firing control circuit. Normally, the susceptance of the SVC (B) is varied to maintain the mid-bus voltage, V_m, within its pre-specified tolerance. The supplementary stabilising signal is added to the output of the voltage regulator. The variation of the susceptance (B) can be related through the differential equation:

$$\Delta\dot{B}_l = \left[-\Delta B_l + B_{l0} + K_{c_l} V_{s_l}\right] / T_{c_l}, \tag{2.50}$$

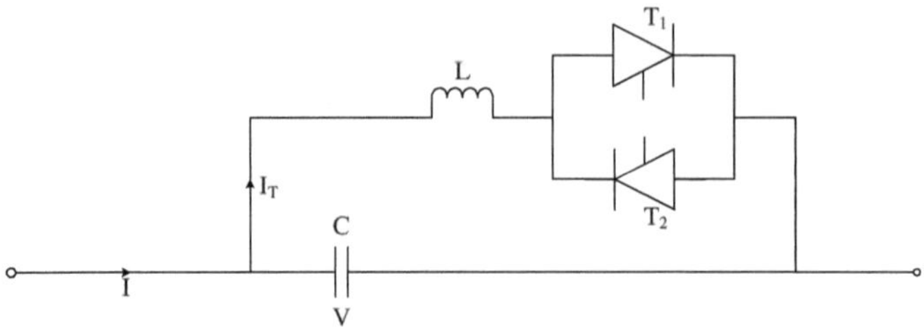

Fig. 2.18 One-line diagram of single-phase TCSC

for $l = 1, \ldots, m$, where m is the number of SVC, K_{C_l} and T_{C_l} the gain and time constants of the SVC's firing angle control circuit, respectively, and V_{s_l} the extra stabilising signal.

2.10.20 Thyristor Controlled Series Capacitor

After the introduction of TCSC in the late 1980s, they have been implemented in several locations around the world. As a controllable series compensation device and with flexible control possibilities, the TCSC has been found to be effective, especially in damping electromechanical and sub-synchronous oscillations. With a properly designed control system, a TCSC can be effectively utilised for enhancing both small signal and transient stabilities of a power system. Therefore, a good understanding of the interaction phenomena between a TCSC and its surrounding network will be necessary in order to design optimal control structures and, at the same time, prevent undesired interactions.

The single-line diagram of a TCSC is shown in Fig. 2.18. The operation of a TCSC involves discrete actions and is periodic in nature, whereby one of its anti-parallel thyristors of the TCSC is turned on during a portion of a half-cycle of the power frequency and is turned-off during the remainder of the cycle. The other anti-parallel thyristor repeats the conduction/non-conduction process during the next half cycle and vice versa. The duration and timing of thyristor conductions are based on the triggering logic and are controlled by the synchronisation system and the higher-level control loops. When a thyristor conducts, a circulating current flows in both the inductor and the capacitor which can either increase or decrease the voltage across the capacitor.

2.10.21 Energy Storage Device

The energy storage system (ESS), as an enabling infrastructure technology, provides ride-through over outages, improves profitability in high-energy applications, increases system reliability and dynamic stability, improves power quality and enhances transmission capacity of the transmission grid in a high power application [44]. For a high power application, the use of short-term (cycles to seconds) energy storage integrated with a power electronics-based controller, well known as a FACTS controller, could offer the following three distinct advantages:

- provide system damping, while maintaining constant voltage following a disturbance;
- provide additional damping in situations where the dynamic reactive power provided by traditional FACTS controllers with similar ratings is inadequate (alternatively, it could provide the same amount of damping at less cost. The damping of oscillation, by repeatedly interchanging small amounts of real power with the system, would be an excellent ESS application); and
- provide energy to maintain the speed of locally connected induction motors during a power system disturbance (This may prevent a voltage collapse in areas where there is a large concentration of induction motors that would otherwise stall).

While the superconductive magnetic energy storage (SMES) technology overcomes many barriers from a technology perspective, and has been commercially available in some specific sizes, the cost and relative complexity of the overall SMES system still makes it an expensive choice for a short-term ESS [47].

Recent advances in supercapacitor (SCAP) technology with an asymmetrical design have brought much excitement to the industry with the hope of a new and better short-term energy storage solution [48]. SCAP, as one type of electrochemical capacitor, stores electrical energy in the electrical double layer using relatively inexpensive materials, provides energy density thousands of times larger than that of conventional electrolytic capacitors, and has miscellaneous advantages over other high-power energy storage devices: high power density, low cost, reliable and long life cycle, fast and deep charge/discharge capability, wide range of operational temperatures, maintenance-free operation and storage, and environmental safety. These characteristics make SCAP a desirable energy storage element for short-term high-power ESS applications [49].

2.10.22 Network Power-Flow Model

The power balance equations pertaining to generator buses are given by:

$$V_i \cos(\delta_i - \theta_i) I_{q_i} - V_i \sin(\delta_i - \theta_i) I_{d_i} - S_{p_i} = 0, \tag{2.51}$$

$$-V_i \sin(\delta_i - \theta_i) I_{q_i} - V_i \cos(\delta_i - \theta_i) I_{d_i} - S_{q_i} = 0, \tag{2.52}$$

where,

$$S_{p_i} = \sum_{k=1}^{n} V_i V_k \left[G_{ik} \cos(\theta_i - \theta_k) + B_{ik} \sin(\theta_i - \theta_k) \right], \tag{2.53}$$

$$S_{q_i} = \sum_{k=1}^{n} V_i V_k \left[G_{ik} \sin(\theta_i - \theta_k) - B_{ik} \cos(\theta_i - \theta_k) \right], \tag{2.54}$$

for $i = 1, 2, \ldots, m$.

The power balance equations for non-generator buses are given by:

$$P_{L_i}(V_i) + \sum_{k=1}^{n} V_i V_k \left[G_{ik} \cos(\theta_i - \theta_k) + B_{ik} \sin(\theta_i - \theta_k) \right] = 0, \tag{2.55}$$

$$Q_{L_i}(V_i) + \sum_{k=1}^{n} V_i V_k \left[G_{ik} \sin(\theta_i - \theta_k) - B_{ik} \cos(\theta_i - \theta_k) \right] = 0, \tag{2.56}$$

for $i = m + 1, m + 2, \ldots, n$, where , n is the total number of buses in the system and $Y_{ik} = G_{ik} + jB_{ik}$ the element of the ith row and k^{th} column of the bus admittance matrix Y.

2.10.23 Power System Modeling

Power system modeling requires the modeling of all system components including, generators, transmission lines, transformers, loads and other control devices/systems, as discussed above. A complete power system modeling approach involves forming the overall system equations in the form of DAEs as:

$$\dot{x} = f(x, z, p), \tag{2.57}$$

$$0 = g(x, z, p), \tag{2.58}$$

where x is the vector of state variables, z the vector of algebraic variables and p the vector of system parameters. The differential equation set includes the dynamics of generators, excitation systems, load dynamics, and the algebraic equation set includes load flow equations and other algebraic relationship among the system components.

In power system modeling studies, the parameter values are chosen as either fixed values or within a certain range because the measurement of actual system parameters is very difficult. In particular, the load parameter values are difficult to obtain due to the large number of load components, the inaccessibility of certain customer loads, load compensation variations and the uncertainties of many load component characteristics.

2.11 Chapter Summary

The basic ideas about voltage instability and the importance of voltage instability analysis are explained from a fundamental as well as practical point of view in this chapter. The main focus of the chapter is to provide underlying causes of voltage instability, and the identification of different categories of stability behaviours that are important in power system stability analysis. The methods of improving voltage stability are also outlined.

In addition, this chapter discusses the dynamic modeling of a large power system. To provide a reliable model for implementation in a standard simulation tool, several factors must be taken into account. The first important process is to clearly define the purpose of the study. Each type of power system study requires a particular frequency bandwidth and a simulation time-frame depending on how fast the system dynamics needs to be investigated. Subsequently, the nature of the system being modelled must be carefully understood and the simulation tool used to simulate the models must be appropriately utilised.

Linear feedback control in power systems has a long history in terms of research and application. Linear controllers are preferred over nonlinear controllers because they impose lower requirements on practical implementation. However, as power system components are nonlinear in order to design a linear controller, linear models of a nonlinear system are needed. Different linearisation and modal analysis techniques will be discussed in the next chapter.

References

1. Vassell GS (1991) Northeast blackout of 1965. IEEE Power Eng Rev 11(1): 4–8
2. Lu W, Besanger Y, Zamai E, Radu D (2006) Blackouts: description, analysis and classification. In: 6th WSEAS international conference on power systems, pp 429–434
3. Custem TV, Vournas CD (1998) Voltage stability of the electric power systems. Kluwer Academic, Norwell
4. Berizzi A (2004) The Italian 2003 blackout. In: IEEE power engineering society general meeting, Denver, CO, pp 1673–1679
5. Ohno T, Imai S (2006) The 1987 Tokyo blackout. In: IEEE PES power systems conference and exposition, Atlanta, GA, pp 314–318
6. Andersson G, Donalck P, Farmer R, Hatziargyriou N, Kamwa I, Kundur P (2005) Causes of the 2003 major grids blackouts in North America and Europe, and recommendationed means to improve system dynamic performance. IEEE Trans Power Syst 20(4): 1922–1928
7. U.S. Canada Power System Outage Task Force (2004) Final report on the Aug 14 2003 blackout in the United States and Canada: causes and recommendations
8. Gao B, Morison GK, Kundur P (1996) Towards the development of a systematic approach for voltage stability assessment of large-scale power systems. IEEE Trans Power Sys 11(3): 1314–1324
9. Hossain MJ, Pota HR, Ugrinovski V (2008) Short and long-term dynamic voltage instability. In: 17th IFAC World Congress, Seoul, Korea, pp 9392–9397
10. IEEE (1990) Special publication 90TH0358-2-PWR: voltage stability of power systems: concepts, analytical tools, and industry experience

11. Vournas CD, Sauer PW, Pai MA (1996) Relationships between voltage and angle stability of power systems. Int J Electr Power Energy Syst 18(8): 493–500
12. Chi Y, Liu Y, Wang W, Dai H (2006) Voltage stability analysis of wind farm integration into transmission network. In: International conference on power system technology, pp 1–7
13. Lei Y, Mullane A, Lightbody G, Yacamini R (2006) Modeling of the wind turbine with a doubly-fed induction generator for grid integration studies. IEEE Trans Energy Convers 21(1): 257–264
14. Morison GK Gao B, Kundur P (1993) Voltage stability analysis using static and dynamic approaches. IEEE Trans Power Syst 8(3): 1159–1171
15. Taylor CW (1994) Power system voltage stability. McGraw-Hill, New York
16. Cutsem TV, Vournas C (1998) Voltage stability of electrical power system. Kluwer Academic, Norwell
17. Cutsem TV (2000) Voltage instability: phenomenon, countermeasures and analysis methods. Proc IEEE 88(2): 208–227
18. Kundur P (1994) Power system stability and control. Mcgraw-Hill, New York
19. Cutsem TV, Mailhot R (1997) Validation of a fast voltage stability analysis method on the Hydro-Quebec system. IEEE Trans Power Syst 12(1): 282–292
20. Ainsworth JD, Gavrilovic A, Thanawala HL (1980) Static and synchronous compensators for HVDC transmission convertors connected to weak AC systems. In: 28th session CIGRE, pp 31–41
21. Molinas M, Suul JA, Undeland T (2008) Low voltage ride through of wind farms with cage generators: STATCOM versus SVC. IEEE Trans Power Electr 23(3): 1104–1117
22. Molinas M, Suul JA, Undeland T (2006) Wind farms with increased transient stability margin provided by a STATCOM. In: IEEE 5th international conference on power electronics and motion control, pp 1–7
23. Pal B, Chaudhuri B (2005) Robust control in power systems. Springer, USA
24. Bergen AR (1986) Power system analysis. Prentice-Hall, New Jersey
25. Das JC, Casey J (1999) Effects of excitation controls on operation of synchronous motors. In: IEEE industrial and commercial power systems technical conference, Sparks, NV, pp 1–5
26. Klein M, Rogers GJ, Kundur P (1991) A fundamental study of inter-area oscillations in power systems. IEEE Trans Power Syst 6(3): 914–921
27. IEEE Committee Report (1981) Excitation system models for power system stability studies. IEEE Trans Power Apparatus Syst PAS 100(2): 494–509
28. Rogers G (2000) Power system oscillations. Kluwer Academic Publishers, Boston
29. Hill DJ (1993) Nonlinear dynamic load models with recovery for voltage stability studies. IEEE Trans Power Syst 8(1): 166–176
30. Ackermann T (2005) Wind power in power systems. Wiley, England
31. Li H, Chen Z (2008) Overview of different wind generator systems and their comparisons. IET Renew Power Gener 2(2):123–138
32. Morren J, de Haan SWH (2005) Ride-through of wind turbines with doubly-fed induction generator during a voltage dip. IEEE Trans Energy Convers 20(2): 435–441
33. Permitting of Wind Energy Facilities (2002) A handbook prepared by National Wind Coordinating Committee (NWCC) Siting Subcommittee, Aug 2002
34. Nandigam K, Chowdhury BH (2004) Power flow and stability models for induction generators used in wind turbines. In: IEEE power engineering society general meeting, Denver, CO, pp 2012–2016
35. Abdin ES, Xu W (2000) Control design and dynamic performance analysis of a wind turbine-induction generator unit. IEEE Trans Energy Convers 15(1): 91–96
36. Akhmatov V, Knudsen H (2002) An aggregate model of a grid-connected, large-scale, offshore wind farm for power stability investigations-importance of windmill mechanical system. Int J Electr Power Energy Syst 24(9): 709–717
37. Akhmatov V, Knudsen H, Nielsen A (2000) Advanced simulation of windmills in the electrical power supply. Int J Electr Power Energy Syst 22(6): 421–434

38. Feijo A, Cidrs J, Carrillo C (200) A third order model for the doubly-fed induction machine. Electric Power Syst Res 56(2): 121–127
39. Ledesma P, Usaola J, Rodriguez JL (2003) Transient stability of a fixed speed wind farm. Renew Energy 28(9): 1341–1355
40. Mahmud MA, Pota HR, Hossain MJ (201)2 Dynamic stability of three-phase grid-connected photovoltaic system using zero dynamic design approach. IEEE J Photovoltaics 12(4): 564–571
41. Mahmud MA, Pota HR, Hossain MJ (2011) Modeling guidelines and a benchmark for power system simulation studies of three-phase single-stage photovoltaic systems. IEEE Trans Power Deliv 26(2): 1247–1264
42. Tan YT, Kirschen DS, Jenkins N (2004) A model of PV generation suitable for stability analysis. IEEE Trans Energy Convers 19: 748–755
43. Nosrat A, Pearce, J.M.: Dispatch strategy and model for hybrid photovoltaic and tri-generation power systems. Appl Energy 88(9): 3270–3276
44. Hingorani MG, Gyugyi L (2000) Understanding FACTS: concepts and technology of flexible AC transmission systems. IEEE Press, New York
45. Divya KC, Ostergaard J (2009) Battery energy storage technology for power systems—an overview. Electric Power Syst Res 79(4): 511–520
46. Acha E, Fuerte-Esquivel CR, Ambriz-Perez H, Angles-Camacho C (2004) FACTS modeling and simulation in power networks. Wiley, London
47. Karasik V, Dixon K, Weber C, Batchelder B, Campbell G, Rebeiro P (1999) SMES for power utility applications: a review of technical and cost considerations. IEEE Trans Appl Supercond 9(2): 541–546
48. Lasseter RH, Jalali SG (1991) Power conditioning systems for superconductive magnetic energy storage. IEEE Trans Energy Convers 6(3): 381–387
49. Varakin IN, Klementov AD, Litvinenko SV, Starodubtsev NF, Stepanov AB (1997) Application of ultracapacitors as traction energy sources. In: Proceedings of 7th international seminar on double layer capacitors and similar energy storage devices, Florida, USA

Chapter 3
Linearisation and Modal Analysis

Abstract This chapter briefly reviews the conventional linearisation and modal analysis technique. The analytical tools that are commonly used in small-signal stability analyses are presented. The proposed linearisation technique using the mean-value theorem with Cauchys remainder is introduced and compared with the conventional linearisation technique.

3.1 Introduction

Many important components of a power system, such as conventional generators, wind generators, dynamic loads and FACTS devices, have nonlinear dynamics. The theory of nonlinear systems can be used to analyse these nonlinearities; however, its application is restricted to small and simple systems [1]. The concept of energy functions has been applied as a powerful tool to assess system security, stability limits and regions of attraction for the post-fault equilibrium state [2]. Suitable energy-like functions have been constructed and examined to see whether their values diminish with time in the post-disturbance period. Construction of the energy function is easy as long as the classical generator model is considered with a constant impedance load [1]. In the presence of large-order model complexities, such as excitation control, wind turbines, dynamic load, FACTS devices and a network with transfer conductances, suitable energy functions are difficult to obtain.

However, the theory of linear system analysis provides useful insights into the operating behaviour of an interconnected power system although the dynamic behaviour of the system must be assumed to be linear for such tools to be applicable. A better understanding of the nature of system dynamics helps to plan the control strategies necessary for the secure operation of the system. The linearisation technique is used throughout this book to gain an idea of the control problems which are at the heart of controller design. This chapter provides a general coverage of the conventional and proposed linearisation techniques and modal analysis used in this research.

J. Hossain and H. R. Pota, *Robust Control for Grid Voltage Stability: High Penetration of Renewable Energy*, Power Systems,
DOI: 10.1007/978-981-287-116-9_3, © Springer Science+Business Media Singapore 2014

3.2 Conventional Linearisation

Multi-machine power system dynamic behaviour is usually expressed as a set of nonlinear differential algebraic equations (DAEs). The algebraic equations result from the network power balance and generator stator current equations. The initial operating state of the algebraic variables, such as bus voltages and angles, are obtained through a standard power-flow solution. The initial values of the dynamic variables are obtained by solving the differential equations through the simple substitution of algebraic variables into the set of differential equations. The set of DAEs is then linearised around the equilibrium point.

A complete power system modeling approach involves forming the overall system equations in the form of DAEs, as:

$$\dot{\bar{x}} = f'(\bar{x}, u, z), \tag{3.1}$$

$$0 = g(\bar{x}, u, z), \tag{3.2}$$

$$y = h'(\bar{x}, u, z), \tag{3.3}$$

where f' and g are the vectors of differential and algebraic equations, respectively, and h' a vector of the output equations. The inputs are normally reference values, such as speed and voltage at individual units, and can be the voltage, reactance and power flow as set in FACTS devices. The output can be unit power output, bus frequency, bus voltage, line power or current, etc. The notations $\bar{x} \in R^n$, $z \in R^m$, $u \in R^p$, and $y \in R^q$ denote the vector of the state and algebraic variables, and the inputs and outputs, respectively. In power systems, the algebraic equation (3.2) is used to eliminate the variables in vector z. By representing $z = g^{-1}(\bar{x}, u)$, the overall power system model can be rewritten as:

$$\dot{x} = f(x, u), \tag{3.4}$$

$$y = h(x, u), \tag{3.5}$$

where

$$f(x, u) = f'(\bar{x}, u, g^{-1}(\bar{x}, u)). \tag{3.6}$$

Setting Eq. (3.4) equal to the zero vector, we get:

$$\dot{x} = f(x, u) = 0, \tag{3.7}$$

the system is said to be at rest or at an equilibrium point since all variables are constant and do not vary with time. Let x_0 be the state vector and u_0 the input vector corresponding to the system at rest, so that:

$$f(x_0, u_0) = 0. \tag{3.8}$$

Let assume x be a point which is different from x_0 by Δx:

$$x = x_0 + \Delta x, \tag{3.9}$$

$$u = u_0 + \Delta u, \tag{3.10}$$

in (3.7). The prefix Δ in Eqs. (3.9) and (3.10) denotes a small deviation. The new state (and every state) must satisfy Eq. (3.7). Hence:

$$\dot{x} = f(x + \Delta x, u + \Delta u). \tag{3.11}$$

By time-differentiating both sides of Eq. (3.9) we get

$$\dot{x} = \dot{x}_0 + \Delta\dot{x}. \tag{3.12}$$

From Eqs. (3.12) and (3.7), we get:

$$\dot{x}_0 + \Delta\dot{x} = f(x + \Delta x, u + \Delta u). \tag{3.13}$$

For small deviations, the non-linear function, $f(x, u)$, in equation (3.13) can be expressed in terms of a Taylor expansion. A Taylor expansion for a general scalar function, $f(x)$, as a function of one variable, x, in a close interval around x_0 is defined in [3] as:

$$f(x) = f(x_0) + \frac{f'(x_0)}{1!}(x - x_0) + \frac{f''(x_0)}{2!}(x - x_0)^2 + \cdots + \frac{f^n(x_0)}{n!}(x - x_0)^n + \cdots. \tag{3.14}$$

If we omit the second and higher order terms in expression (3.14), we get:

$$f(x) = f(x_0) + f'(x_0)(x - x_0). \tag{3.15}$$

The closer the interval we choose around x_0 as shown in Fig. 3.1, the better the linear approximation in (3.15).

So, neglecting the higher order terms, we can write for n order system, $i = 1, 2, \ldots, n$:

$$\dot{x}_{i0} + \Delta\dot{x}_i = f_i(x + \Delta x, u + \Delta u), \tag{3.16}$$

$$= f_i(x_0, u_0)) + \frac{\partial f_i}{\partial x_1}\Delta x_1 + \cdots + \frac{\partial f_i}{\partial x_n}\Delta x_n + \frac{\partial f_i}{\partial u_1}\Delta u_1 + \cdots + \frac{\partial f_i}{\partial u_n}\Delta u_n. \tag{3.17}$$

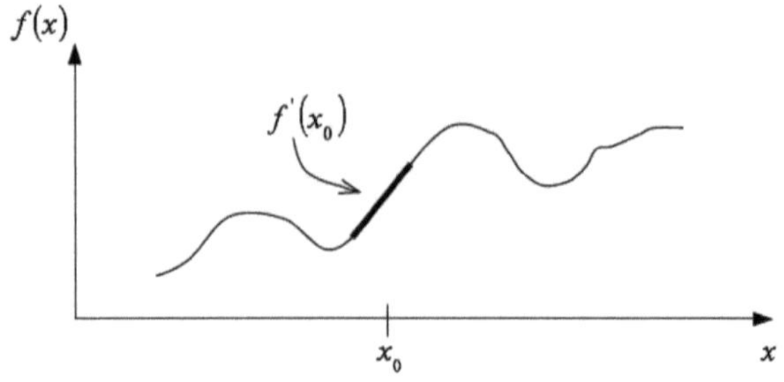

Fig. 3.1 An approximation of f(x) at (x, f(x))

Since $x_{i0} = f_i(x_0, u_0)) = 0$, we get:

$$\Delta \dot{x_i} = f_i(x + \Delta x, u + \Delta u), \tag{3.18}$$

$$= \frac{\partial f_i}{\partial x_1}\bigg|_{\substack{x=x_0\\u=u_0}} \Delta x_1 + \cdots + \frac{\partial f_i}{\partial x_n}\bigg|_{\substack{x=x_0\\u=u_0}} \Delta x_n + \frac{\partial f_i}{\partial u_1}\bigg|_{\substack{x=x_0\\u=u_0}} \Delta u_1 + \cdots + \frac{\partial f_i}{\partial u_n}\bigg|_{\substack{x=x_0\\u=u_0}} \Delta u_n. \tag{3.19}$$

Similarly, for output signal y_j, we can write:

$$\Delta y_j = \frac{\partial g_j}{\partial x_1}\bigg|_{\substack{x=x_0\\u=u_0}} \Delta x_1 + \cdots + \frac{\partial g_j}{\partial x_n}\bigg|_{\substack{x=x_0\\u=u_0}} \Delta x_n + \frac{\partial g_j}{\partial u_1}\bigg|_{\substack{x=x_0\\u=u_0}} \Delta u_1 + \cdots + \frac{\partial g_j}{\partial u_n}\bigg|_{\substack{x=x_0\\u=u_0}} \Delta u_n. \tag{3.20}$$

Finally, in matrix form, we can rewrite:

$$\Delta \dot{x} = A\Delta x + B\Delta u, \tag{3.21}$$

$$\Delta y = C\Delta x + D\Delta u, \tag{3.22}$$

where

$$A = \begin{bmatrix} \frac{\partial f_1}{\partial x_1} & \cdots & \frac{\partial f_1}{\partial x_n} \\ \vdots & \cdots & \vdots \\ \frac{\partial f_n}{\partial x_n} & \cdots & \frac{\partial f_n}{\partial x_n} \end{bmatrix}\Bigg|_{\substack{x=x_0\\u=u_0}}, \quad B = \begin{bmatrix} \frac{\partial f_1}{\partial u_1} & \cdots & \frac{\partial f_1}{\partial u_n} \\ \vdots & \cdots & \vdots \\ \frac{\partial f_n}{\partial u_n} & \cdots & \frac{\partial f_n}{\partial u_n} \end{bmatrix}\Bigg|_{\substack{x=x_0\\u=u_0}}, \tag{3.23}$$

$$C = \begin{bmatrix} \frac{\partial g_1}{\partial x_1} & \cdots & \frac{\partial g_1}{\partial x_n} \\ \vdots & \cdots & \vdots \\ \frac{\partial g_n}{\partial x_n} & \cdots & \frac{\partial g_n}{\partial x_n} \end{bmatrix}\Bigg|_{\substack{x=x_0\\u=u_0}}, \quad D = \begin{bmatrix} \frac{\partial g_1}{\partial u_1} & \cdots & \frac{\partial g_1}{\partial u_n} \\ \vdots & \cdots & \vdots \\ \frac{\partial g_n}{\partial u_n} & \cdots & \frac{\partial g_n}{\partial u_n} \end{bmatrix}\Bigg|_{\substack{x=x_0\\u=u_0}}, \tag{3.24}$$

where Δx is the state vector of dimension n, Δu the input vector of dimension r, Δy the output vector of dimension m, A the state matrix of size $n \times n$, B the input matrix of size $n \times r$, C the output matrix of size $m \times n$ and D the feed-forward matrix of size $m \times r$.

3.2.1 Linearisation by Perturbation

In many situations in which the partial derivatives are difficult or inconvenient to calculate, perturbation analysis can be used to obtain a linearised model. The (i, j)th element of matrix A in (3.21) can be obtained numerically as:

$$a_{ij} = \frac{f_i(x_0 + \varepsilon_j, u_0) - f_i(x_0, u_0)}{\varepsilon}, \tag{3.25}$$

where ε_j is a vector the same size as x with all its elements zero except that its jth element is a small number, ε. The (i, j)th element of matrix B can be obtained numerically as:

$$b_{ij} = \frac{f_i(x_0, u_0 + \varepsilon_j) - f_i(x_0, u_0)}{\varepsilon}, \tag{3.26}$$

where ε_j is a vector the same size as u with all its elements zero except for its jth element.

3.3 Proposed Linearisation

The method of conventional linearisation described above has been widely used to design controllers for power system stability. However, it has some serious limitations which cannot be ignored. As the operating point of the system drifts away from the selected equilibrium point, the state equations obtained by an approximate linearised method will not represent the original nonlinear system. It would not be surprising if the controllers, well-designed according to the mathematical model obtained through conventional linearisation, are only able to improve the system's stability under small disturbances, but reveal their inability to do so under large disturbances. Therefore, it would be desirable to have a robust controller that could ensure stability of the system for a wider operating region, comprising of operating conditions that could be much more distant from the equilibrium point than the ones covered by the conventional linearisation method.

In a typical robust control design the nominal system is considered with unmodeled dynamics in a feedback arrangement and the *size* of the unmodeled dynamics is used in robust control design. In the technique developed in this chapter, in addition to the linear term, Cauchy remainder is used which includes all the other terms. The Cauchy remainder term is an existence result and does not specify the point in the interval at which the system Jacobian needs to be evaluated. This term can also be linear but it has to be evaluated not at the system equilibrium point but on a point lying on the segment joining the equilibrium point and the current operating point. Clearly we do not know which point is this and it is not computationally feasible to evaluate that point at each integration step. Instead we obtain the largest value of the

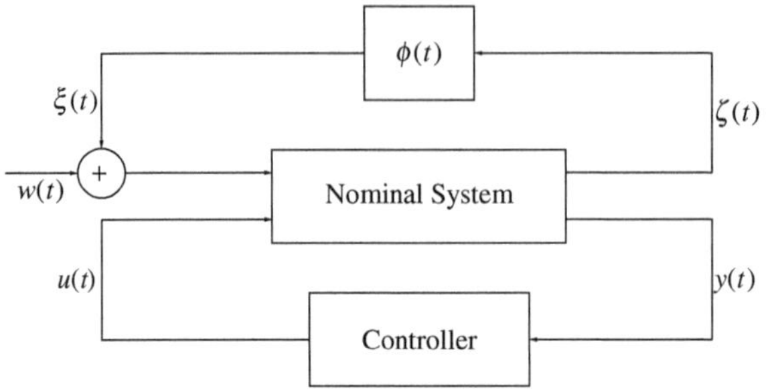

Fig. 3.2 Block diagram of uncertain system

size of the unmodeled dynamics, $\phi(t)$, shown in Fig. 3.2, over a region of interest (how to choose this region is the strength of our method) and use that in the design of the controller.

In this book, a robust linear control method is used to design stabilising control for large disturbances in power systems. To do so, the range of the validity of linearised models needs to be quantified. This can be done by the use of the Cauchy remainder formula for the Taylor series [4]. In this book, in the design of the linear controller, the Cauchy remainder is incorporated as an uncertain term in the robust control design process, thus, quantifying the deviations from the equilibrium point. Next, the mean-value theorem and linearisation process are described.

3.3.1 Mean-Value Theorem

Assume that $f\colon R^n \to R$ is continuously differentiable at each point, x, of an open set, $S \in R^n$. Let x and y be two points of S such that the line segment $L(x, y) \in S$. Then, there exists a point, z, of $L(x, y)$ such that [3]:

$$f(y) - f(x) = \left(\frac{\partial f}{\partial x}\right)'\Big|_{x=z}(y - x). \tag{3.27}$$

Theorem 1 *First, recall the Taylor series with a remainder. Let $f = f(x)$ be defined for $x \in R^n$. Then i) if $f(x)$ is continuously differentiable in the vicinity of x_0;*

$$f(x) = f(x_0) + \left(\frac{\partial f}{\partial x}\right)'\Big|_{x=\xi^*}(x - x_0). \tag{3.28}$$

(ii) if $f(x)$ has continuously iterated second partial derivatives in the vicinity of x_0;

$$f(x) = f(x_0) + \left(\frac{\partial f}{\partial x}(0)\right)'(x - x_0) + \frac{1}{2}(x - x_0)'\left(\frac{\partial^2 f}{\partial x^2}\right)\Big|_{x=\xi^*}(x - x_0); \quad (3.29)$$

for some $\xi^* \in R^n$ *such that* $\| \xi^* - x_0 \| < \| x - x_0 \|$*; here,* $\frac{\partial^2 f}{\partial x^2} = \left[\frac{\partial^2 f}{\partial x_i \partial x_j}\right]$ *is the Hessian of* f.

Proof Introduce the function $g(t) = f(tx + (1-t)x_0)$, $t \in [0, 1]$. This function has as many continuous derivatives in t as the number of continuous iterated partial derivatives of f. We will apply Taylor's theorem by stating that, if $g(t)$ is k times continuously differentiable, then:

$$g(t) = g(0) + \sum_{s=1}^{k} \frac{g^s(0)}{k!} t^k + R_k(t), \quad (3.30)$$

$$R_k(t) = \int_0^t \frac{(t-\tau)^k}{k!} g^{(k+1)}(\tau) d\tau = \frac{g^{(k+1)}(\lambda)}{(k+1)!} t^{k+1}, \quad (3.31)$$

for some $\lambda \in [0, t]$ and $R_k(t)$ is the remainder. The first (integral) expression is called the Cauchy remainder and the second the Lagrange remainder. Note that:

$$g'(t) = \left(\frac{\partial f}{\partial x}(tx + (1-t)x_0)\right)'(x - x_0), \quad (3.32)$$

$$g''(t) = \sum_{i=1}^{n}\sum_{j=1}^{n} \frac{\partial^2 f}{\partial x_i \partial x_j}(tx + (1-t)x_0)(x_i - x_{i0})(x_j - x_{j0}), \quad (3.33)$$

$$= (x - x_0)' \frac{\partial^2 f}{\partial x_i^2}(tx + (1-t)x_0)(x_i - x_{i0}). \quad (3.34)$$

(i) Let $k = 0, t = 1$ and define $\xi(\lambda) = \lambda x + (1-\lambda)x_0$ for each $\lambda \in [0, 1]$. Then, for some $\lambda^* \in [0, 1]$, we have:

$$f(x) = g(1), \quad (3.35)$$

$$= g(0) + g'(\lambda^*)t, \quad \text{(essentially, we apply the mean value theorem)} \quad (3.36)$$

$$= f(x_0) + \left(\frac{\partial f}{\partial x}(0)\right)'(x - x_0) + \frac{1}{2}(x - x_0)'\left(\frac{\partial^2 f}{\partial x^2}\right)\Big|_{x=\xi^*}(x - x_0), \quad (3.37)$$

where $\xi^* = \xi(\lambda^*)$. Note that $\|\xi^* - x_0\| = \|\lambda^*(x - x_0)\| \leq |\lambda^*|.\|x - x_0\| \leq \|x - x_0\|$, are required. (ii) In the same fashion, let $k = 1, t = 1$ and define $\xi(\lambda) = \lambda x + (1-\lambda)x_0$ for each $\lambda \in [0, 1]$. Then, for some $\lambda^* \in [0, 1]$, we have:

$$f(x) = g(1), \tag{3.38}$$

$$= g(0) + g'(0) + \frac{1}{2}g''(\lambda^*)t \text{ Taylor's formula for } k = 1,\ t = 1, \tag{3.39}$$

$$= f(x_0) + \left(\frac{\partial f}{\partial x}(0)\right)'(x - x_0) + \frac{1}{2}(x - x_0)'\left(\frac{\partial^2 f}{\partial x^2}\right)\Big|_{x=\xi^*}(x - x_0). \tag{3.40}$$

3.3.2 Reformulation Technique

Let (x_0, u_0) be an arbitrary point in the control space; using the mean-value theorem, the test system dynamics can be rewritten as [3, 5]:

$$\dot{x} = f(x_0, u_0) + L(x - x_0) + M(u - u_0), \tag{3.41}$$

where

$$L = \left[\frac{\partial f_1}{\partial x}\Big|_{\substack{x=x^{*1}\\u=u{*1}}}, \ldots, \frac{\partial f_n}{\partial x}\Big|_{\substack{x=x^{*n}\\u=u{*n}}}\right]^T, \quad \text{and } M = \left[\frac{\partial f_1}{\partial u}\Big|_{\substack{x=x^{*1}\\u=u{*1}}}, \ldots, \frac{\partial f_n}{\partial u}\Big|_{\substack{x=x^{*n}\\u=u{*n}}}\right]^T.$$

Here, (x^{*p}, u^{*p}) and $p = 1, \ldots, n$, denote points lying on the line segment connecting points (x, u) and (x_0, u_0), $f = [f_1, \ldots, f_n]^T$ denotes the vector function on the right-hand side of the differential equations used to represent the dynamics of the system. Equation (3.41) is an exact reformulation of the system equations. The nonlinearity of the system is captured through the nonlinear dependencies $x^{*p} = \phi(x, u, x_0, u_0)$ and $u^{*p} = \psi(x, u, x_0, u_0)$, $p = 1, \ldots, n$. It should be noted that the exact form of the functions ϕ and ψ are not available and, therefore, instead of the exact expressions for L and M in (3.41), their bounds are used in the control design.

Letting (x_0, u_0) be the equilibrium point about which the trajectory is to be stabilised and defining $\Delta x \triangleq x - x_0$ and $\Delta u \triangleq u - u_0$, it is possible to rewrite (3.41) as:

$$\Delta\dot{x} = \dot{x} - \dot{x_0},$$
$$= L(x - x_0) + M(u - u_0),$$
$$= \left[\frac{\partial f_1}{\partial x}\Big|_{\substack{x=x_0\\u=u_0}}, \ldots, \frac{\partial f_n}{\partial x}\Big|_{\substack{x=x_0\\u=u_0}}\right]^T \Delta x + \left(\left[\frac{\partial f_1}{\partial x}\Big|_{\substack{x=x^{*1}\\u=u^{*1}}}, \ldots, \frac{\partial f_n}{\partial x}\Big|_{\substack{x=x^{*n}\\u=u^{*n}}}\right]^T\right.$$
$$\left. - \left[\frac{\partial f_1}{\partial x}\Big|_{\substack{x=x_0\\u=u_0}}, \ldots, \frac{\partial f_n}{\partial x}\Big|_{\substack{x=x_0\\u=u_0}}\right]^T\right)\Delta x + \left[\frac{\partial f_1}{\partial u}\Big|_{\substack{x=x_0\\u=u_0}}, \ldots, \frac{\partial f_n}{\partial u}\Big|_{\substack{x=x_0\\u=u_0}}\right]^T \Delta u,$$

$$+\left(\left[\frac{\partial f_1}{\partial u}\bigg|_{\substack{x=x^{*1}\\u=u^{*1}}},\ldots,\frac{\partial f_n}{\partial u}\bigg|_{\substack{x=x^{*n}\\u=u^{*n}}}\right]^T-\left[\frac{\partial f_1}{\partial u}\bigg|_{\substack{x=x_0\\u=u_0}},\ldots,\frac{\partial f_n}{\partial u}\bigg|_{\substack{x=x_0\\u=u_0}}\right]^T\right)\Delta x$$

$$= A\Delta x + (L - A)\,\Delta x + B_1\Delta u + (M - B_1)\,\Delta u, \tag{3.42}$$

where $A = \frac{\partial f}{\partial x}\Big|_{\substack{x=x_0\\u=u_0}}$, $B_1 = \frac{\partial f}{\partial u}\Big|_{\substack{x=x_0\\u=u_0}}$, and Δx is the state vector. Since x^{*p} and $p = 1, \ldots, m$ are not known, it is difficult to obtain the exact value of $(L - A)$, but it is possible to obtain a bound on $\|(L - A)\|$.

In most systems $(L - A)\Delta x$ term does not depend on all the state variables but only on a subset, ξ, of the state vector, Δx. System (3.42) is shown as a block diagram in Fig. 3.2. We introduce a signal ξ such that

$$(L - A)\,\Delta x + (M - B_1)\,\Delta u = B_2\xi, \tag{3.43}$$

and

$$\xi = \tilde{\phi}(\tilde{C}_1\Delta x) + \tilde{\psi}(\tilde{D}_1\Delta u),\ \phi = (1/\sqrt{\beta})\left[\tilde{\phi}\ \ \tilde{\psi}\right], \tag{3.44}$$

where $\xi(t)$ is known as the uncertainty input and $\tilde{\phi}(t)$ is a uncertain gain matrix. In a typical robust control design the block $\phi(t)$ in Fig. 3.2 contains unmodeled dynamics and the size of the block $\phi(t)$ is used in control design. If the maximum value of $\phi(t)$ is evaluated over the entire region than the value is so large that the performance of the designed controller is limited. In this chapter the size of $\phi(t)$ is evaluated over the region of interest which is estimated from detailed simulations.

There are several choice for B_2 and $\tilde{C}_1$ such that the equality in (3.43) is satisfied. In general, x^{*p}, $p = 1, \ldots, m$, are not known beforehand, it is difficult to obtain the exact value of $(L - A)$, but it is possible to obtain a bound on $\tilde{\phi}$ and $\tilde{\psi}$ over the operating range and parameter β is chosen to ensure,

$$\|\phi\|^2 \leq 1, \tag{3.45}$$

where β is the scaling parameter. From this, we have

$$\|\xi(t)\|^2 \leq \beta\|\left(\tilde{C}_1\Delta x + \tilde{D}_1\Delta u\right)\|^2. \tag{3.46}$$

and we recover the norm bound constraints [6],

$$\|\xi(t)\|^2 \leq \|\zeta(t)\|^2. \tag{3.47}$$

The expressions for obtaining $\tilde{\phi}$ and $\tilde{\psi}$ can be determined for any power system model.

The system can now be rewritten as:

$$\Delta \dot{x} = A\Delta x + B_1 \Delta u + B_2 \xi. \tag{3.48}$$

The output matrix, C_2 depends on the measured output of a power system.

Finally the value of β is chosen such that the uncertainty, $\phi(t)$, shown in Fig. 3.2 satisfies:

$$\|\phi\|^2 \leq 1. \tag{3.49}$$

From this, we have:

$$\|\xi\|^2 \leq \beta \|\tilde{C}_1 \Delta x\|^2, \tag{3.50}$$

and we recover the following IQC (integral quadratic constraint) [7]:

$$\|\xi\|^2 \leq \|\varsigma\|^2. \tag{3.51}$$

To facilitate control design, the power system model is summarised as:

$$\Delta \dot{x}(t) = A\Delta x(t) + B_1 \Delta u(t) + B_2 \xi(t), \tag{3.52}$$

$$y(t) = C_2 \Delta x(t) + D_2 \xi(t) + D_2 w(t), \tag{3.53}$$

$$\varsigma(t) = C_1 \Delta x(t), \tag{3.54}$$

where ς is known as the uncertainty output, $y(t)$ the measured output and C_2 the output matrix.

Equations (3.52) to (3.54) provide a new representation of the power system model with one part being linear and another having higher-order terms. The new formulation presented in this section is used to design a robust controller for the nonlinear power system.

3.3.3 Application of the Proposed Technique to a Simple System

A single-line diagram of a single wind farm infinite bus system is shown in Fig. 3.3 and parameters for this system is given in Table 3.1. A simple block diagram of a wind turbine is shown in Fig. 3.4.

A wind farm and a STATCOM connected to an infinite bus, shown in Fig. 3.3, can be represented with the following equations [8, 9]:

$$\dot{\omega}_m = (1/2H_m)\left[T_{ae_i} - K_s\gamma - D_m\omega_m\right], \tag{3.55}$$

$$\dot{\omega}_G = (1/2H_G)\left[K_s\gamma - T_e - D_G\omega_G\right], \tag{3.56}$$

$$\dot{\gamma} = 2\pi f(\omega_m - (1/N_g)\omega_G) \tag{3.57}$$

$$\dot{s} = (1/2H_G)\left[T_m - T_e\right], \tag{3.58}$$

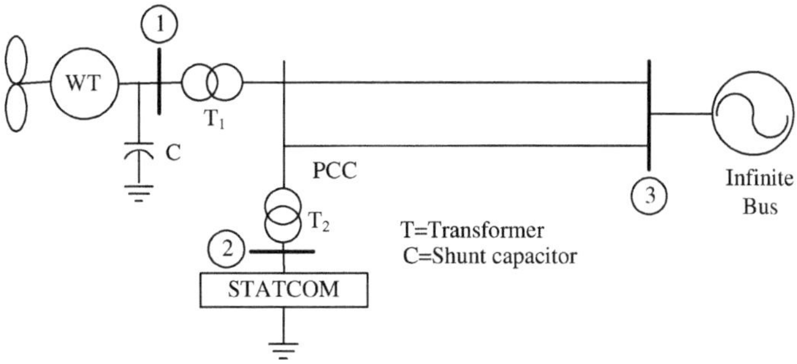

Fig. 3.3 Single wind farm infinite bus system

Table 3.1 Machines and grid parameters of single wind farm infinite bus system

Asynchronous machines	
Power: 2 MW	$R_s = 0.0121$ pu
Voltage: 690 V	$X_s = 0.0742$ pu
Frequency, $f = 50$ Hz	$X_m = 2.7626$ pu
Self damping, 0.008 pu	$R_r = 0.008$ pu
Rated slip: 0.02	$X_r = 0.1761$ pu
Two mass model	**STATCOM**
$H_m = 2.6$ s, $H_G = 0.22$ s	Capacity: 10 kVA
$D_m = 3$ pu, $K_s = 141$ pu	$R_C = 0.01$ pu
Gearbox ratio: 23.75	$C = 300$ μF
Grid and line	**Turbine parameters**
$x_g = 0.1126$ pu	Hub height: 30 m
$r_g = 0.01126$ pu	Rotor diameter: 23.2 m
$x_{13} = 0.75$ pu	Rated speed: 42 rpm
$r_{13} = 0.075$ pu	
Shunt compensator	**Load**
Capacitor: 25 kVAr	No load

$$\dot{E}'_{qr} = -(1/T'_o)\left[E'_{qr} - (X - X')i_{ds}\right] - s\omega_s E'_{dr}, \tag{3.59}$$

$$\dot{E}'_{dr} = -(1/T'_o)\left[E'_{dr} + (X - X')i_{qs}\right] + s\omega_s E'_{qr}, \tag{3.60}$$

$$\dot{v}_{dc}(t) = -P_s/Cv_{dc} - v_{dc}/(R_C C), \tag{3.61}$$

$$V_{ds} = R_s i_{ds} - X' i_{qs} + E'_{dr}, \tag{3.62}$$

$$V_{qs} = R_s i_{qs} + X' i_{ds} + E'_{qr}, \tag{3.63}$$

$$V_t = \sqrt{V_{ds}^2 + V_{qs}^2}, \tag{3.64}$$

where $X' = X_s + X_m X_r/(X_m + X_r)$ is the transient reactance, $X = X_s + X_m$ the rotor open-circuit reactance, $T'_o = (L_r + L_m)/R_r$ the transient open-circuit time

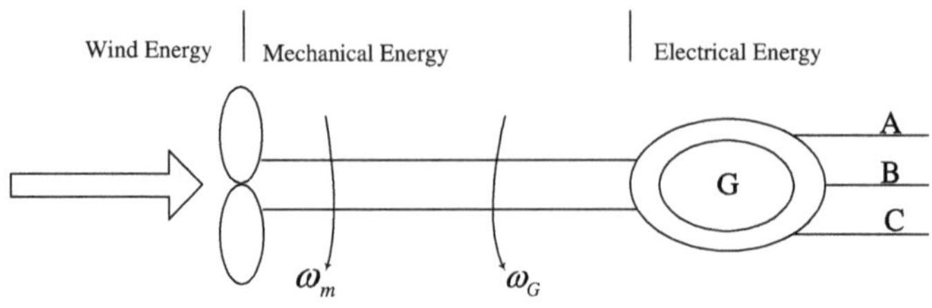

Fig. 3.4 Block diagram of a wind turbine

constant, V_t the terminal voltage of the IG, s the slip, E'_{dr} the direct-axis transient voltages, E'_{qr} the quadrature-axis transient voltages, V_{ds} the d-axis stator voltage, V_{qs} the q-axis stator voltage, T_m the mechanical torque, $T_e = E_{dr}i_{ds} + E_{qr}i_{qs}$ the electrical torque, X_s the stator reactance, X_r the rotor reactance, X_m the magnetizing reactance, R_s the stator resistance, R_r the rotor resistance, H_G the inertia constant of the IG, i_{ds} and i_{qs} the d- and q-axis components of the stator current, respectively, v_{dc} the capacitor voltage, C the DC capacitor, R_C the internal resistance of the capacitor and P_s the power supplied by the system to the STATCOM to charge the capacitor, given by

$$\begin{aligned} P_s = |E|^2 G_{22} + |E||V_\infty|\,[B_{23}\sin\alpha + G_{23}\cos\alpha] + |E||E'_{dr}|\,[B_{21}\sin(\delta-\alpha) \\ +G_{21}\cos(\delta-\alpha)] + |E||E'_{qr}|\,[B_{21}\cos(\delta-\alpha) - G_{21}\sin(\delta-\alpha)]\,. \end{aligned} \tag{3.65}$$

The d- and q-axes stator current are given by:

$$\begin{aligned} i_{ds} &= E'_{dr}G_{11} + E'_{qr}B_{11} - V_\infty(G_{13}\sin\delta + B_{13}\cos\delta) + kv_{dc}\,[G_{12}\sin(\alpha-\delta) + B_{13}\cos(\alpha-\delta)]\,, \\ i_{qs} &= E'_{dr}B_{11} + E'_{qr}G_{11} + V_\infty(G_{13}\cos\delta + B_{13}\sin\delta) + kv_{dc}\,[G_{12}\cos(\alpha-\delta) - B_{13}\sin(\alpha-\delta)]\,. \end{aligned}$$

The above system (3.55)–(3.61) can be rewritten in a compact form as follows:

$$\dot{x} = A_1 x + f_1(\zeta) + B_1 u \tag{3.66}$$

where A_1x is the linear part and $f_1(\zeta)$ the nonlinear part.

For this test system:

$$A_1 = \begin{bmatrix} 0 & 0 & 0 & 0 & 0 & 0 & 0 \\ 0 & -\frac{1-B_{11}(X-X')}{T'_o} & \frac{G_{11}(X-X')}{T'_o} & 0 & 0 & 0 & 0 \\ 0 & -\frac{B_{11}(X-X')}{T'_o} & -\frac{1+G_{11}(X-X')}{T'_o} & 0 & 0 & 0 & 0 \\ 0 & 0 & 0 & -\frac{D_m}{2H_m} & 0 & -\frac{K_s}{2H_m} & 0 \\ 0 & 0 & 0 & 0 & -D_G & 0 & 0 \\ 0 & 0 & 0 & 0 & 2\pi f & -\frac{1}{N_G} & 0 \\ 0 & 0 & 0 & 0 & 0 & 0 & -\frac{1}{R_c C} \end{bmatrix} \tag{3.67}$$

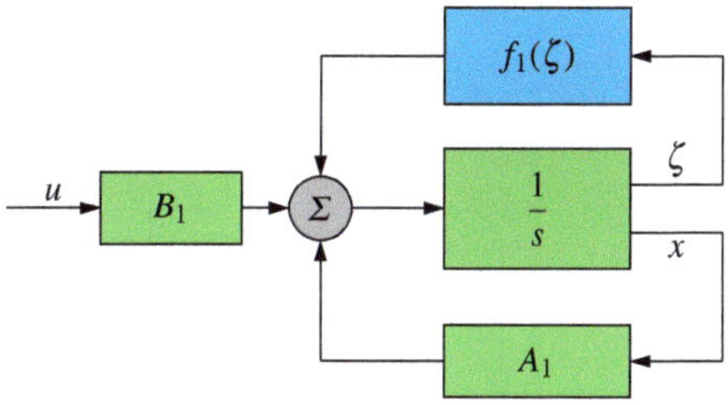

Fig. 3.5 Linear system with feedback nonlinearity

$x = [s, E'_{qr}, E'_{dr}, \omega_m, \omega_G, \gamma, v_{dc}]$; ζ is a subset of the state-vector x; and the vector $f_1(\zeta)$ includes all the nonlinear terms in Eqs. (3.55)–(3.61). The system (3.66) in terms of the block diagram is shown in Fig. 3.5. This representation can be used to either design nonlinear controllers [10, 11] or treat $f_1(\zeta)$ as modeling uncertainty and design linear robust controllers [12, 13].

In many linear control methods a nonlinear system is first linearised about an equilibrium point and this model is then used for controller design [14, 15]. In the approach presented in this book we quantify the region of the validity of the linear controller and also significantly reduce the conservativeness of the controllers designed using the formulation in (3.66).

The first step in the design of robust controller using the proposed method is to linearise the system in the regions given by x_0 and $\bar{x}$. The desired equilibrium point is x_0 and $\bar{x}$ is a vector made up of the end-points of the region of interest. The region defined by x_0 and $\bar{x}$ is a "polytope" region Ω with the centre at x_0 and corner points given by $\bar{x}$. The mean-value theorem is used to obtain the following linear model [3]:

$$\Delta\dot{x} = L\Delta x + B_1\Delta u \tag{3.68}$$

where L and B_1 are the state and control Jacobian matrices evaluated at a point in the region Ω.

The linearised system (3.68) is rewritten in the following form [16]:

$$\Delta\dot{x} = A\Delta x + (L - A)\Delta x + B_1\Delta u \tag{3.69}$$

where A is the Jacobian matrix evaluated at the equilibrium point x_0. The idea behind writing the system in the form (3.69) is to be able to use linear robust control methods where $\Delta\dot{x} = A\Delta x + B_1\Delta u$ is the system model and $(L - A)$ is the modeling uncertainty. The system (3.69) in terms of the block diagram is shown in Fig. 3.6, where

$$(L - A)\,\Delta x = B_2\phi(\zeta). \tag{3.70}$$

The system can now be written as

$$\Delta\dot{x} = A\Delta x + B_1\Delta u + B_2\phi(\zeta). \tag{3.71}$$

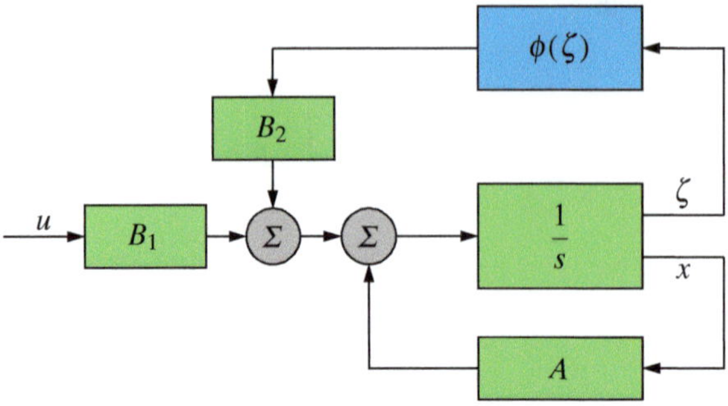

Fig. 3.6 Linearised system with Cauchy residue

For the wind farm system, with parameters in Table 3.1, the state Jacobian matrix around the desired equilibrium point $s_0 = 0.10$ pu, $E'_{dr_0} = 0.925$ pu, $E'_{qr_0} = 0.96$ pu, $\omega_{m_0} = 1.1$ pu, $\omega_{G_0} = 1.1$ pu, $\gamma_0 = 8.5°$, $v_{dc_0} = 1$ pu, is

$$A = \begin{bmatrix} 0 & 4.950 & 329.5 & 0 & 0 & 0 & -0.4491 \\ -4.978 & -19.72 & 89.96 & 0 & 0 & 0 & 6.595 \\ -3.381 & -0.01185 & -0.7888 & 0 & 0 & 0 & -0.3736 \\ 0 & 0 & 0.000 & -18.92 & 4.947 & -27.15 & 0 \\ 0 & 0 & -4.975 & -19.71 & -89.95 & 6.596 & 0 \\ 0 & 0 & 0.000 & -3.382 & 377 & -0.04 & 0 \\ 0 & -7.171 & -1.016 & -7.172 & 0 & 0 & -39.82 \end{bmatrix}. \tag{3.72}$$

Based on the simulation of various fault conditions we determine the region of operation Ω in which the controller needs to be effective. The region Ω is given by the corner points $\left[\bar{s}, \bar{E}'_{dr}, \bar{E}'_{qr}, \bar{\omega}_m, \bar{\omega}_G, \bar{\gamma}, \bar{v}_{dc}\right]^T$ and $\left[\underline{s}, \underline{E}'_{dr}, \underline{E}'_{qr}, \underline{\omega}_m, \underline{\omega}_G, \underline{\gamma}, \underline{v}_{dc}\right]^T$ which have the following values:

$\bar{s} = s_0 + 0.225$ pu, $\underline{s} = s_0 - 0.225$ pu, $\bar{E}'_{dr} = E'_{dr_0} + 0.242$ pu, $\underline{E}'_{dr} = E'_{dr_0} - 0.242$ pu, $\bar{E}'_{qr} = E'_{qr_0} + 0.225$ pu, $\underline{E}'_{qr} = E'_{qr_0} - 0.225$ pu, $\bar{\omega}_m = \omega_{m_0} + 0.395$ pu, $\underline{\omega}_m = \omega_{m_0} - 0.395$ pu, $\bar{\omega}_G = \omega_{G_0} + 0.328$ pu, $\underline{\omega}_G = \omega_{G_0} - 0.328$ pu, $\bar{\gamma} = \gamma_0 + 25°$, $\underline{\gamma} = \gamma_0 - 25°$, $\bar{v}_{dc} = v_{dc_0} + 0.334$ pu and $\underline{v}_{dc} = v_{dc_0} - 0.334$ pu.

The bounds on the unmodeled nonlinear parts in representation (3.66) and (3.69), over the region Ω are:

$$\|f_1(\zeta)\|^2 \le 1.75\|\zeta\|^2, \tag{3.73}$$

$$\|\phi(\zeta)\|^2 \le 0.69\|\zeta\|^2. \tag{3.74}$$

From the above two bounds it is clearly seen that the size of $\phi(\zeta)$ is about one-third of the size of $f_1(\zeta)$ and this is one of the reasons for obtaining high performance controllers using the formulation in this work. Also using this linearisation method it is certain that the designed controller has a validity over the entire region of interest Ω.

There do exist control design methods where the A matrix is evaluated at different operating points, e.g., for different load conditions, and the difference between the A matrices, commonly called ΔA, is used to design robust controllers [17–19]. The matrix ΔA is evaluated from end-point A matrices. The above bound for $\phi(\zeta)$ is obtained at a point interior to the region, i.e., $s^* = 0.175$ pu, $E_{dr}^{'*} = 0.825$ pu, $E_{qr}^{'*} = 0.765$ pu, $\omega_m^* = 1.28$ pu, $\omega_G^* = 1.325$ pu, $\gamma^* = 17.5°$, $v_{dc}^* = 0.725$ pu. This clearly indicates that one has to be very careful when using the usual ΔA for the purposes of the design of robust controllers.

3.4 Modal Analysis for Power Systems

In this section some basics of modal analysis, which are necessary to understand the controller design methods, are introduced. By linearising the nonlinear power system model, about an operating point, the total linearised system model can be represented by:

$$\Delta \dot{x} = A\Delta x + B\Delta u, \tag{3.75}$$

$$\Delta y = C\Delta x + D\Delta u, \tag{3.76}$$

where Δx the state vector of length equal to the number of states n, Δy the output vector of length m, Δu the input vector of length r, A the n by n state matrix, B the control or input matrix with dimensions n by the number of inputs r, C the output matrix of size m by n and D the feed forward matrix of dimensions m by r.

By taking the Laplace transform of the above equations, we obtain:

$$s\Delta x(s) - \Delta x(0) = A\Delta x(s) + B\Delta u(s), \tag{3.77}$$

$$\Delta y(s) = C\Delta x(s) + D\Delta u(s). \tag{3.78}$$

A formal solution of the state equations results in:

$$\Delta y(s) = C(sI - A)^{-1}\left[\Delta x(0) + B\Delta u(s)\right] + D\Delta u(s), \tag{3.79}$$

where I represents the identity matrix. The equation

$$\det(sI - A) = 0, \tag{3.80}$$

is referred to as the characteristic equation of matrix A and the values of s which satisfy the characteristic equation are the eigenvalues of matrix A. The natural modes of the system response are related to the eigenvalues. Analysis of the eigenproperties of A provides valuable information regarding the stability characteristics of the system [20].

The system matrix, A is a n by n matrix and has n solutions of the eigenvalues given as:

$$\lambda = \lambda_1, \lambda_2 \ldots, \lambda_n. \tag{3.81}$$

For any eigenvalue, λ_i, the n-column vector, ϕ_i, which satisfies (3.82), is called the right eigenvector of A associated with the eigenvalue λ_i [20] as:

$$A\phi_i = \lambda_i \phi_i. \tag{3.82}$$

Similarly, the n-row vector, ψ_i, which satisfies:

$$\psi_i A = \lambda_i \psi_i, \tag{3.83}$$

is called the left eigenvector associated with the eigenvalue λ_i.

Physically, the right eigenvector describes how each mode of oscillation is distributed among the systems states and is called the mode shape. The left eigenvector, together with the input coefficient matrix and the disturbance determines the amplitude of the mode in the time-domain solution for a particular case [20].

In order to express the eigenproperties of A succinctly, the modal matrices are also introduced:

$$\Phi = [\phi_1, \phi_2, \ldots, \phi_n], \tag{3.84}$$

$$\Psi = [\psi_1, \psi_2, \ldots, \psi_n]. \tag{3.85}$$

If we define a transformed vector, z, as $x = \Phi z$ (since $\Phi \Psi = I$, we have $z = \Psi x$) then, for ($u = 0$):

$$\dot{z} = \Phi^{-1} A \Phi z = \Lambda z. \tag{3.86}$$

This means $z_i(t) = e^{\lambda_i t} z(0) = e^{\lambda_i \psi_i x(0)}$ and, finally:

$$x(t) = \sum_{i=1}^{n} \phi_i \psi_i x(0) e^{\lambda_i t}. \tag{3.87}$$

- The ith element of $z(t)$ is called the ith mode of the system corresponding to the eigenvalue λ_i.
- The ith right eigenvector, ϕ_i, is the mode shape corresponding to the eigenvalue λ_i.
- The jth element of the left eigenvector, ψ_i, ψ_{ij}, gives the contribution of the jth state in the ith mode.

- For a complex eigenvalue, $\lambda_i = a_i + jb_i$ and its eigenvector, $\phi_i = U_i + jV_i$, we have:

$$AU_i = a_i U_i - b_i V_i, \tag{3.88}$$

$$AV_i = a_i V_i + b_i U_i. \tag{3.89}$$

3.5 Eigenvalue Sensitivity

Eigenvalue sensitivity is applied to determine the sensitivity of the eigenvalues to changes in the elements of the state matrix A. The sensitivity of the eigenvalue λ_i to the element a_{kj} of the state matrix is equal to the product of the left eigenvector element, ψ_{ik}, and the right eigenvector element, ϕ_{ji}:

$$A\phi_i = \lambda_i \phi_i, \tag{3.90}$$

$$\frac{\partial A}{\partial a_{kj}}\phi_i + A\frac{\partial \phi_i}{\partial a_{kj}} = \frac{\partial \lambda_i}{\partial a_{kj}}\phi_i + \lambda_i \frac{\partial \phi_i}{\partial a_{kj}}, \tag{3.91}$$

$$\psi_i \frac{\partial A}{\partial a_{kj}}\phi_i = \frac{\partial \lambda_i}{\partial a_{kj}}, \tag{3.92}$$

$$\frac{\partial \lambda_i}{\partial a_{kj}} = \psi_{ik}\phi_{ji}. \tag{3.93}$$

Since we know that $\phi_i \psi_i = I$ and $\psi_i(A - \lambda_i) = 0$ and, finally, only the (k, j)th element of A depends on a_{kj}.

From the above sensitivity formula, let's see what happens if we change the (k, k)th element of the A matrix, i.e., provide a feedback in the state equation for $\dot{x}_k$ from the state variable, x_k:

$$\frac{\partial \lambda_i}{\partial a_{kk}} = \psi_{ik}\phi_{ki} = P_{ki}. \tag{3.94}$$

This tells us the best way to change the ith mode is to apply a control to the state variable, k, such that p_{ki} is the largest participating factor.

3.6 Participation Matrix

The participation matrix, P, which combines the right and left eigenvectors as a measure of the association between the state variables and the modes, denotes the eigenvalue sensitivity with respect to the diagonal element of the state matrix.

The dynamics of the $x_j = \sum_{i=1}^{n} \phi_i \psi_i x(0) e^{\lambda_i t}$ state are made up of the dynamics of various modes and we see that ϕ_{ji} gives the participation of mode z_i in state x_j.

We have $z_i = \sum_{i=1}^{n} \psi_{ij} x_j$, which means that $\phi_{ji}\psi_{ij}$ is the participation of mode i in state x_j and vice versa. The participation matrix, P, is made up of elements

$$P_{ki} = \phi_{ki}\psi_{ki} = P_{ki}, \tag{3.95}$$

- p_{ki} is the participation factor of the ith state in the kth mode and vice versa.
- The states of a system are not unique; they can be scaled and eigenvectors may also be scaled by an arbitrary constant.
- The participation factors are scale independent.
- The participation factors are useful in the placement of exciters and STATCOMs.

3.7 Residues

The transfer function of a system with poles at $p_1, \ldots, p_n$ can be written as:

$$G_s = \frac{R_1}{(s - p_1)} + \frac{R_2}{(s - p_2)} + \cdots + \frac{R_n}{(s - p_n)}, \tag{3.96}$$

where $R_i, i = 1, \ldots, n$, are the system. Let us see what is the relationship between R_i and the modal matrices. Let the state-space representation of the same system be:

$$\Delta\dot{x} = A\Delta x + B\Delta u, \tag{3.97}$$

$$\Delta y = C\Delta x + D\Delta u. \tag{3.98}$$

Define $x = \phi z$, giving $\dot{z} = \phi^{-1}A\phi z + \phi^{-1}Bu$ and $y = C\phi z$.

$$G_s = C\phi(sI - A)^{-1}\phi^{-1}B = \sum_{i=1}^{n} \frac{C\psi_i\phi_i B}{s - p_i} \rightarrow R_i = C\psi_i\phi_i B. \tag{3.99}$$

Residues give the sensitivity of the corresponding eigenvalue to feedback of the transfer function output to its input. They are useful in finding the feedback signal which exerts the largest influence on the researched mode.

3.8 Bus Participation Factors, Eigenvalues and Voltage Stability

The bus participation factors give the areas associated with each mode (eigenvalue) of the system and the relative participation of a bus to a particular mode. This analysis is carried out at the maximum power transfer level to find weak points and areas susceptible to voltage instability based on the load flow equations rather than the dynamic equations. In order to assess the voltage stability margin and voltage weak points, modal analysis is performed.

The modal analysis involves the computation of the eigenvalues of a reduced system steady-state Jacobian matrix (J_R) which retains the Q–V relationship in the network. It should be noted that (J_R) represents the linearised relationship between the incremental changes in bus voltage magnitude and bus reactive power injection, and does not represent a dynamic system. A positive eigenvalue indicates that the modal voltage and modal reactive power are in the same direction and, thus, the system is voltage-stable. On the other hand, a negative eigenvalue indicates that the modal voltage and modal reactive power are in opposite direction and, thus, the system is voltage-unstable.

The linearised steady-state power voltage equations are given by [21]:

$$\begin{bmatrix} \Delta P \\ \Delta Q \end{bmatrix} = \begin{bmatrix} J_{P\Theta} & J_{PV} \\ J_{Q\Theta} & J_{QV} \end{bmatrix} \begin{bmatrix} \Delta\Theta \\ \Delta V \end{bmatrix}. \tag{3.100}$$

Let $\Delta P = 0$, then $\Delta Q = J_R \Delta V$ and $\Delta V = J_R^{-1} \Delta Q$ where $J_R = [J_{QV} - J_{Q\Theta} J_{P\Theta}^{-1} J_{PV}]$. Let

$$J_R = \zeta \Lambda \eta, \tag{3.101}$$

where ζ and η are the right left eigenvector, respectively, and $\Lambda = \text{diag}\{\lambda_1, \ldots, \lambda_n\}$ the eigenvalues of J_R so:

$$J_R^{-1} = \zeta \Lambda^{-1} \eta. \tag{3.102}$$

From (3.102) we have

$$\Delta V = \zeta \Lambda^{-1} \eta \Delta Q, \tag{3.103}$$

or

$$\Delta V = \sum_i \frac{\zeta_i \eta_i}{\lambda_i} \Delta Q, \tag{3.104}$$

where ζ_i is the ith column right eigenvector and η_i the ith row left eigenvector of J_R and, from the basic definition of left and right eigenvectors, ζ_i and η_i are orthogonal.

In this analysis, we wish to find the sensitivity to voltage of a change in reactive power consumption. Let a change in reactive power be such that:

$$\Delta Q_{mi} = K_i \zeta_i, \tag{3.105}$$

where the normalising constant, K_i is given by:

$$K_i^2 \sum_j \zeta_{ji}^2 = 1, \tag{3.106}$$

where ζ_{ji} is the jth element of the eigenvector ζ_i. The corresponding ith modal voltage variation is:

$$\Delta V_{mi} = \frac{1}{\lambda_i} \Delta Q_{mi}. \tag{3.107}$$

The smaller the smallest eigenvalue λ_i the more sensitive is the voltage to changes in reactive power.

Next, in (3.104), let $\Delta Q = \eta_k$ where η_k has all its elements as zero except for the kth which is 1. Then:

$$\Delta V = \sum_i \frac{\eta_{ik} \zeta_i}{\lambda_i}, \tag{3.108}$$

with η_{ik} is the kth element of the eigenvector η_i. The V–Q sensitivity of bus k:

$$\begin{aligned} \frac{\partial \Delta V_k}{\partial \Delta Q_k} &= \sum_i \frac{\zeta_{ki} \eta_{ik}}{\lambda_i}, \\ &= \sum_i \frac{p_{ki}}{\lambda_i}, \end{aligned} \tag{3.109}$$

where $p_{ki} = \zeta_{ki} \eta_{ik}$ are the bus participation factors.

Branch and generator participation factors are calculated for a given reactive power injection ΔQ_{mi}. Then $\Delta V_{mi} = J_R^{-1} \Delta Q_{mi}$ and $\Delta Q_{mi} = -J_{P\Theta}^{-1} J_{PV} \Delta V_{mi}$. This information can be used to calculate the reactive power flowing over all the transmission lines and supplied by the generators. The branch and generator participation factors comprise the ratio of the individual reactive power transmitted or generated divided by the maximum taken over the entire system [21].

3.9 Chapter Summary

Techniques for conventional and proposed linearisation and modal analysis techniques are presented in this chapter. Inspection of the eigenvalues of the state matrix provides sufficient information regarding the small-disturbance voltage stability of a power system in some neighbourhood of a given operating point. This will be used throughout this book to obtain an idea of the critical modes and participation factors required to design the proposed controllers.

Before designing a controller, it is essential to gain a clear idea as to how voltage instability occurs in a power system. The next chapter will present a number of possible voltage collapse mechanisms to provide a deeper insight into the dynamic mechanisms of the voltage instability phenomenon.

References

1. Pal B, Chaudhuri B (2005) Robust control in power systems. Springer, NewYork
2. Pavella M, Murthy PG (1994) Transient stability of power systems: theory and practice. Wiley, Chichester
3. Khalil HK (1992) Nonlinear systems. Prentice-Hall, New York
4. Leith DJ, Leithead WE (2000) Survey of gain-scheduling analysis and design. Int J Control 73(11):1001–1025
5. Hossain MJ, Pota HR, Mahmud MA, Ramos RA (2011) New approach of modelling power systems for robust control. In: IEEE power and energy society general meeting, Michigan, pp 1–8
6. Ugrinovskii VA, Petersen IR (2001) Minimax LQG control of stochastic partially observed uncertain systems. SIAM J Control Optim 40(4):1189–1226
7. Petersen IR, Ugrinovskii VA, Savkin AV (2000) Robust control design using H_∞ methods. Springer, London
8. Ackermann T (2005) Wind power in power systems. Wiley, London
9. Nandigam K, Chowdhury BH (2004) Power flow and stability models for induction generators used in wind turbines. In: IEEE power engineering society general meeting, Denver, CO, pp 2012–2016
10. Gao L, Chen L, Fan Y, Ma H (1992) A nonlinear control design for power systems. Automatica 28(5):975–979
11. Wang Y, Guo G, Hill DJ (1997) Robust decentralized nonlinear controller design for multi-machine power systems. Automatica 33(9):1725–1733
12. Kwang MS, Park JK (2000) On the robust LQG control of TCSC for damping power system oscillations. IEEE Trans Power Syst 15(4):1306–1312
13. Boukarim GE, Wang S, Chow JH, Taranto GN, Martins N (2000) A comparison of classical, robust, and decentralized control designs for multiple power system stabilizers. IEEE Trans Power Syst 15(4):1287–1292
14. Rao P, Crow ML, Yang Z (2000) STATCOM control for power system voltage control applications. IEEE Trans Power Delivery 15(4):1311–1317
15. Ko HS, Lee KY, Kim HC (2004) An intelligent based LQR controller design to power system stabilization. Electric Power Syst Res 71(1):1–9
16. Hossain MJ, Pota HR, Ugrinovski V, Ramos RA (2009) Robust STATCOM control for the enhancement of fault ride-through capability of fixed-speed wind generators. In: IEEE multi-conference on systems and control. Saint Petersburg, pp 1505–1510
17. Ramos RA, Martins ACP, Bretas NG (2005) An improved methodology for the design of power system damping controllers. IEEE Trans Power Syst 20(4):1938–1945
18. Ramos RA, Alberto LFC, Bretas NG (2004) A new methodology for the coordinated design of robust decentralized power system damping controllers. IEEE Trans Power Syst 19(1):444–454
19. Li L, Ugrinovskii VA, Orsi R (2007) Decentralized robust control of uncertain markov jump parameter systems via output feedback. Automatica 43(11):1932–1944
20. Kundur P (1994) Power system stability and control. Mcgraw-Hill, New York
21. Klein M, Rogers GJ, Moorty S, Kundur P (1992) Analytical investigation of factors influencing power system stabilizers performance. IEEE Trans Energy Convers 7(3):383–390

Chapter 4
Dynamic Voltage Instability Analysis with Wind Generators and FACTS Devices

Abstract The complexity of power systems has increased in recent years due to the operation of existing transmission lines closer to their limits, using flexible AC transmission system devices (FACTS), and also due to the increased penetration of new types of generators that have more intermittent characteristics and lower inertial response, such as wind generators. This changing nature of a power system has considerable effect on its dynamic behaviours resulting in power swings, dynamic interactions between different power system devices and less synchronized coupling. This chapter presents some analyses of this changing nature of power systems and their dynamic behaviours to identify critical issues that limit the large-scale integration of wind generators and FACTS devices. In addition, this chapter addresses some general concerns towards high compensations in different grid topologies. The studies in this chapter are conducted on the New England and New York power system model under both small and large disturbances. From the analyses, it can be concluded that high compensation can reduce the security limits under certain operating conditions, and the modes related to operating slip and shaft stiffness are critical as they may limit the large-scale integration of wind generation.

4.1 Introduction

Power systems are complex systems that evolve in response to economic growth and continuously increasing power demands. With growing population and the industrialisation of the developing world, more energy is required to satisfy basic needs and to attain improved standards of human welfare [1]. The structure of the modern power system is becoming highly complex in order to make energy available economically with reduced carbon emissions and the use of renewable energy.

The challenges of providing reliable and efficient supplies of electricity to residential and commercial users in the digital age are great. Regulatory uncertainty, cost, and lengthy delays in transmission line construction are just a few of the reasons

J. Hossain and H. R. Pota, *Robust Control for Grid Voltage Stability: High Penetration of Renewable Energy*, Power Systems,
DOI: 10.1007/978-981-287-116-9_4,

for the serious deficiencies in power transmission capacities currently prevailing in many countries [2]. Solving these issues requires innovative thinking. Increasing numbers of electricity stake-holders now recognise that low environmental impact technologies, such as flexible AC transmission system (FACTS) devices and DC links, are appropriate for enhancing reliability and upgrading transmission capacity on a long-term and cost-effective basis [2].

In recent years, power demand has increased substantially while the expansion of power transmission lines has been severely limited due to inadequate resources and environmental restrictions. As a consequence, some transmission lines are heavily loaded and system stability becomes a power transfer-limiting factor. FACTS controllers have been used to solve various power system steady-state control problems, therefore, enhancing power system stability in addition to their main function of power-flow control [3].

Dynamic reactive devices, such as thyristor-controlled series capacitors (TCSCs), mechanically switched capacitors (MSCs), static VAr (volt-ampere-reactive) compensators (SVCs) and static synchronous compensators (STATCOMs) do not require extensive amounts of land nor are they especially visible when compared with major new EHV (extra-high voltage) transmission lines [4]. These characteristics make them much more acceptable to government agencies and the public and, thus, to transmission system operators who are responsible for providing reliable electricity delivery services.

In many cases, these dynamic reactive devices have become less expensive to build than the equivalent number of new transmission lines that may otherwise be necessary. Employed in moderation, such devices are useful additions to the set of tools that system planners and operators should use to relieve voltage or VAr problems and provide flexibility. However, if used to excess, such devices will likely increase the risk of uncontrolled system collapse and significantly increase the complexity of system design and operation. This complexity may introduce new failure modes into the system and reduce its overall reliability in unexpected ways [5].

Following the issuance of the renewable energy regulations in recent years to give impetus to the development of renewable energy by governments in *Denmark, Germany, USA, China, Ireland, Australia* and *India*, a large number of wind farms are currently interconnected into transmission networks at the 220 kV voltage level with higher installed capacities. Being connected to a higher voltage level, their impact is becoming more widespread. The European Wind Energy Association (EWEA) projects that there will be 230 GW and 300 GW of total installed wind power capacity in *Europe* in 2020 and 2030, respectively. This will result in wind power generation of the same order of magnitude as the contributions from conventional technologies developed over the past century. An overview of the historical development of wind energy technology and the current world-wide status of grid-connected, as well as stand-alone, wind power generation is given in [6]. The present and progressive scale of integration has brought to a head serious concern about the impact of such a scale of wind penetration on the future safety, stability, reliability and security of the electricity systems [7].

Voltage instability is also influenced due to the integration of induction generators into grids. There are several technical constraints, including steady-state or dynamic stability, that may limit wind power integration into a power system. A majority of large wind farms, including proposed large wind projects, are geographically far away from load centres and connected into relatively weak transmission networks [8]. The presence of wind farms in such weak transmission networks raises serious concerns about system security and stability. Concerns regarding power system utilities are shifting focus from power quality issues to stability problems caused by wind power integration. In the grid impact studies of wind power integration, the voltage stability issue is a key problem because a large proportion of existing wind farms are based on fixed-speed wind turbines (FSWTs) equipped with simple squirrel-cage induction generators (SCIGs) [9]. The impact of wind turbines in the voltage profile of distribution systems is discussed in [10].

SCIGs consume reactive power and behave similarly to IMs during a system contingency and deteriorate the local grid voltage stability [11]. Presently, variable-speed wind turbines (VSWTs) equipped with doubly-fed induction generators (DFIGs) are becoming more widely used due to their reactive power and voltage control capability. DFIGs make use of power electronic converters and are, thus, able to regulate their own reactive power in order to operate at a given power factor or to control grid voltage. However, because of the limited capacity of the PWM (pulse-width modulation) converter, the voltage control capability of a DFIG cannot match that of a synchronous generator. When the voltage control requirement is beyond the capability of a DFIG, the voltage stability of a grid is also affected.

The application of FACTS devices to confront some major operating problems in voltage regulation, power-flow control, transient stability and the damping of power oscillations have been investigated widely in recent years [3, 12–14]. The objective of the current research is to investigate the applications, benefits and locations of FACTS devices and to meet new challenges, such as the increasing penetration of renewable energy sources, in particular wind generation, growing demands, limited resources and competitive electricity markets. However, the standard power system FACTS controllers are local non-coordinated linear controllers [15]. Possible interactions of these local controllers might lead to adverse effects and cause inappropriate control efforts by the different controllers. As the density of FACTS devices has increased rapidly, it is necessary to take into account interactions amongst controllers in future power systems [15].

Recently, a lot of work has been done on analysing the effects of the large-scale integration of wind generators on dynamic stability. The effect of wind power on oscillations and damping has been investigated by gradually replacing the power generated by synchronous generators in a system by power from either constant or variable-speed wind generators [16, 17]. Only the impact on electromechanical modes are investigated in [16] despite the voltage modes also being influenced by the penetration of wind generators. The limits for voltage stability at different wind power integration levels and grid alternatives are illustrated in [9] for situations both with and without extra stabilising controls in the system and for different characteristics of wind turbine generators. Detailed dynamic load modeling and fault analysis are not treated in [9]. However, load characteristics are closely related to voltage instability [18, 19].

The issues of interconnecting large wind parks to an electrical power network are discussed and possible solutions to anticipated problems using AC and DC transmission technologies and FACTS devices with energy storage are presented in [20]. The impact of grid-connected large DFIG-based wind farms on power system transient stability is studied in detail, including on a full generator model, as there is a lack of this generator model in the most common transient simulation software packages [21]. It is shown in [21] that power system transient stability can be improved to some extent when the specified synchronous generator is replaced by a wind farm of the DFIG type. The effect of short-circuiting the rotor when the fault current exceeds the converter rating is not considered in this chapter [21]. However, during large disturbances, DFIGs behave as conventional squirrel-cage induction generators (SCIGs) with an increased rotor resistance [22].

The challenge of accommodating increasingly larger amounts of wind energy in a system and its impact on system operation is discussed in [23] while those regarding the incorporation of dispersed power generation, and particularly large offshore wind power, into the Danish power system are presented in [24]. The latter article focuses mainly on investigations into short-term voltage stability as one of the main concerns has been to evaluate the response of a power grid to a short-circuit fault in the transmission system. Technical issues relating to the integration of large wind power into weak grids with long transmission lines, together with the most significant challenges for wind generation facilities, including voltage control, reactive power management, dynamic power-swing stability and behaviour following disturbances in a power grid, are addressed in [25]. It is shown in [25] that the decoupled characteristics of variable wind power plants with grids ensure stability performances that can exceed those of conventional synchronous generations with the same ratings and installed at the same locations.

Large wind generation facilities are evolving to look more and more like conventional generating plants in terms of their abilities to interact with a transmission network, other generating units and FACTS devices in a way that does not compromise performance or system reliability. But many challenges and new breakthroughs are needed before their use becomes a reality.

The high penetration of wind power is foreseen in many countries and regions globally. FACTS devices will mainly be used to integrate the large-scale wind energy into power systems. The dynamics of power systems is likely to be dominated by the dynamics of wind turbines and FACTS devices in the near future. Therefore, the impact of wind power and FACTS devices on power system stability and reliability needs to be thoroughly analysed before they are integrated into existing power system networks. This chapter will analyse the changing nature of systems and their dynamic behaviours to identify future issues that need to be either resolved or to have limits defined to a degree that can be tolerated. The following cases are considered in this chapter:

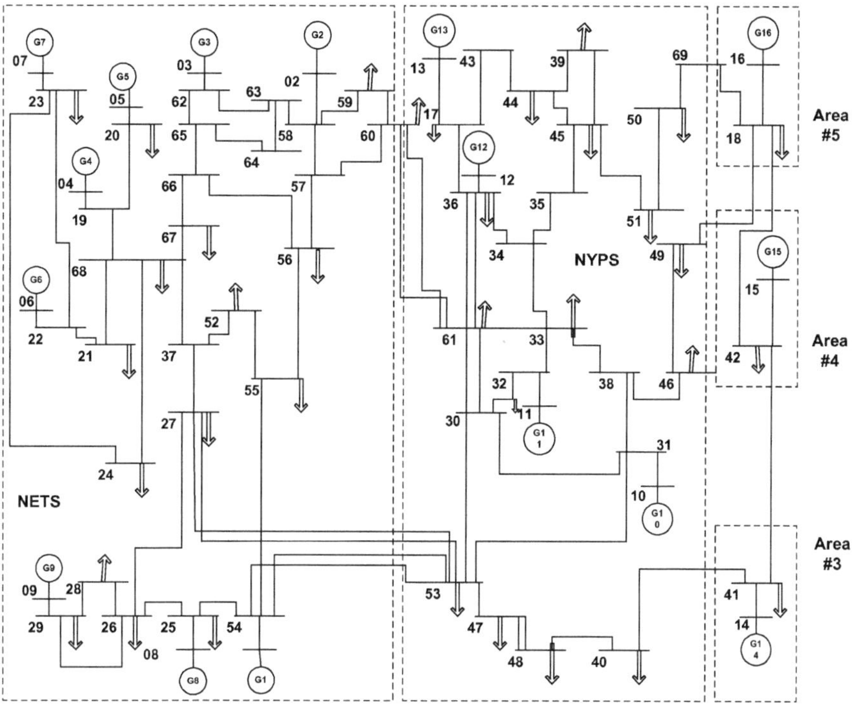

Fig. 4.1 16 machine 69 bus test system

(i) the possible effects of higher FACTS density on the dynamic performance of a distributed power system;
(ii) the critical interactions amongst FACTS devices; and
(iii) whether there is a level of wind generation capacity and FACTS devices above which the system dynamic behaviour is unstable and has poor controllability.

4.2 Case Studies

A 16-machine, 69-bus system is considered in this section [26]. The single line diagram of the system is shown in Fig. 4.1 and its parameters are given in Appendix-X. This is the reduced order equivalent of the interconnected New England (NETS) and the New York power system (NYPS). There are five geographical regions, of which NETS and NYPS are represented by a group of generators whereas imports from each of the other three neighbouring areas 3, 4 and 5, are approximated by generator equivalent models. The generators, loads and imports from other neighbouring areas are representative of operating conditions in the early 1970s. The total load on the system is $P_L = 17,620.65$ MW, $Q_L = 1,971.76$ MVAr and generation

$P_G = 18,408.00$ MW. The line loss in the system is 152.2 MW. Generators G_1 and G_9 are the equivalent representations of the NETS generators whilst machines G_{10} to G_{13} represent those of the NYPS. Generators G_{14} to G_{16} are the dynamic equivalents of the three neighbouring areas connected to the NYPS. There are three major transmission corridors between the NETS and the NYPS connecting buses 60–61, 53–54 and 27–53. All these corridors have double-circuit tie-lines. In steady-state, the tie-line power exchange between the NETS and the NYPS is 700 MW in total. The NYPS is required to import 1,500 MW from area 5.

All the generators of the test system (G_1 to G_{16}) are represented by a sub-transient model [26]. The mechanical input power to the generators is assumed to be constant during the disturbance. Generators G_1 to G_8 are equipped with slow excitation systems (IEEE-DC1A) whilst G_9 is equipped with a fast-acting static excitation system (IEEE-ST1A) [26] which has a PSS to provide supplementary damping control for the local modes. The rest of the generators are under manual excitation control [26]. The load is modelled as (i) 20 % large IM load [19], (ii) 25 % small IM load [19], and (iii) 55 % static load. The active components of static loads were represented by constant current models and the reactive components by constant impedance models, as recommended in [27] for dynamic simulations.

In this section, FSWTs and DFIGs are considered for analysis. The modeling of IGs for power flow and dynamic analysis is discussed in [28, 29]. A general model for representation of VSWTs in power system dynamics simulations is presented in [30]. Among different FACTS devices, an SVC [26], a TCSC [26] and a STATCM [11] are used in this section.

4.2.1 Case Studies

A number of cases are considered to get a deeper insight into these complex issues.

4.2.1.1 Load Area Fed by High Import

A large amount of compensation is needed within the load area of this test system. This is due to the high demand for reactive power by the loads, the need for a lot of reactive power by the feeding lines and the lack of generation units within the load area which could deliver reactive power as a by-product. A large amount of reactive compensation results in a P–V curve that is flat to a certain point from there it falls very steeply whereas in a moderately compensated system this P–V curve has a slow gradient and it curves towards the critical point. The effect is that the voltages are on a good level for a long time and nobody can see a problem. However, in reality the security margin for voltage instability is very small and a small event can pull the system down.

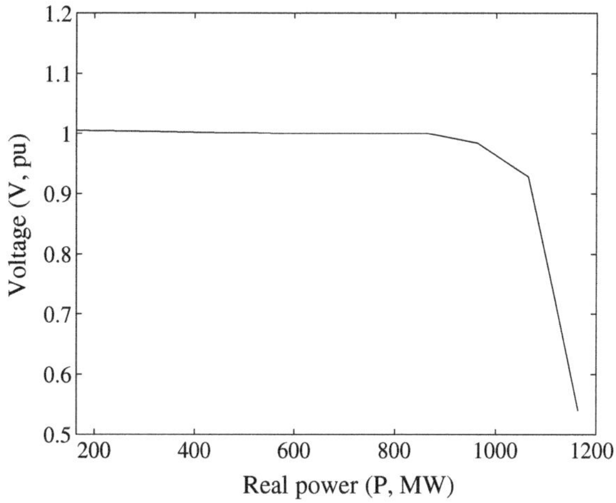

Fig. 4.2 P–V relationship at bus 49

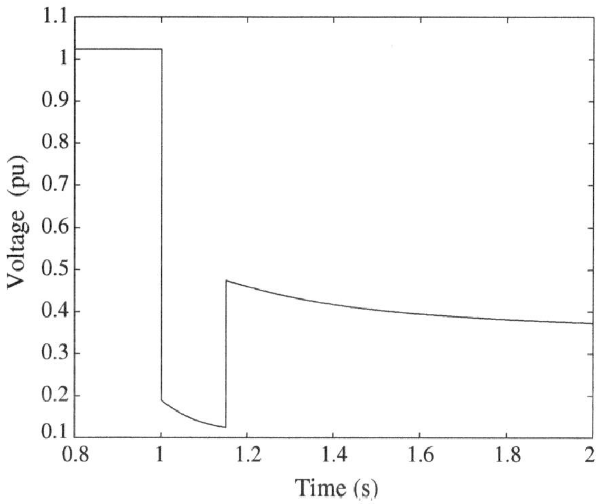

Fig. 4.3 Voltage at bus 49 for three-phase fault at middle of one of lines 60–61

The P–V curve for bus 49 in area NYPS which is highly compensated (750 MVAr) and the load ($P = 1{,}350$ MW and $Q = 29$ MVAr) is supplied from the NETS, as shown in Fig. 4.2. The system is operated with an 8, 64 MW load at bus 49 and the voltage remains constant for this loaded condition. A three-phase fault is applied at the middle of one of the lines 60–61 and the resulting voltage at bus 49 is shown in Fig. 4.3 in which it is clear that, although the system is highly compensated and operated at a constant voltage, instability occurs due to the small security margin.

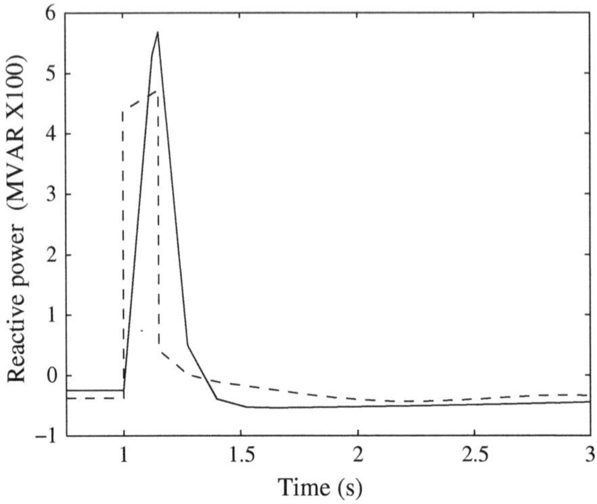

Fig. 4.4 Reactive power for three-phase fault at middle of one of lines 60–61 (*solid line* generator and *dashed line* DFIG)

4.2.1.2 Lots of DFIG Type Wind Farms and a Few Synchronous Generators

Statically, DFIGs often do not deliver as much reactive power as do synchronous generators and, dynamically, they cannot produce the same short-circuit current. The post-fault voltage support provided by the feeding reactive power is normally worse for a DFIG than for a synchronous machine. Although, recent work shows that an improvement in the area of voltage support may be possible. Due to the inferior behaviour of DFIGs compared to the synchronous generators, a system dominated by them behaves worse than one with synchronous generation. Normally, the effect is that more reactive compensation is needed in such a system. During deep voltage sags, a synchronous generator feeds in more reactive current than does a DFIG-based wind farm and, thus, provides a stronger support to the grid voltage. Also, DFIGs consume reactive power when they behave as SCIGs during transients and can reduce the voltage stability limit.

Figure 4.4 shows the reactive power supplied from a synchronous generator (G_{10}) and a DFIG of the same capacity during a three phase fault at the middle of one of the lines 60–61. Voltage transients for a power system with only synchronous generators (SGs) and one with a combination of 60 % DFIGs and 40 % synchronous generators for the same fault are shown in Fig. 4.5. The synchronous generators supply more reactive power and, thus, provide better performances in contrast to DFIG when recovering post-fault voltage. In some countries the grid-codes are so rigid that the DFIGs must be combined with static synchronous compensators (STATCOMs) in order to achieve similar behaviour to that of synchronous generators. In these cases, wind generation can be considered to be equivalent to conventional power generation.

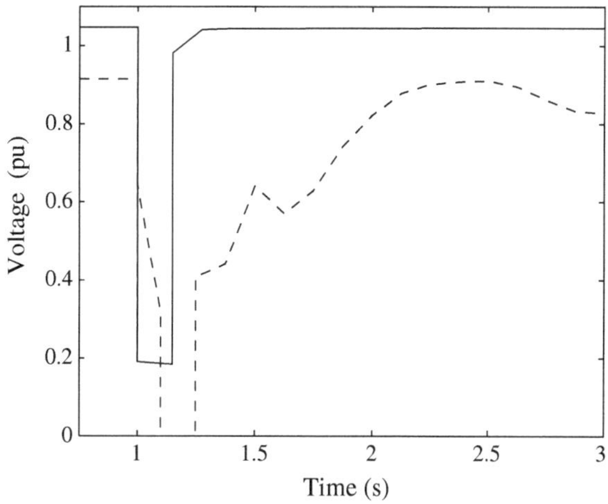

Fig. 4.5 Voltage at bus 49 for three-phase fault at middle of one of lines 60–61 (*solid line* synchronous generator, and *dashed line* 60 % DFIGs and 40 % SGs)

4.2.1.3 Interactions Amongst Different FACTS Devices

Interactions amongst the FACTS devices in a multi-machine system can adversely influence the damping properties due to individual FACTS devices. In some critical cases, it may even amplify power swings or increase voltage deviations. Due to local, uncoordinated control strategies used in many power systems, destabilising interactions amongst FACTS controls are possible. This problem may occur especially after the clearance of a critical fault, if shunt and series connected devices, e.g., SVC or STATCOM and TCSC, are present in the same area. Interactions amongst FACTS controls can adversely influence the rotor damping of generators and for weakly interconnected system conditions it can even cause dynamic instability and restrict the operating power range of the generators. The following two case studies are conducted: (i) a STATCOM at bus 31; and (ii) a STATCOM at bus 31, an MSC at bus 53 and a TCSC in the middle of the line 30–31. Figure 4.6 shows the rotor angle for both the cases with three-phase fault at the middle of one of the lines between buses 60–61. It is clear that the addition of TCSC and MSC nearest to STATCOM increases the rotor angle oscillations. To improve overall system dynamic performance, interactions amongst FACTS controls must be minimised or prevented.

4.2.1.4 A Few Bulk Transmission Lines with Series Compensations

Series compensation can be optimised so that the electrical length of the lines can be reduced. If the compensation is equipped with thyristor control, it can even be used for the damping of power swings. Case studies are carried out for: (i) a base case power flow (200 MW) through one of the lines 60–61; and (ii) an extra power flow

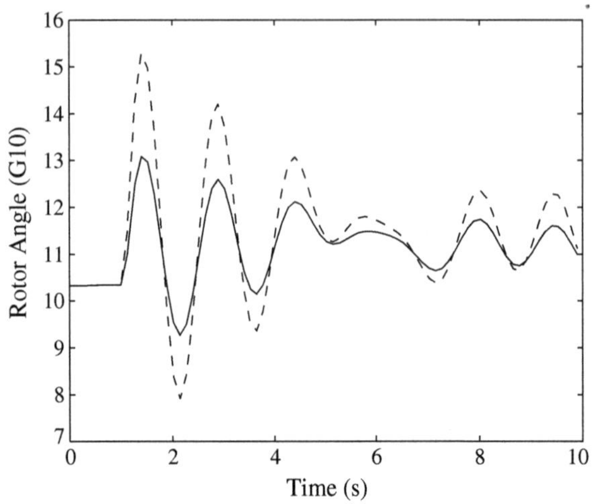

Fig. 4.6 Rotor angle G_{10} for three-phase fault at middle of one of lines 60–61 (*solid line* STATCOM and *dashed line* STATCOM+TCSC+MSC)

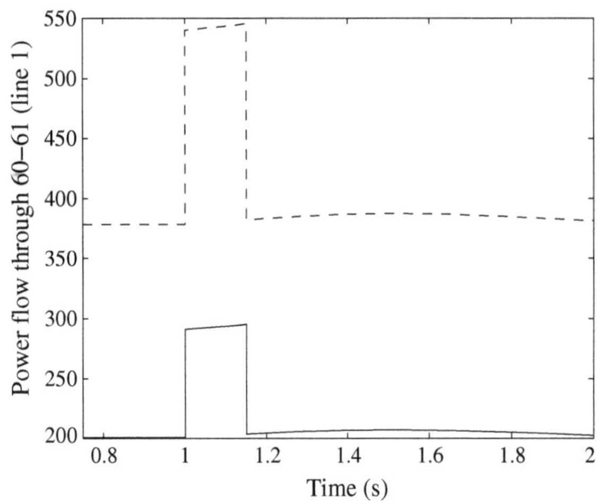

Fig. 4.7 Power flow (lines 60–61(2)) for outage of one of transmission lines 60–61 (*solid line* base case power flow and *dashed line* extra power flow with TCSC)

with a TCSC of 380 MW (almost double) through the same line. Figure 4.7 shows the power flow through the line 60–61(2) for both cases which raise no concerns regarding that configuration. The series compensation makes a stable transmission possible.

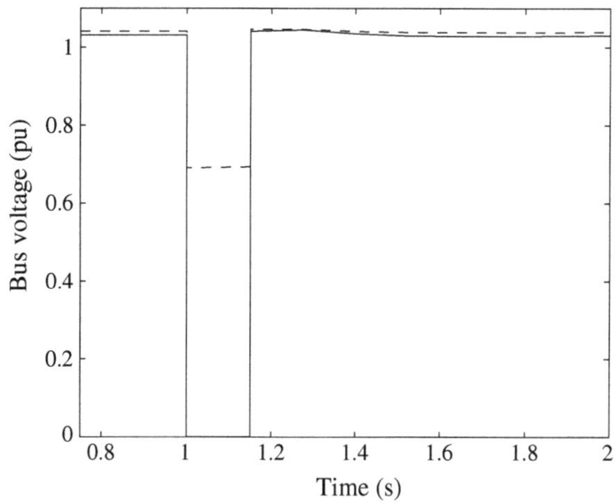

Fig. 4.8 Voltages at buses 60 and 61 for three-phase fault at middle of one of lines 60–61 (*solid line* bus 60 and *dashed line* 61)

4.2.1.5 Longitudinal System with Shunt Compensation at Middle

If the distances are not too great, line reactive power requirement can be met by compensation in the middle of the system which minimises the distances from the compensation device to the nodes of the system. For this case, a STATCOM is placed in the middle of one the of lines 60–61. Figure 4.8 shows the voltage profiles at buses 60 and 61 for a three-phase fault at one of the lines for 150 ms. From Fig. 4.8, it is evident that the post-fault voltage remains in good condition at both buses.

4.2.1.6 Comparison of Different Compensation Devices

A possible way of assessing the 'quality' of a compensation device concerning voltage stability may be that the reactive power output is observed at minimum voltage that is accepted in the grid and not at its rated voltage. The voltage at which the performance is observed could be the voltage at which power plants trip due to under-voltage (or other under-voltage criteria).

An SVC has constant impedance and its reactive power output decreases in proportion to the square of the voltage. It is an active fast reacting device, if it is not yet at its limit before the fault, it can improve the voltage recovery after fault by activating the reserve. A STATCOM is a constant current source and its reactive power output decreases linearly with the voltage. It is also an active fast reacting device with limited energy storage. Because it can inject a constant current, it can better improve the voltage recovery after fault. Figure 4.9 shows the bus voltage with a STATCOM and a SVC for a three phase fault in the middle of line 60–61.

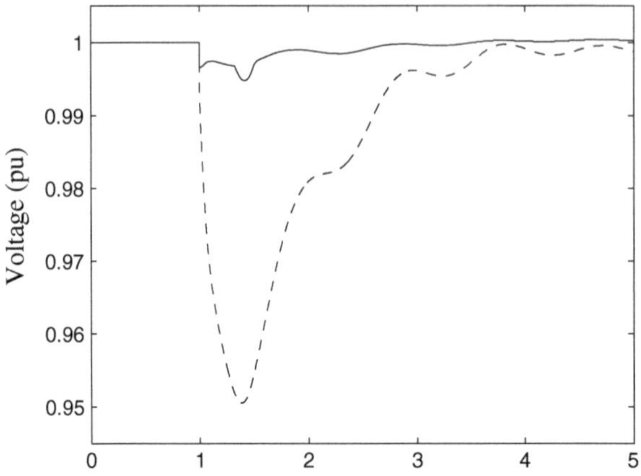

Fig. 4.9 Voltage at bus 60 for three-phase fault at middle of one of lines 60–61 (*solid line* STATCOM and *dashed line* SVC)

It is clear that during low voltage the STATCOM provides better response and the lower voltage with the STATCOM is much higher than that of the SVC.

4.2.1.7 Classical Generation Close to Load Centres

When generators are close to the load centre, only a small amount of compensation is required because the reactive power needed by the lines is relatively low and the consumption of reactive power by the loads can be covered by the generators. A synchronous generator offers dynamic compensation of high quality, i.e., in the case of a decreasing voltage at the connecting point, it does not reduce its output of reactive power. The job of additional compensation is to deliver a base compensation in order to create an adequate security margin for the generator action.

A study is conducted by increasing the load by 10 % in the NYPS (New York) which, in the first case, is supplied from local generation. In the last case, the extra load is supplied from a remote area wind turbines (New England). Figure 4.10 shows the voltage at bus 49 for a three phase fault in the middle of one of the lines 60–61. Voltage is not recovered for the last case due to high transmission losses in the connecting long lines and the different behaviours of wind turbines compared with those of synchronous generators.

4.2.1.8 Effects of Integration of Large-Scale FSWTs

A grid-connected wind turbine generation system consists of both mechanical and electrical systems which are connected to the distribution system to form part of the existing utility network. The analyses described in this section show that, for an

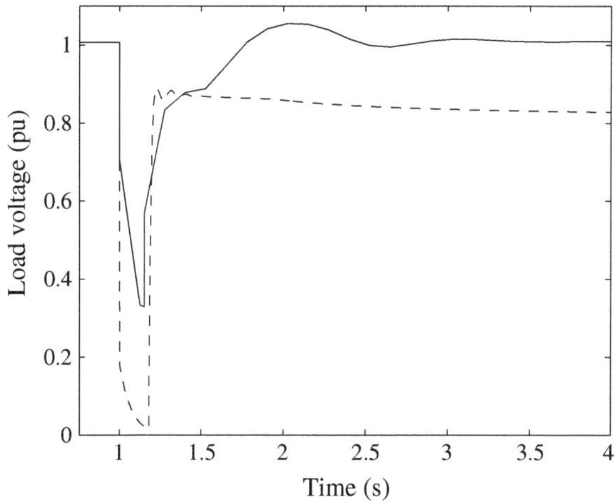

Fig. 4.10 Voltage at bus 49 for three-phase fault at middle of one of lines 60–61 (*solid line* classical local generator and *dashed line* far FSWT)

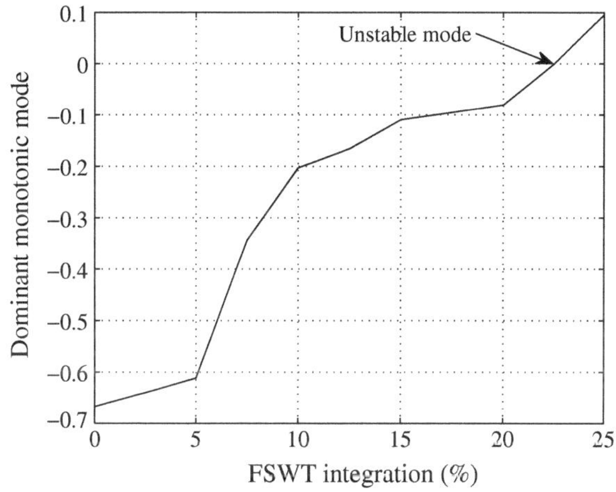

Fig. 4.11 Monotonic mode as function of FSWT integration

interconnected power system, stability is likely to be affected by various factors contributed to by the constituent distribution system and the wind turbine's mechanical and electrical properties. Modal analysis is conducted on the test system by replacing the synchronous generator by fixed speed wind turbines. Figure 4.11 shows the monotonic mode under different FSWT integration levels. Figure 4.12 shows the damping and frequency of the mechanical modes as a function of the shaft stiffness. From the participation factors, we find that the instability is caused mainly by shaft stiffness and the operating slip.

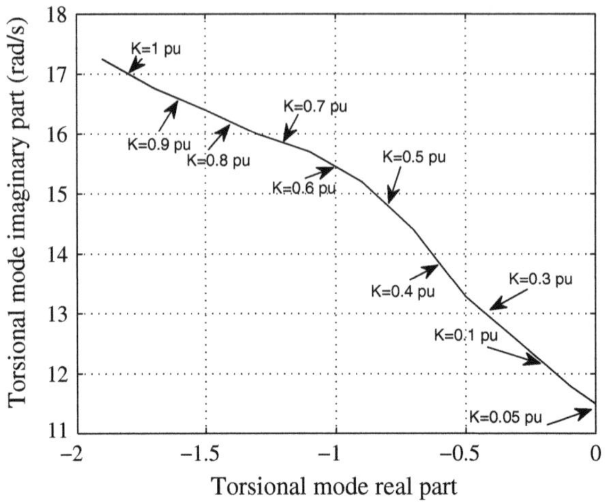

Fig. 4.12 Mechanical mode as function of shaft stiffness

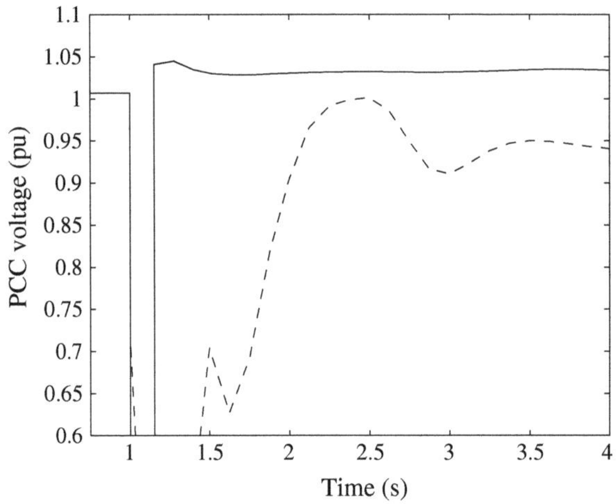

Fig. 4.13 PCC voltage for three-phase fault at middle of one of lines 60–61 (*solid line* with LVRT and *dashed line* without LVRT)

It is essential that wind generators have the fault ride-through capability. If this is not enforced, the generators trip, based on a certain under-voltage criteria (e.g. 80 %). This means that, in the case of a fault, many units trip and there is a great imbalance in the system. If this amount exceeds the primary control reserve, the system can collapse. This scenario is analysed by replacing G_3 with a wind farm. Figure 4.13 shows the voltage at bus 3 having a wind farm (i) with a low-voltage-ride-through (LVRT) capability, and (ii) without a LVRT capability. From Fig. 4.13, it is clear that the wind farm without a LVRT capability may cause instability which is also visible

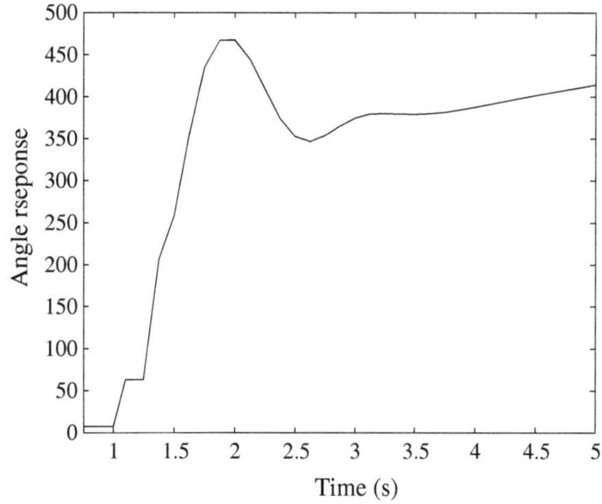

Fig. 4.14 Angle response for three-phase fault at middle of one of lines 60–61

in Fig. 4.14. In order to obtain an optimal integration of wind energy in a system, large wind farms must be able to withstand network disturbances.

4.2.1.9 Integration of FSWTs by Using STATCOMs

In the case of renewable energy, FACTS devices are especially advantageous when integrated with wind generators. As wind farms become a larger part of total power generation and as their penetration levels increase, issues related to integration, such as transient stability and voltage control are becoming increasingly important. For wind generation applications, FACTS can be implemented for voltage control in the form of shunt-connected static VAr compensators (SVCs) or STATCOMs configurations.

FSWTs and STATCOMs are integrated into the system repeatedly to find a certain level at which the system becomes unstable. The upper limits for compensations, such as STATCOM and shunt capacitor, are shown in Table 4.1. It is found that a 100 MVA STATCOM with a 500 MVAr capacitor is required to integrate a 24.5 % level of FSWTs in different areas of the test system. For identifying the nature of instability during transients at the upper limit, the method of eigenvalue tracking is used [31]. In this method, the system is repeatedly linearised at selected time instants during the simulation and the system eigenvalues are computed at each snapshot. With a fixed compensation level, the system becomes unstable after the integration of a certain level of FSWTs. Online linearisation and eigenvalue tracking show that, after fault clearing, the monotonic mode shown in Fig. 4.15 is correlated to generator slip and the mechanical state is the first to become unstable. The conclusions, drawn from Fig. 4.15, are an indication of potential instability.

Table 4.1 FSWT integration and STATCOM

FSWT (MW)	4,500	6,300	7,325	8,120	8,600
STATCOM (MVAr)	100	200	300	400	500
Capacitor (MVAr)	500	1,000	1,500	2,000	2,500

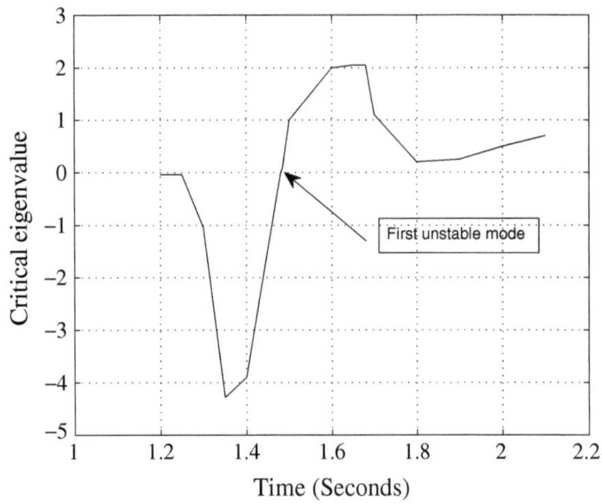

Fig. 4.15 Evolution of critical eigenvalue in unstable case

From the above analyses, the following conclusions can be drawn:

- accurate dynamic load modeling is crucial for both short- and long-term dynamic voltage instability analyses;
- although both static and dynamic compensations enhance the security level, high compensation reduces the security limits under certain operating conditions and, because of the local, uncoordinated control strategies used in many power systems, destabilising interactions amongst FACTS controls are possible,
- although interactions among local shunt FACTS controls can adversely influence the rotor damping of generators, series compensations, such as TCSCs do not have adverse effects on power flow;
- new transmission lines may be needed where wind generation is situated at a substantial distance from existing networks; and
- the modes related to operating slip and shaft stiffness are critical as they may limit the large-scale integration of wind generation.

4.3 Chapter Summary

In this chapter, different aspects of the voltage instability problem through both static (P–V and Q–V) and dynamic analyses are discussed. We explore the effects of including dynamics of a conventional generator, a wind generator, an AVR, a PSS, an OXL, an IM, an OLTC and FACTS devices on the voltage stability characteristics of a power system using a time domain analysis during low-voltage transients. In particular, in this chapter, the mechanism of the voltage collapse phenomenon is analysed from the physical rather than the mathematical point of view, and some meaningful physical interpretations are given.

The effect of load representation on voltage stability is studied. Under the same conditions in a system, load representation affects the location of the operating point in the P–V curves, leading the system either closer to or further away from the collapse point. Since load behaviour is critical for the stability of a system, more accurate models are necessary. As the traditional static models are not sufficient to represent load dynamics, dynamic load models are introduced.

For the IM load, after large disturbances take place, the IM can still maintain voltage stability if the disturbances are removed in time. Otherwise, the system voltage will become unstable or even collapse. The IM at a given operating state, and subject to a given disturbance, undergoes voltage collapse if post-disturbance equilibrium voltages are below acceptable limits. IM instability may develop into voltage instability or even voltage collapse unless effective control measures are taken against it.

Voltage stability can be enhanced by the use of an exciter or compensating devices, among which the exciter is the cheapest and it is effective in a distributed power system where the load is not far away from the generator. Of the different control methods, the robust control technique is getting more attention recently and can ensure stability against uncertainty. A novel robust exciter control technique will be given in the next chapter.

References

1. Anderson PM, Fouad AA (2002) Power system control and stability. Wiley, New York
2. Reed G, Paserba J, Salavantis P (2004) The key to resolving transmission gridlock: the case for implementing power electronics control technologies. In: Conference on electricity transmission in deregulated markets: challenges, opportunities, and necessary R&D Agenda, Pittsburgh PA, USA, pp 1–5
3. Qiao W, Harley RG, Venayagamoorthy GK (2006) Effects of FACTS devices on a power system which includes a large wind farm. In: IEEE PES power systems conference and exposition, Atlanta, GA, pp 2070–2076
4. Grunbaum R, Halvarsson B, Wilk-Wilczynski A (1999) FACTS and HVDC light for power system interconnections. In: Power delivery confrerence, Madrid, Spain, pp 1–18
5. Hossain MJ, Pota HR, Mahmud MA, Ramos RA (2012) Investigation of the impacts of large-scale wind power penetration on the transient and voltage stability of power systems. IEEE Syst J 6(1):76–84
6. Ackermann T, Soder L (2002) An overview of wind energy-status 2002. Renew Sustain Energy Rev 6(1–2):67–128
7. Hossain MJ, Pota HR, Mahmud MA, Ramos RA (2011) Impacts of large-scale wind generators penetration on the voltage stability of power systems. In: IEEE power and energy society general meeting, Michigan, USA, pp 1–8

8. Rathi MR, Mohan N (2005) A novel robust low voltage and fault ride through for wind turbine application operating in weak grids. In: 31st annual conference of IEEE industrial electronics society, pp 2481–2486
9. Palsson MP, Toftevaag T, Uhlen K, Tande JOG (2002) Large-scale wind power integration and voltage stability limits in regional networks. In: IEEE power engineering society summer meeting, Chicago, IL, USA, pp 762–769
10. Boulaxis NG, Papathanassiou SA, Papadopoulos MP (2002) Wind turbine effect on the voltage profile of distribution networks. Renew Energy 25(3):401–415
11. Hossain MJ, Pota HR, Ugrinovski V, Ramos RA (2009) Robust STATCOM control for the enhancement of fault ride-through capability of fixed-speed wind generators. In: IEEE multi-conference on systems and control, Saint Petersburg, Russia, pp 1505–1510
12. Hossain MJ, Pota HR, Ramos RA (2010) Investigation of the dynamic performance of compensation devices for voltage stability during low voltages. In: Australasian Universities power engineering conference, Christchurch, New Zealand, pp 1–6
13. Galiana FD, Almeida K, Toussaint M, Griffin J, Atanackovic D, Ooi BT, McGillis DT (1996) Assessment and control of the impact of FACTS devices on power system performance. IEEE Trans Power Syst 11(4):1931–1936
14. Salehi V, Afsharnia S, Kahrobaee S (2006) Improvement of voltage stability in wind farm connection to distribution network using FACTS devices. In: 32nd annual conference on IEEE industrial electronics, Paris, pp 4242–4247
15. Qiao W, Venayagamoorthy GK, Harley RG (2007) DHP-based wide-area coordinating control of a power system with a large wind farm and multiple FACTS devices. In: Proceedings of international joint conference on neural networks, Orlando, FL, pp 2093–2098
16. Slootweg JG, Kling WL (2003) The impact of large scale wind power generation on power system oscillations. Electr Power Syst Res 67(1):9–20
17. Hagstrom J, Norheim I, Uhlen K (2005) Large-scale wind power integration in Norway and impact on damping in the Nordic grid. Wind Energy 8(3):375–384
18. Hossain MJ, Pota HR, Ugrinovskii V, Ramos RA (2009) Excitation control for large disturbances in power systems with dynamic loads. In: IEEE power and energy society general meeting, Calgary, Canada, pp 1–8
19. Hossain MJ, Pota HR, Ugrinovski V (2009) Short and long-term dynamic voltage instability. In: 17th IFAC World Congress, Seoul, Korea, pp 9392–9397
20. Enslin J, Knijp J, Jansen C, Bauer P (2003) Integrated approach to network stability and wind energy technology for on-shore and offshore applications. In: International conference for power electronics, intelligent motion and power, quality, pp 1–8
21. Shi L, Dai S, Ni Y, Yao L, Bazargan M (2009) Transient stability of power systems with high penetration of DFIG based wind farms. In: IEEE power and energy society general meeting, Calgary, Canada, pp 1–6
22. Hansen AD, Michalke G (2007) Fault ride-through capability of DFIG wind turbines. Renew Energy 32(9):1594-1610
23. Smith JC, Milligan MR, DeMeo EA, Parsons B (2007) Utility wind integration and operating impact state of the art. IEEE Trans Power Syst 22(3):900–908
24. Akhmatova V, Knudsenb H (2007) Large penetration of wind and dispersed generation into Danish power grid. Electr Power Syst Res 77(9):1228–1238
25. Piwko R, Miller N, Sanchez-Gasca J, Yuan X, Dai R, Lyons J (2006) Integrating large wind farms into weak power grids with long transmission lines. In: IEEE 5th international conference on power electronics and motion, control, pp 1–7
26. Pal B, Chaudhuri B (2005) Robust control in power systems. Springer, USA
27. IEEE Task Force (1993) Load representation for dynamic performance analysis. IEEE Trans Power Sys 8(1):472–482
28. Ackermann T (2005) Wind power in power systems. Wiley, England
29. Nandigam K, Chowdhury BH (2004) Power flow and stability models for induction generators used in wind turbines. In: IEEE power engineering society general meeting, Denver, CO, pp 2012–2016

30. Slootweg JG, de Haan SWH, Polinder H, Kling WL (2003) General model for representing variable speed wind turbines in power system dynamics simulations. IEEE Trans Power Syst 18(1):144–151
31. Potamianakis EG, Vournas CD (2006) Short-term voltage instability: effects on synchronous and induction machines. IEEE Trans Power Syst 21(2):791–798

Chapter 5
Control for Voltage Stability with Dynamic Loads

Abstract This chapter presents a novel modeling and excitation control design to enhance large-disturbance voltage stability in power systems with significant induction motor (IM) loads. The excitation controller is designed using minimax linear quadratic Gaussian (LQG) controller synthesis method. The nonlinear power system model is reformulated with a linear and a nonlinear term. The nonlinear term is the Cauchy remainder in the Taylor series expansion and its bound is used in a robust control design. An advantage of this approach over the existing linearisation scheme is the treatment of the nonlinear dynamic load model in a rigorous framework for excitation control design. The performance of the designed controller is demonstrated by large disturbance simulations on a benchmark power system for various types of loads.

5.1 Introduction

During the last few decades, power systems have been operated under increasingly stressed conditions. This is due to transmission expansion, increased electricity consumption and new loading patterns due to deregulation of the power market. Under these stressed conditions, slow voltage drops or even voltage collapses have become a serious operating concern and, therefore, the dynamic analysis and control design of power systems in terms of voltage stability issues have become more critical [1–3]. The trend has been towards the imposition of voltage stability requirements for power systems due to increases in load from air conditioners, refrigerators, heaters, etc., which have voltage characteristics that approach constant power. The expansion of 275 kV cable systems in urban areas, and the emergence of long-distance power transmission systems with concentrated large-scale power sources, also play significant roles in causing voltage instability.

J. Hossain and H. R. Pota, *Robust Control for Grid Voltage Stability: High Penetration of Renewable Energy*, Power Systems,
DOI: 10.1007/978-981-287-116-9_5,

The assessment of voltage stability in power systems has recently gained increasing attention because voltage instability has been responsible for several major network collapses [4]. In contrast to the traditional machine instability problem, which deals with the rotor dynamics, voltage instability is closely related to the load behaviour [5]. Different studies have shown the importance of load representation in voltage stability analyses [4, 6, 7]. Currently, although static load models are commonly used in the power industry to model dynamic behaviours of reactive loads, these models do not adequately represent the loads [8]. Induction machines play a crucial role in voltage dynamics [9, 10]. Motor behaviour has been a major contributing factor in a number of documented voltage instability problems and collapses [11, 12]. The problem of voltage instability is likely to increase because of the growing use of dynamic motor loads for air conditioning, heat pumps, refrigeration, etc. [13, 14]. Therefore, dynamic load models are needed in order to analyse voltage instability and to design controllers for enhancing stability.

The active power consumption of an induction motor (IM) remains constant even after a voltage drop [9]. Moreover, if the disturbance is such that the electrical torque cannot balance the mechanical load, the motor can decelerate, absorbing higher reactive current, thus inducing a further voltage drop and occasionally a voltage collapse. Induction machines are usually shunt-capacitor compensated to improve their power factor. However, the reactive support provided by shunt capacitors varies with the square of the voltage, and consequently, during a voltage drop it is greatly reduced. Therefore, in order to avoid induction machine instability, dynamic and fast reactive compensations may become necessary.

The control of voltage levels is accomplished by controlling the production, absorption and flow of reactive power at all levels in a power system. There are two basic approaches to controlling voltage. One is to adjust the voltage at the generating plants by the use of excitation controllers. However, when the transmission lines are long and the generators are far away from major load centres, this method is not effective in improving voltage at the load end [5]. Still there are many situations in which excitation controllers can be used to stabilise the voltage in modern distributed power systems.

High-gain voltage regulators are often used but they deteriorate the small-signal stability of a system by reducing of the electromechanical modes of oscillation [3, 15]. Power system stabilisers (PSSs), supplementary to the main excitation control loop, are required to damp the electromechanical oscillations. Another approach is to use flexible AC transmission system (FACTS) devices which act directly over the power flowing across the transmission lines. However, these devices are quite expensive [16] and, therefore, where possible, the use of generator excitation control for enhancement of voltage stability is much more desirable.

A power system is highly nonlinear and a mathematical model obtained via linearisation may only satisfy the security requirement under small disturbances. Small disturbance voltage stability can be effectively studied with steady-state approaches that use linearisation of the system dynamic equations at a given operating point. However, large-disturbance voltage stability is closely related to the nonlinear behaviour of power systems [3] and controller designed using a linear model cannot ensure stability during severe disturbances. Therefore, it would be desirable to have a robust controller that could ensure the stability of a system for a wider operating region under operating conditions which could be much more distant from the equilibrium point. With this motivation, a detailed investigation into the need to include higher-order terms for small signal (modal) analysis in power systems was recently conducted by an IEEE Task Force [17]. It is shown in [18] that the addition of higher-order terms accurately predict system behaviour for large system perturbations.

Conventional excitation controllers are usually designed based on approximately linearised models obtained about a given operating condition and, therefore, may not work properly when the operating conditions change [19]. A linear optimal excitation controller based on the linear quadratic regulator (LQR) theory has been developed to improve power system stability [20]. It is more effective, with better damping for the enhancement of power system stability, under small disturbances than a conventional excitation controller. But this approach is often difficult to implement since not all the state variables are directly measurable. To overcome this difficulty, an output feedback linear quadratic Gaussian (LQG) excitation control scheme is proposed in [21–23].

The LQG controller is more realistic because it is designed using only the measurable outputs and the state variables estimated from them [24]. It is known that it provides good robustness in terms of gain margin and phase margin [25]. However, LQR and LQG controllers are unable to provide robustness against uncertainties in the operating conditions [26]. Robust LQG controllers to damp power system oscillations for thyristor-controlled series capacitors (TCSCs) are proposed in [27] and [28] in which loop transfer recovery (LTR) method is used to design them. Robust excitation control for large uncertain power systems using the H_∞ method is proposed in [29]. Controllers in [28] and [29], designed using linearised models, are only able to damp the oscillations due to electromechanical modes.

Feedback linearisation schemes are also widely used in power system robust control designs to linearise nonlinear power system models [30, 31]. A feedback linearising controller (FBLC) is used in the design of a controller for a synchronous generator connected to an infinite bus in [32]. The direct feedback linearisation (DFBL) theory is easier to understand for power engineers but, so far, its application has been limited to single-input single-output systems. Another type of FBLC to damp electromechanical oscillations in power systems is proposed in [33], but this is achieved at the expense of reducing the voltage regulation ability of the excitation system.

In [34], it is indicated that the implementation of an exciter control based on the FBLC theory requires a fixed and known equilibrium point of the power system which is usually not well-defined in practice. Furthermore, the controllers designed using a FBLC require information about the power system topology and the states must be measurable. In practice, it is very difficult to measure all the states of a power system. In addition, feedback linearisation schemes need exact plant parameters to cancel the inherent system nonlinearities. Although many developments have been proposed in the area of feedback linearisation theory, it has not yet been applied to a system with dynamic loads.

A great deal of attention has been given to the control of power systems using recently developed nonlinear control design tools to improve transient stability [35–37]. Rather than using a linearised model, as in the design of the usual PSS, nonlinear models are used and nonlinear feedback linearisation techniques are employed to linearise them which has the advantage that they are not dependent on the operating conditions as linearised models. Using nonlinear controllers, power system transient stability can be improved significantly. However, nonlinear controllers usually have a more complicated structure and are harder to implement in practice.

This chapter presents a novel large-disturbance controller for power systems with significant induction motor load. Unlike the standard linearisation technique, a linearisation is used where the Cauchy remainder is included in the design process as a bounded uncertainty. Details of the process to obtain a bound on Cauchy remainder are given in the chapter; the bound is used to design the robust minimax linear quadratic Gaussian (LQG) large-disturbance controller. The efficacy of the proposed control strategy is evaluated through simulations on a benchmark four bus power system for various types of loads and large disturbances. The organisation of the rest of this chapter is as follows: Sect. 5.2 provides a short description of the background to power system stability and exciter control; Sect. 5.3 discusses the mathematical modeling of the power system devices under consideration. Section 5.4 presents the test system and control task; Sect. 5.5 describes the linearisation technique and the technique to obtain a bound for uncertainties; Sect. 5.6 contains the essential details of the minimax LQG controller design technique as applied to our problem; Sect. 5.7 outlines the designed control algorithm and performance of the controller; and Sect. 5.8 presents the conclusions.

5.2 Power System Stability and Exciter Control

Power system stability involves (a) voltage stability in which a constant voltage can be restored and maintained even when changes in load occur and (b) power stability in which the power perturbation that arises between generators is quickly suppressed and a constant power can be maintained. It is necessary to guarantee both types of stability taking the most severe operating conditions into consideration.

Approaches to improving power system stability include improving the main circuits by increasing the system voltage, constructing additional power transmission lines, installing series capacitors and static VAr (volt-ampere-reactive) compensators (SVCs), etc., and the method of generator exciter control. Although the main circuit improvement approach is a fundamental measure, the scale of reconstruction is very large. The control approach, on the other hand, makes it possible to extract the maximum capability from the generator by improving the control algorithm.

The commonly used method for stabilising voltage against increases in load is to install a power capacitor or synchronous phase modifier at the load termination. However, local synchronous generators in modern distributed power systems can be used to control the dynamic load voltage. Synchronous generators can generate or absorb reactive power depending on the excitation. When over-excited they supply reactive power, and when under-excited they absorb reactive power. Synchronous generators are normally equipped with AVRs which continually adjust the excitation so as to control the phase of the armature current.

The excitation control of generators is one of the most effective and economical techniques for improving dynamic voltage performance and the voltage stability of power systems. Early in the 1950s, a number of scholars emphasised the importance regulating of synchronous generator's magnetic field to improve power system stability [38]. Since then, a great deal of attention has been paid to research in this field which covers two areas: amelioration of the main excitation system and improvement of the excitation control strategy [39, 40].

The advances in excitation control systems over the last 20 years have been influenced by developments in solid-state electronics. Developments in analog-integrated circuitry have made it possible to easily implement complex control strategies. In recent times, static excitation technology has also seen great advances. Since the static exciter with a thyristor-controlled self-shunt excitation has the advantages of a simple structure, high reliability and low cost, it is adopted by many large power systems around the world. The design principle and approach of the robust excitation control, including dynamic loads, to be discussed in this chapter will be aimed mainly at this type of excitation.

Although the problem of excitation control has been approached using classic control and linear control techniques with good results, it is only locally valid. Due to the nonlinearities of various components of power systems and the inherent characteristics of changing loads, the operating points of a power system may change during a daily cycle. As a result, a conventional excitation control design around an operating point may not work under large disturbances.

5.3 Power System Model

Power system operation can be modelled at several different levels of complexity, depending on the intended application of the model. In this design, a benchmark power system model, with a large IM, is considered due to its suitability for voltage stability analysis [9].

The actual transient behaviour of a synchronous generator in a real power system is very complex and difficult to accurately model unless some simplifications are made [41]. It is pointed out in [41] that the classical third-order single-axis generator dynamic model can be reliably used when designing an excitation controller as the dynamic phenomena with much smaller time constants (e.g., the effects of the damper windings) or much larger ones (e.g., the governor dynamics) can be neglected. The transmission network is considered as a passive circuit and modelled by a set of algebraic constraints representing interconnections among the dynamic models of generators and IMs. Under typical assumptions, the synchronous generator can be modelled by the following set of nonlinear differential equations [42]:

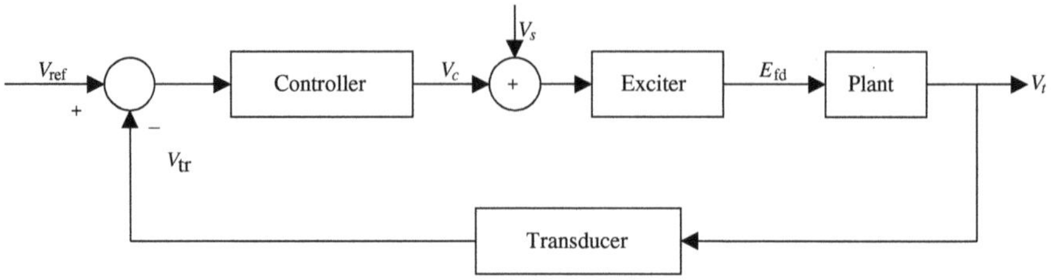

Fig. 5.1 Excitation control scheme

$$\dot{\delta} = \omega\omega_s - \omega_s, \tag{5.1}$$

$$\dot{\omega} = \frac{1}{2H}\left[P_m - E_q' I_q - D\omega\right], \tag{5.2}$$

$$\dot{E}_q' = \frac{1}{T_{do}'}\left[E_{\rm fd} - E_q' - (X_d - X_d')I_d\right], \tag{5.3}$$

where $E_{\rm fd}$ is the equivalent emf in the exciter coil, δ the power angle of the generator, ω the rotor speed with respect to a synchronous reference, E_q' the transient emf due to field flux linkage, ω_s the absolute value of the synchronous speed in radians per second, H the inertia constant of the generator, D the damping constant of the generator, T_{do}' the direct-axis open-circuit transient time constant of the generator, X_d the synchronous reactance, X_d' the transient reactance, I_d and I_q the direct and quadrature axis components of stator current, respectively. The mechanical input power P_m to the generator is assumed to be constant.

The excitation system is a high-gain static system and the terminal voltage is measured using a transducer with first-order dynamics:

$$E_{\rm fd} = K_a V_c,$$
$$\dot{V}_{tr} = \frac{1}{T_r}[-V_{tr} + V_t], \tag{5.4}$$

where V_{tr} and T_r are the output and time constant of the voltage transducer, respectively, K_a the gain of the exciter amplifier, $V_t = [(E_q' - X_d' I_d)^2 + (X_d' I_q)^2]^{\frac{1}{2}}$ the generator terminal voltage and V_c the input to the exciter (output of the designed controller).

The output-feedback controller shown in Fig. 5.1 is represented as:

$$\dot{\hat{x}}_c = A_c\hat{x}_c(t) + B_c V_{tr}, \quad V_c = C_c\hat{x}_c(t),$$

where A_c, B_c and C_c are the appropriate matrices of the controller in (5.25)–(5.26).

The transient model of an IM is described by the following algebraic-differential equations written in a synchronously-rotating reference frame [4]:

$$
\begin{aligned}
\dot{s} &= \frac{1}{2H_m}\left[T_e - T_L\right], \\
T'_{dom}\dot{e}'_{qm} &= -e'_{qm} + (X - X')i_{dm} - T'_{dom} S\omega_s e'_{dm}, \\
T'_{dom}\dot{e}'_{dm} &= -e'_{dm} - (X - X')i_{qm} + T'_{dom} S\omega_s e'_{qm}, \\
(v_{ds} + jv_{qs}) &= (R_s + jX')(i_{dm} + ji_{qm}) + j(e'_{qm} - je'_{dm}),
\end{aligned}
$$

where $X' = X_s + X_m X_r / (X_m + X_r)$ is the transient reactance, $X = X_s + X_m$ the rotor open-circuit reactance, $T'_{dom} = (L_r + L_m)/R_r$ the transient open-circuit time constant, $T_e = e'_{qm} i_{qm} + e'_{dm} i_{dm}$ the electrical torque, s the slip, e'_{dm} the direct-axis transient voltage, e'_{qm} the quadrature-axis transient voltages, T_L the load torque, X_s the stator reactance, X_m the magnetizing reactance, R_s the stator resistance, H_m the inertia constant of the motor, v_{ds} the d-axis stator voltage, v_{qs} the q-axis stator voltage, i_{dm} and i_{qm} the d- and q-axis components of stator current, respectively. However, these equations represent the induction machine in its own direct and quadrature axes, which are different from the d- and q-axis of the generator. A transformation is used to represent both dynamic elements with respect to the same reference frame, as given in Sect. 11.2. Then, the modified third-order induction machine model can be rewritten as:

$$
(V_d + jV_q) = -(R_s + jX')(I_{dm} + jI_{qm}) + jE'_{qm},
$$

$$
\dot{s} = \frac{1}{2H_m}\left[T_m - E'_m I_{qm}\right], \tag{5.5}
$$

$$
\dot{E}'_m = -\frac{1}{T'_{dom}}\left[E'_m + (X - X')I_{dm}\right], \tag{5.6}
$$

$$
\dot{\delta}_m = s\omega_s - \omega_s - \frac{X - X'}{T'_{dom}E'_m} I_{qm}. \tag{5.7}
$$

To complete the model, the d- and q-axis components of the currents for both the generator and motor are given by the following network interface equations:

$$
\begin{aligned}
I_d = -(E'_q B_{11} - V_\infty(G_{13}\sin\delta - B_{13}\cos\delta) - E'_m(G_{12}\sin(\delta_m - \delta) \\
- B_{12}\cos(\delta_m - \delta))),
\end{aligned} \tag{5.8}
$$

$$
\begin{aligned}
I_q = E'_q G_{11} + V_\infty(G_{13}\cos\delta + B_{13}\sin\delta) + E'_m(G_{12}\cos(\delta_m - \delta) \\
- B_{12}\sin(\delta_m - \delta)),
\end{aligned} \tag{5.9}
$$

$$
\begin{aligned}
I_{dm} = -(E'_m B_{22} - V_\infty(G_{23}\sin\delta_m - B_{23}\cos\delta_m) - E'_q(G_{21}\sin(\delta - \delta_m) \\
- B_{21}\cos(\delta - \delta_m))),
\end{aligned} \tag{5.10}
$$

$$
\begin{aligned}
I_{qm} = E'_m G_{22} + V_\infty(G_{23}\cos_m + B_{23}\sin\delta_m) + E'_q(G_{21}\cos(\delta - \delta_m) \\
- B_{23}\sin(\delta - \delta_m)).
\end{aligned} \tag{5.11}
$$

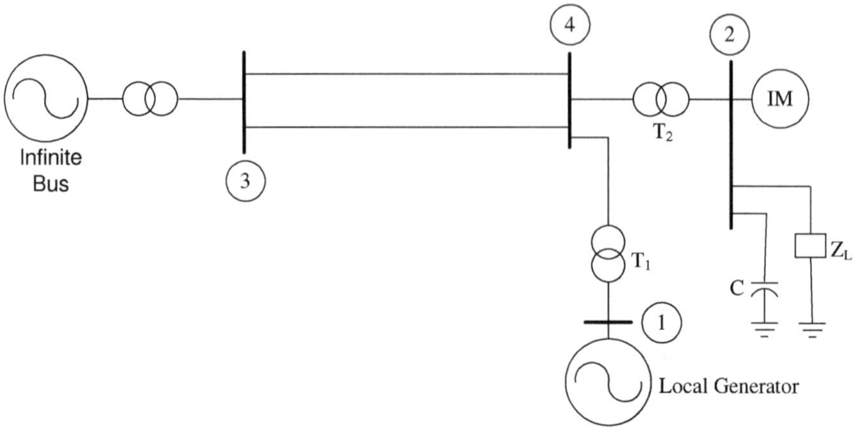

Fig. 5.2 Test system with large IM

where parameters G_{ij} and B_{ij} are the real and imaginary parts of the equivalent transfer impedances of the reduced network between the ith and jth buses, respectively. The complete model is given by Eqs. (5.1)–(5.5) with the direct and quadrature axes currents given by Eqs. (5.8)–(5.11).

5.4 Test System and Control Task

The test system used in this chapter is made up of a long transmission line connecting an infinite bus to a bus with a synchronous machine and a load ($P_L = 1{,}500$ MW, $Q_L = 150$ MVAr) [9]. Most of this load power is supplied by the remote system ($P = 1{,}200$ MW, $Q = 185$ MVAr) through a long double-circuit transmission line, while the remaining power (approximately $P = 300$ MW, $Q = 225$ MVAr) is supplied by the local generator at bus 1. The nominal transmission voltage is 380 kV and the length of the transmission line between buses 3 and 4 is 534 km (twice the length of the same line given in [9]).

The single-line diagram of the test system is shown in Fig. 5.2 and the numerical values of the parameters are given in Sect. 11.1. Bus 2 represents a distribution feeder. The power consumed at this bus may correspond to a large number of individual loads fed through medium-voltage (MV) distribution lines, shunt capacitors, etc. The load at bus 2 is made up of three parts: (i) a constant impedance load; (ii) an equivalent IM; and (iii) a 250 MVAr shunt capacitor for compensation purposes. The load selection is such that 50 % is static and 50 % dynamic (IM). With these load values, the operating point for the test system is: $\delta_0 = 19.788°$, $\omega_0 = 1$ pu, $E'_{q0} = 1.1672$ pu, $s_0 = 0.05$ pu, $E'_{m0} = 0.758$ pu, $\delta_{m0} = 22.38°$, $V_{tr0} = 1$ pu.

Table 5.1 Participation factors

States	$\Delta E_q'$	$\Delta E_m'$	Δs	ΔV_o	$\Delta\omega$	$\Delta\delta$	$\Delta\delta_m$
Part. factor for mode 1	1.0	0.846	0.313	0.277	0.148	0.108	0.021
Part. factor for mode 2	1.0	0.701	0.081	0.058	0.047	0.008	0.001

PSSs are designed to damp electromechanical oscillations due to large generator inertia and low damping. The objective in PSS design is to increase damping of the electromechanical mode by adding an auxiliary signal to the AVR. PSSs are designed to have very low gains in the frequency range outside of a narrow band centred around the resonant mode frequency. This necessitates the design of controllers to maintain system stability for other unstable or lightly damped modes.

The control problem considered in this case is different from the problem considered in the design of a PSS [3]. The control objective here is to stabilise unstable voltage modes which lead to voltage instability during large disturbances. The test system in this section, with a 50 % dynamic load, has two unstable modes corresponding to the two positive eigenvalues at 3.6507 and 0.36118. The participation vectors corresponding to these two modes are given in Table 5.1 [3]. The participation vectors indicate that the states $\Delta E_q'$ and $\Delta E_m'$ make the most significant contributions in these unstable modes. These modes cannot be damped by a PSS.

A controller based on an approximate linearised model might fail to stabilise the system during recovery from large disturbances. To deal with this problem, a robust stabilisation approach to the design of an excitation controller is proposed, with the objective of stabilising the response after relatively large disturbances. Using participation vector and eigenvector analyses, it is shown that such a system becomes unstable due to the voltage mode. In this design, the controlled variable and controller input-output are selected by the residue method which ensures that the voltage mode is stabilised by the designed controller.

The block diagram of the proposed controller is given in Fig. 5.1 and, unlike for the PSS, the terminal voltage is used as the feedback signal. A coordinated PSS and voltage stability controller can be designed as suggested by the auxiliary input, V_s in Fig. 5.1, but the focus of this work is the design of a robust voltage stability controller for large disturbances.

5.5 Linearisation and Uncertainty Modeling

The test system nonlinear model (5.1)–(5.5) is represented, using mean-value theorem, as the following form:

$$\Delta\dot{x} = A\Delta x + (L - A)\,\Delta x + B_1\Delta u, \tag{5.12}$$

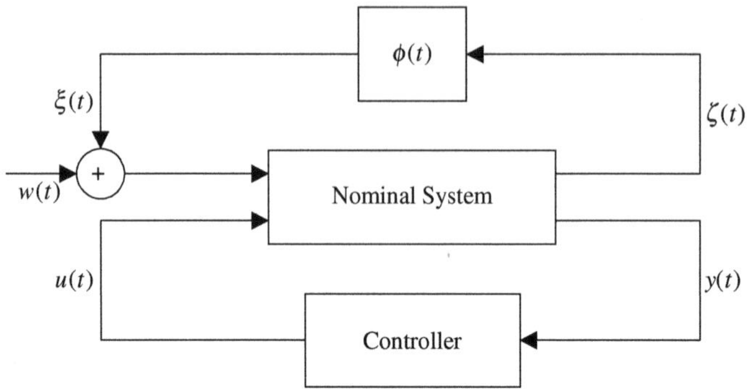

Fig. 5.3 Block diagram of robust control scheme

where $A = \frac{\partial f}{\partial x}\Big|_{\substack{x=x_0\\u=u_0}}$, $B_1 = \frac{\partial f}{\partial u}\Big|_{\substack{x=x_0\\u=u_0}}$, and $\Delta x = \left[\Delta\delta, \Delta E'_q, \Delta\omega, \Delta V_{tr}, \Delta\delta_m, \Delta E'_m,\right.$ $\Delta s]^T$. Equations (5.1)–(5.5) is linear with respect to the control vector.

The system (5.12) in terms of the block diagram is shown in Fig. 5.3, where

$$(L - A)\,\Delta x = B_2\xi, \quad \text{and} \quad \xi = \tilde{\phi}(\tilde{C}_1\Delta x), \tag{5.13}$$

and $\xi(t)$ is known as the uncertainty input.

There are several choice for B_2 and $\tilde{C}_1$ such that the equality in (5.13) is satisfied. Noting the nonlinearities in this system are only due to δ, E'_q, δ_m and E'_m, we choose

$$B_2 = \text{diag}\left(0, \frac{X_d - X'_d}{T'_{do}}, \frac{1}{2H}, \frac{1}{T_r}, \frac{X - X'}{T'_{dom}}, \frac{1}{2H_m}, \frac{X - X'}{T'_{dom}}\right), \tag{5.14}$$

$$\tilde{C}_1 = \begin{bmatrix} 1\,0\,0\,0\,0\,0\,0 \\ 0\,1\,0\,0\,0\,0\,0 \\ 0\,0\,0\,0\,1\,0\,0 \\ 0\,0\,0\,0\,0\,1\,0 \end{bmatrix}. \tag{5.15}$$

The expressions for obtaining $\tilde{\phi}$ are given in Sect. 11.3.

To facilitate control design, the power system model is summarised as:

$$\Delta\dot{x}(t) = A\Delta x(t) + B_1\Delta u(t) + B_2\xi(t), \tag{5.16}$$
$$y(t) = C_2\Delta x(t) + D_2\xi(t) + D_2 w(t), \tag{5.17}$$
$$\zeta(t) = C_1\Delta x(t). \tag{5.18}$$

where ζ is known as the uncertainty output and $y(t)$ is the measured output. The output matrix, C_2, is defined as:

$$C_2 = \left[0\,0\,0\,1\,0\,0\,0\right]. \tag{5.19}$$

Equations (5.16)–(5.18) provide a new representation of the power system model with a linear part, and another part with higher-order terms. The new formulation presented in this section is used with the minimax LQG control theory to design a voltage controller for the nonlinear power system.

5.6 Minimax LQG Control

In this chapter, we use a simplified version of the main results in [43] and [44] concerning the minimax LQG control problem. A complete description of this method can be found in [45]. In this case, the minimax LQG method is applied to uncertain systems of the form shown in Fig. 5.3.

In association with the uncertain system (5.16)–(5.18), we consider a cost functional, J, of the form

$$J = \lim_{T\to\infty} \frac{1}{2T} E \int_0^T (x(t)^T R\Delta x(t) + u(t)^T G\Delta u(t))dt, \tag{5.20}$$

where $R \geq 0$, $G > 0$, $R \in R^{n\times n}$, $G \in R^{m\times m}$ and E is the expectation.

The quadratic cost (5.20) is particularly suited to the design of an excitation controller for a power system. Every generator has an over-excitation limiter which limits the field voltage based on the time integral of the voltage. This means that one can apply large voltages as long as they are for short durations. Thus, the quadratic cost optimisation is much more suitable in this situation than are H_∞-norm based designs.

The minimax optimal control finds the controller which minimises J over all admissible uncertainties. The cost function, J, satisfies the following relationship [45]:

$$\sup_{\|\xi\|^2 \leq \|\zeta\|^2} J(u^*) \leq \inf_{\tau} V_\tau \tag{5.21}$$

where V_τ is given by:

$$V_\tau = \frac{1}{2}\text{tr}[Y_\infty R_\tau + (Y_\infty C_2^T + B_2 D_2^T)(D_2 D_2^T)^{-1}(C_2 Y_\infty + D_2 B_2^T) X_\infty (I - \frac{1}{\tau} Y_\infty X_\infty)^{-1}], \tag{5.22}$$

where τ is a free parameter and the matrices X_∞ and Y_∞ are the solution to the following pair of parameter-dependent algebraic Riccati equations [45]:

$$\begin{aligned}(A - B_2 D_2^T (D_2 D_2^T)^{-1} C_2) Y_\infty + Y_\infty (A - B_2 D_2^T (D_2 D_2^T)^{-1} C_2)^T \\ - Y_\infty (C_2^T (D_2 D_2^T)^{-1} C_2 - \frac{1}{\tau} R_\tau) Y_\infty + B_2 (I - D_2^T (D_2 D_2^T)^{-1} D_2) B_2^T = 0,\end{aligned} \tag{5.23}$$

and

$$X_\infty(A - B_1 G_\tau^{-1}\gamma_\tau^T + (A - B_1 G_\tau^{-1}\gamma_\tau^T)X_\infty$$
$$+ (R_\tau - \gamma_\tau G_\tau^{-1}\gamma_\tau^T) - X_\infty(B_1 G_\tau^{-1} B_1^T - \frac{1}{\tau} B_2 B_2^T)X_\infty = 0. \qquad (5.24)$$

The solutions are required to satisfy the following conditions: $Y_\infty > 0$, $X_\infty > 0$, the spectral radius of the matrix, $X_\infty Y_\infty$ is $\rho(X_\infty Y_\infty) < \tau$, $R_\tau - \gamma_\tau^T G_\tau^{-1}\gamma_\tau \geq 0$, $R_\tau = R + \tau C_1^T C_1$, $G_\tau = G + \tau D_1^T D_1$ and $\gamma_\tau = \tau C_1^T D_1$.

To obtain the minimax LQG controller, the parameter $\tau > 0$ is chosen to minimise V_τ. A line search is carried out to find the value of $\tau > 0$ which attains the minimum value of the cost function, V_τ. This line search involves solving the Riccati equations (5.23) and (5.24) for different values of τ and finding the value which gives the smallest V_τ.

The minimax LQG optimal controller is given by the equations:

$$\dot{\hat{x_c}} = \left(A - B_1 G_\tau^{-1}\gamma_\tau^T\right)\hat{x_c} - \left((B_1 G_\tau^{-1} B_1^T - \frac{1}{\tau} B_2 B_2^T)X_\infty\right)\hat{x_c} + \left(I - \frac{1}{\tau} Y_\infty X_\infty\right)^{-1}$$
$$\left(Y_\infty C_2^T + B_2 D_2^T\right) \times (D_2 D_2^T)^{-1}\left(y - (C_2 + \frac{1}{\tau} D_2 B_2^T X_\infty)\hat{x_c}\right), \qquad (5.25)$$
$$u = -G_\tau^{-1}(B_1^T X_\infty + \gamma_\tau^T)\hat{x_c}. \qquad (5.26)$$

In the next section, we design a minimax LQG controller for the test power system and evaluate its performance.

5.7 Controller Design and Performance Evaluation

Prior to the controller design, we carry out several large disturbance simulations to gain an idea of the region of interest. The maximum value of $\|\phi(t)\|$ is obtained over this region and not globally. If the maximum value of $\|\phi(t)\|$ is evaluated over the entire uncertainty region, the calculation burden will be very high and lead to a conservative controller. This method of evaluating the maximum value of $\|\phi(t)\|$ is different from the global maximum method used in [46]. The controller is then designed as follows,

Step 1 From the simulations of the faulted system, obtain the range of variations of all state variables and form a volume, Ω, with corner points given by $(x0_p - x_{f_p})$ and $(x0_p + x_{f_p})$, $p = 1, \ldots, 7$, where $2x_{f_p}$ is the largest variation of the pth state variable about its equilibrium value, $x0_p$. Formally $x \in \Omega$ if $|x - x0_p| \leq |x_{f_p} - x0_p|$.

Step 2 Obtain

$$\alpha^* = \max_{x^{*p} \in \Omega} \left\{\alpha : ||\phi(t)||^2 < 1\right\}.$$

The process to obtain α^* involves obtaining the maximum value of $||\tilde{\phi}(t)||$ over the volume Ω.

Step 3 Check if there exists a feasible controller with $\alpha = \alpha^*$, i.e., if there is a scalar τ such that there is a feasible solution to the coupled Riccati equations (5.23) and (5.24).

Step 4 If we obtain a feasible controller in the above step, either enlarge the volume, Ω, i.e., increase the operating region of the controller or if we have arrived at the largest possible volume, perform an optimal search over the scalar parameter, τ, to get the infimum of V_τ. If there is no feasible solution with the chosen $\alpha = \alpha^*$, reduce the volume, Ω, and go to step 2.

The term $x(t)'Rx(t)$, in the cost function (5.20), corresponds to the norm-squared value of the nominal system output and $u'Gu$ is treated as a design parameter affecting controller gain. Although a certain amount of trial-and-error is needed, it is determined that the value of $G = 10^{-4}$ and the matrix $R = \mathrm{diag}(1,\ 3,\ 1,\ 1,\ 1,\ 2,\ 1)$ in the cost function (5.20) ensure the satisfactory performance of the controller. It can be seen that the state variable representing the transient emf is given higher weight compared with those of other state variables since we want a quick controller response to compensate for the reactive power mismatch. Furthermore, the theory requires that $D_2 D_2' > 0$ [43], so we choose $D_2 = 10^{-6}\,[1,\ 1,\ 1,\ 1]$.

The process described above enables the selection of the largest range for which a feasible controller is obtained. For the given power system model, we obtain the optimum value of the parameter $\tau = 3.42$ and $\alpha^* = 0.64$, for the polytope region Ω whose corner points $\left[\bar{\delta},\, \bar{\omega},\, \bar{E}_q',\, \bar{V}_{tr},\, \bar{\delta}_m,\, \bar{s},\, \bar{E}_m'\right]$ and $\left[\underline{\delta},\, \underline{\omega},\, \underline{E}_q',\, \underline{V}_{tr},\, \underline{\delta}_m,\, \underline{s},\, \underline{E}_m'\right]$ are given by: $\bar{\delta} = \delta_0 + 42.38°$, $\underline{\delta} = \delta_0 - 42.38°$, $\bar{\omega} = \omega_0 + 0.345$ pu, $\underline{\omega} = \omega_0 - 0.345$ pu, $\bar{E}_q' = E_{q0} + 0.25$ pu, $\underline{E}_q' = E_{q0} - 0.25$ pu, $\bar{V}_{tr} = V_{tr0} + 0.275$ pu, $\underline{V}_{tr} = V_{tr0} - 0.275$ pu, $\bar{\delta}_m = \delta_{m0} + 41.19°$, $\underline{\delta}_m = \delta_{m0} - 41.19°$, $\bar{s} = s_0 + 0.225$ pu, $\underline{s} = s_0 - 0.225$ pu, $\bar{E}_m' = E_{m0} + 0.25$ pu and $\underline{E}_m' = E_{m0} - 0.25$ pu.

The output-feedback controller is:

$$\dot{x}_c = A_c x_c(t) + B_c y(t), \quad u(t) = C_c x_c(t), \tag{5.27}$$

where

$$A_c = \begin{bmatrix} 0 & 375.96 & 0 & 0 & 0 & 482.85 & 0.3433 \\ -0.25 & -15.32 & -0.13 & 0 & -0.54 & -19.83 & -0.48 \\ -1175.90 & 1.5875e6 & -625.12 & -0.57 & -2592.7 & 4.157e6 & 0.73 \\ -3.96 & 121.09 & 39.14 & -66.67 & -2.174 & 136.9 & 36.70 \\ 0 & 464.7 & 0 & 377 & 0 & 463 & 0.9386 \\ -1057 & 1.223e4 & -953.4 & -1.25 & -579.3 & -40 & 0 \\ -3.712 & 109.82 & 15.45 & -60.00 & -3.9386 & 160.25 & -20.65 \end{bmatrix}, \tag{5.28}$$

Table 5.2 Closed loop eigenvalues

$-2.946 \pm i6.14$	$-4.846 \pm i5.46$	$-1.189 \pm i3.16$	-27.470

$$B_c = \begin{bmatrix} -225.92 \\ 40.64 \\ -356.09 \\ -222.235 \\ -107.02 \\ 20.47 \\ -183.17 \end{bmatrix}, \quad C_c = \begin{bmatrix} -42.76 \\ 5.143e4 \\ -25.24 \\ 0.87 \\ -79.41 \\ 7.72e4 \\ -31.46 \end{bmatrix} \tag{5.29}$$

The closed loop system are shown in Table 5.2. The dominant mode for the closed loop system is $-1.189 \pm j1.8794$ and the damping ratio is 0.12971. From the eigenvalues, it is clear that the closed loop system is well-damped. The performance of the designed robust excitation controller is tested by simulating responses to three contingencies on the test system which are: (I) the outage of one transmission line; (II) a three-phase short-circuit at one of the transmission lines between bus 3 and bus 4; and (III) sudden changes in load power. The simulation sampling time is 2.5 ms for all cases.

5.7.1 Contingency I: Outage of One Transmission Line

A transmission line outage increases line impedance and weakens interconnections. Due to the increase in line reactance, extra reactive power is needed in order to maintain the voltage at the load bus. The AVR responds to this condition by increasing the generator field voltage which has a beneficial effect on voltage stability. The effectiveness of this type of control depends on the ability of the excitation system to quickly increase the field voltage to meet the required reactive power.

Here, a transmission line outage simulation is performed by opening the line at 1 s and reclosing it after five cycles. Figures 5.4 and 5.5 show the generator angle and terminal voltage response, respectively, with the designed excitation controller and an IEEE ST1A exciter. The simulation is repeated for the same contingency with a 25 % dynamic load for which the terminal voltage of the generator is shown in Fig. 5.6. From the responses, it can be concluded that the designed controller can stabilise the voltage as well as the angle of the generator within 2 s of this large disturbance. Although the excitation controller is designed with the load being 50 % dynamic and 50 % static, it also performs well if the composition of the total load includes less than a 50 % dynamic load.

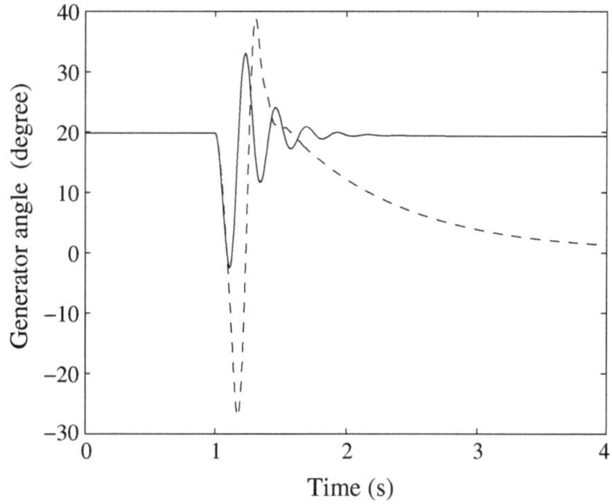

Fig. 5.4 Generator angle—outage of one line (*solid line* designed controller and *dashed line* IEEE ST1A exciter)

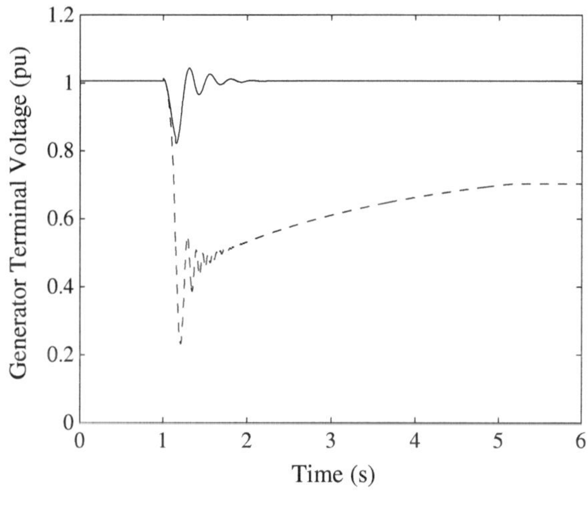

Fig. 5.5 Generator terminal voltage (50 % dynamic load)—outage of one line (*solid line* designed controller and *dashed line* IEEE ST1A exciter)

5.7.2 Contingency II: Three-Phase Short-Circuit

One of the most severe disturbances, perhaps leading to voltage collapse, is a three-phase fault on one of the key transmission circuits. From the power system viewpoint, the excitation system should contribute to the effective control of voltage and the enhancement of system stability. The excitation system should be capable

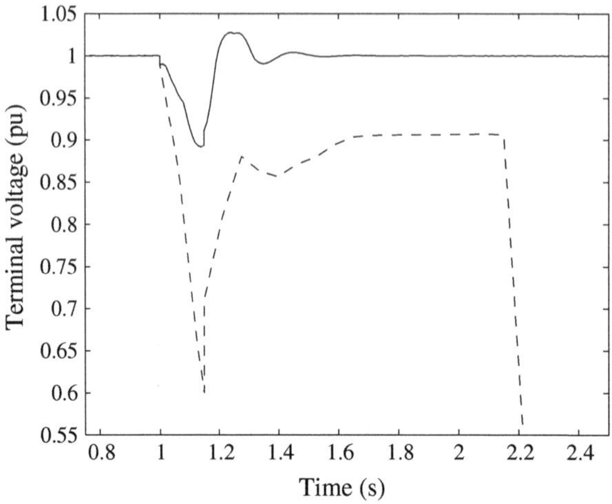

Fig. 5.6 Generator terminal voltage (25 % dynamic load)—outage of one line (*solid line* designed controller and *dashed line* IEEE ST1A exciter)

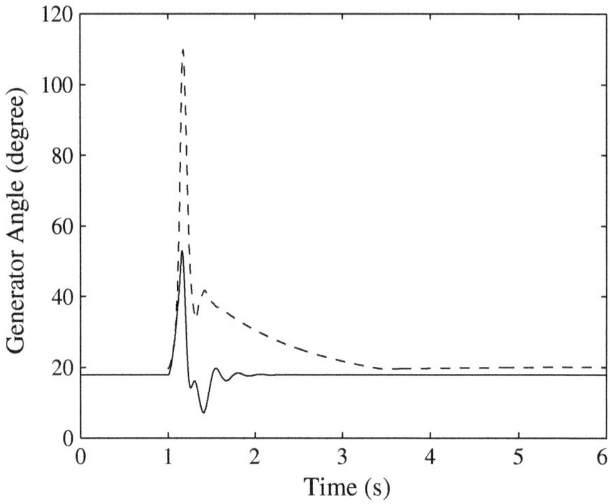

Fig. 5.7 Generator angle—three-phase fault (*solid line* designed controller and *dashed line* IEEE ST1A exciter)

of responding rapidly to a disturbance by modulating the generator field so as to enhance voltage stability.

Here, a simulation is performed for a three-phase fault on one of the transmission lines. The fault is cleared by auto-reclosing the circuit breaker after 0.15 s. Figures 5.7 and 5.8 show the angle and terminal voltage responses of the local generator, respectively, from which it can be seen that the proposed controller stabilises voltage within five cycles of a fault occurrence and damps out the power angle oscillations.

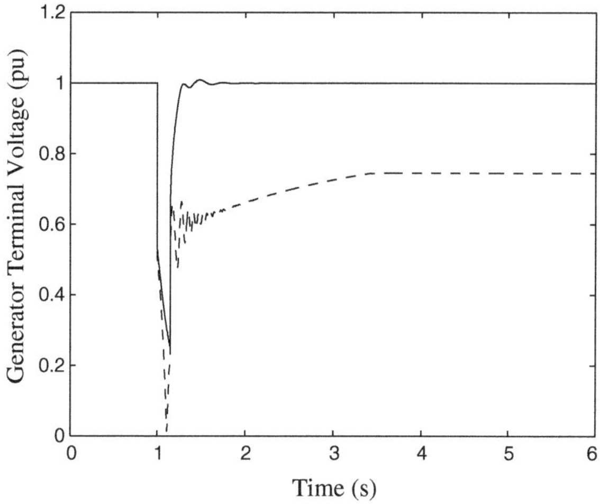

Fig. 5.8 Generator terminal voltage—three-phase fault (*solid line* designed controller and *dashed line* IEEE ST1A exciter)

5.7.3 Contingency III: Sudden Change in Load

The load characteristics and dynamics indicate the dependency between the load and the voltage, and the close coupling of the load dynamics with the voltage stability phenomenon. The stable operation of a power system depends on its ability to continuously match the electrical output of generating units to the electrical load on the system. As loads vary, the reactive power requirements of the transmission system also vary. In practice, a power system experiences continuous changes in load demand which can cause voltage instability unless an appropriate control action is taken.

The performance of the designed controller is also evaluated for a less severe fault, such as a sudden change in load. At 1 s, the load is increased by 5 % and the transient responses of the local generator and the load voltage due to this change are shown in Figs. 5.9 and 5.10, respectively, in which it is clear that both the controllers can stabilise the system against a less severe disturbance but that the proposed controller performs better in terms of settling time, damping and oscillations.

The simulation is repeated for a comparatively severe disturbance, i.e., a 25 % change in load. In this case, although the designed controller requires more settling time to stabilise the system, it still provides good performance. On the other hand, the IEEE ST1A cannot stabilise the system so the voltage collapses and the generator goes out of synchronism.

From the above analyses, it is clear that the proposed controller ensures a wider stable operating region and stabilises the system against large disturbances with improved settling time, damped oscillations, and a smaller steady-state error. The conventional controller can stabilise the power system against less severe disturbances but fails to do so when large-disturbances occur.

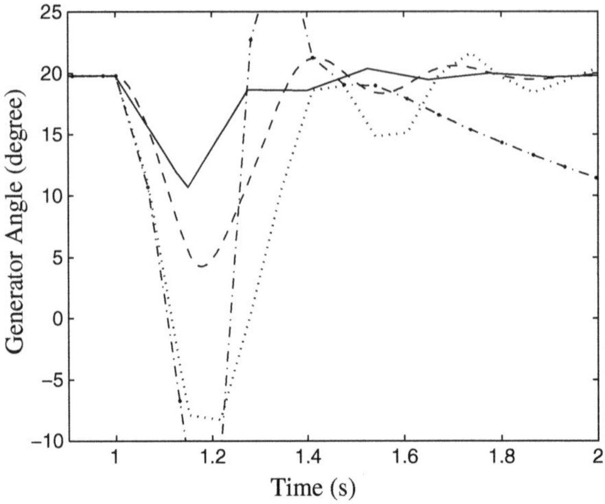

Fig. 5.9 Generator angle—change in load power (for 5 and 25 % change in load, *solid* and *dotted line* designed controller, and the *dashed* and *dashdot line* IEEE ST1A exciter)

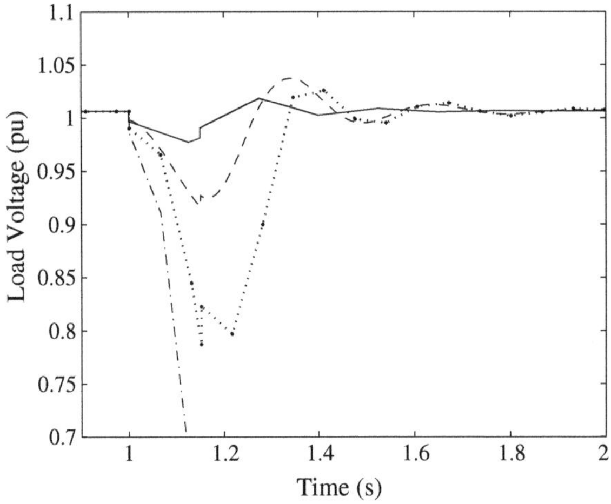

Fig. 5.10 Load voltage—change in load power (for 5 and 25 % change in load, *solid* and *dotted line* designed controller, and the *dashed* and *dashdot line* IEEE ST1A exciter)

5.8 Chapter Summary

In this chapter, an algorithm to design a robust output-feedback excitation controller is proposed. An advantage of this approach over the methods that use the existing linearisation scheme is the treatment of the nonlinear dynamic load model in a rigorous framework for excitation control design. The design method is tested on a benchmark study system. Linear analyses and nonlinear simulations demonstrate the robustness and efficiency of the designed controllers. The performance of the proposed robust excitation controller is compared with that of a standard IEEE exciter and the simulation results confirm the improved efficacy of the proposed controller.

The reformulation used in this design explicitly accounts for the effect of nonlinearities in system dynamics which enables us to more accurately represent the system and also provides guaranteed performance and stability characteristics over a pre-specified region around the equilibrium point. It is demonstrated by simulation results that the proposed controller is very effective under different operating conditions, fault locations and varying load conditions in damping both small and large disturbances in the systems studied.

Today, the focus regarding interactions between an electrical network and wind energy installation has shifted since the loss of such a considerable part of power production (as wind energy constitute in some regions) due to network disturbances can no longer be accepted. Accordingly, the important issue is to avoid a disconnection of a wind energy installation during a network disturbance. The fault ride-through capabilities of FSIGs with STATCOMs will be investigated and a suitable control technique for enhancing them will be outlined in the next chapter.

References

1. Custem TV, Vournas CD (1998) Voltage stability of the electric power systems. Kluwer Academic, Norwell
2. Cutsem TV (February 2000) Voltage instability: phenomenon, countermeasures and analysis methods. Proc IEEE 88(2):208–227
3. Kundur P (1994) Power system stability and control. Mcgraw-Hill, New York
4. Taylor CW (1994) Power system voltage stability. McGraw-Hill, New York
5. Vu KT, Liu CC, Taylor CW, Jimma KM (1995) Voltage instability: mechanism and control strategies [power systems]. Proc IEEE 83(11):1442–1455
6. Roy NK, Pota HR, Mahmud MA, Hossain MJ (2013) Key factors affecting voltage oscillation of distribution networkswith distributed generation and inductionmotor loads. Int J Electr Power Energy Syst 53:515–528
7. Overbye TJ (1994) Effects of load modelling on analysis of power system voltage stability. IEEE Trans Power Syst 16(5):329–338
8. Li Y, Chiang HD, Choi BK, Chen YT, Lauby DHHMG (2008) Load models for modeling dynamic behaviors of reactive loads: evaluation and comparison. Int J Electr Power Energy Syst 30(9):497–503
9. Cutsem TV, Vournas C (1998) Voltage stability of electric power systems. Kluwer Academic, Norwell
10. Hossain MJ, Pota HR, Ugrinovski V (2008) Short and long-term dynamic voltage instability. In: 17th IFAC world Congress, Seoul, Korea, pp 9392–9397
11. Hossain MJ, Pota HR, Ugrinovskii V, Ramos RA (2009) Excitation control for large disturbances in power systems with dynamic loads. In: IEEE power and energy society ceneral meeting, Calgary, Canada, pp 1–8
12. Williams BR, Schmus WR, Dawson DC (1992) Transmission voltage recovery delayed by stalled air conditioners compressors. IEEE Trans Power Syst 7(3):1173–1179
13. Mahmud MA, Hossain MJ, Pota HR (2012) Effects of large dynamic loads on power systems stability. Int J Electr Power Energy Syst 41(1):357–363

14. Leon JADD, Taylor CW (2002) Understanding and solving short term voltage stability problems. In: IEEE power engineering society summer meeting, Chicago, USA, pp 745–752
15. Hossain MJ, Pota HR, Ramos RA (2011) Excitation control for improving transient stability limit and voltage regulation including dynamic loads. In: 18th IFAC world congress, Milano, Italy, pp 4971–4976
16. Niioka S, Yokoyama R, Fujita G, Shirai G (2002) Decentralized exciter stabilizing control for multimachine power systems. Electr Eng Jpn 139(1):35–43
17. Sanchez-Gasca JJ, Vittal V, Gibbard MJ, Messina AR, Vowles DJ, Liu S, Annakkage UD (2005) Inclusion of higher order terms for small-signal (modal) analysis: committee report-task force on assessing the need to include higher order terms for small-signal (modal) analysis. IEEE Trans Power Syst 20(4):1886–1904
18. Martinez I, Messina A, Barocio E (2004) Perturbation analysis of power systems: effects of second- and third-order nonlinear terms on system dynamic behaviour. Electr Power Syst Res 71(2):159–167
19. Cao YJ, Jiang L, Cheng SJ, Chen DD, Malik OP, Hope GS (1994) A nonlinear variable structure stabilizer for power system stability. IEEE Trans Energy Convers 9(3):489–495
20. Lu Q, Sun YZ, Mei S (2001) Nonlinear control systems and power system dynamics. Kluwer Academic Publishers, Boston
21. Hossain MJ, Mahmud MA, Roy NK, Pota HR (2013) Enhancement of transient stability limit and voltage regulation with dynamic loads using robust excitation control. Int J Emerg Electr Power Syst (in press)
22. Hossain MJ, Pota HR, Ugrinovski V, Ramos RA (2010) Voltagemode stabilisation in power systems with dynamic loads. Int J Electr Power Energy Syst 32(8):911–920
23. Seo JC, Kim TH, Park JK, Moon SI (1996) An LQG based PSS design for controlling the SSR in power systems with series-compensated lines. IEEE Trans Energy Convers 11(2):423–428
24. Roy NK, Pota HR, Mahmud MA, Hossain MJ (2013) Voltage control of emerging distribution systems with induction motor loads using robust lqg approach. Int Trans Electr Energy Syst. doi:10.1002/etep.1739
25. Anderson BD, Moore JB (1990) Optimal Control: linear quadratic methods. Prentice-Hall, New Jersey
26. Ray LR (1993) Stability robustness of uncertain LQG/LTR systems. IEEE Trans Autom Control 38(2):304–308
27. Kwang MS, Park JK (2000) On the robust LQG control of TCSC for damping power system oscillations. IEEE Trans Power Syst 15(4):1306–1312
28. Ferreira A, Barreiros JA , Barra W Jr, de Souza JRB (2007) A robust adaptive LQG/LTR TCSC controller applied to damp power system oscillations. Electric Power Syst Res 77(8):956–964
29. Dou CX, Zhang XZ, Guo SL, Mao CC (2010) Delay-independent excitation control for uncertain large power systems using wide-area measurement signals. Int J Electr Power Energy Syst 32(3):210–217
30. Mahmud MA, Pota HR, Aldeen M, Hossain MJ (2013) Partial feedback linearizing excitation controller for multimachine power systems to improve transient stability. IEEE Trans Power Syst. doi:10.1109/TPWRS.2013.2283867
31. Mahmud MA, Pota HR, Hossain MJ (2012) Dynamic stability of three-phase grid-connected photovoltaic system using zero dynamic design approach. IEEE J Photovoltaics 12(4):564–571
32. Mielczarski W, Zajaczkowski AM (1994) Nonlinear field voltage control of a synchronous generator using feedback linearisation. Automatica 30(10):1625–1630
33. Ramos RA, Alberto LFC, Bretas NG (2003) Linear matrix inequality based controller design with feedback linearisation: application to power systems. IEEE Proc-Control Theory Appl 150(5):551–556
34. Chapman JW, Ilic MD, King CA (1993) Stabilizing a multi-machine power system via decentralized feedback linearizing excitation control. IEEE Trans Power Syst 8(3):830–838
35. Gan D, Qu Z, Cai H (2000) Multimachine power system excitation control design via theories of feedback linearisation control and nonlinear robust control. Int J Syst Sci 31(4):519–527

36. Guo Y, Hill DJ, Wang Y (1999) Robust decentralized excitation control of multimachine power systems. In: American control conference, San Diego, California, USA , pp 61–74
37. Xianrong C, Renzhou W, Yihan Y (1993) The nonlinear excitation control of generators. In: IEEE conference on computer, communication, control and power engineering
38. Concordia C (1951) Synchronous machines. Willey, New York
39. Demello FP, Concordia C (1969) Concepts of synchronous machine stability as affected by excitation control. IEEE Trans Power Apparatus Syst PAS-88(4):316–329
40. Wang Y, Guo G, Hill DJ (1997) Robust decentralized nonlinear controller design for multimachine power systems. Automatica 33(9):1725–1733
41. Lu YN (1983) Electric power system dynamics. Academic Press, London
42. Bergen AR (1986) Power system analysis. Prentice-Hall, New Jersey
43. Ugrinovskii VA, Petersen IR (2001) Minimax LQG control of stochastic partially observed uncertain systems. SIAM J Control Optim 40(4):1189–1226
44. Ugrinovskii VA, Petersen IR (1999) Finite horizon minimax optimal control of stochastic partially observed time varying uncertain systems. Mathe Control Signals Syst 12(1):1–23
45. Petersen IR, Ugrinovskii VA, Savkin AV (2000) Robust control design using H_∞ methods. Springer, London
46. Hossain MJ, Pota HR, Ugrinovski V, Ramos RA (2009) Robust STATCOM control for the enhancement of fault ride-through capability of fixed-speed wind generators. In: IEEE multiconference on systems and control. Saint Petersburg, Russia, pp 1505–1510

Chapter 6
Control for Dynamic Transfer Capability Enhancement

Abstract This chapter presents an algorithm to design a decentralized robust controller for STATCOMs (static synchronous compensators) using minimax linear quadratic (LQ) output-feedback control design approach. There is an increase of the available (dynamic) transfer capability (ATC) of power systems with fixed-speed wind generators (FSWGs) due to the designed decentralized controllers. The effects of the integration of various types of wind generators into power systems based on transfer limit has also been analyzed in this chapter. The effectiveness of the suggested control strategy is validated by simulations on a benchmark two area power system. The performance of the designed controller is also compared with a conventional PI (proportional-integral)-based STATCOM controller. Simulation results show that both the dynamic voltage stability and the transient stability can be improved by the use of the robust STATCOM control presented in this chapter.

6.1 Introduction

Power system transfer capability indicates the possible increase in the inter-area power transfer transfer without compromising system security. The accurate identification of this capability provides vital information for both the planning and operation of the bulk power market. Planners need to know the system bottlenecks, and system operators must not implement transfers which exceed the calculated transfer capability. Repeated estimates of transfer capabilities are needed to ensure that the combined effects of power transfers do not cause an undue risk of system overload, equipment damage and blackouts [1]. However, an overly conservative estimate of the transfer capability unnecessarily limits the power transfers and is a costly and inefficient use of the network. Power transfers are increasing in both number and variety as deregulation proceeds. Indeed, they are necessary for a competitive electric power market. There is a very strong economic incentive to improve

J. Hossain and H. R. Pota, *Robust Control for Grid Voltage Stability: High Penetration of Renewable Energy*, Power Systems,
DOI: 10.1007/978-981-287-116-9_6,

the accuracy and effectiveness of transfer capability computations for use by system operators, planners and power marketers.

With growing consumer demands and geographically separated energy sources, large power exchanges over long transmission lines play a vital role in the secure and economic operation of modern power systems. As a power system becomes more stressed with increasing loads, the need to transfer power over long transmission lines increases. Deregulation of the power supply has introduced new power-flow patterns to bulk transmission systems. The net result is that power systems operate much closer to their transfer limits and operate for longer times than was previously necessary. With political, economical and environmental restrictions on the development of new transmission and generation facilities, it is critical to utilise the total capability of existing transmission lines while also maintaining adequate system reliability.

The generation of electricity using wind power sources has received considerable attention in recent years. It has been reported that targets have been set to generate ten to fifteen percent of the world's electricity from wind power by 2020 [2]. Wind farms are generally erected in remote areas and it is difficult to control their voltages at these distant places by the use of synchronous generators located at substantial distances from them. As more and more attention is being paid to the increased use of wind farms, a number of complex issues need to be investigated in more detail.

Voltage control assessments and reactive power compensations play an increasingly important roles during planning and development for determining secure transfer limits for large-scale wind power plants in areas distant from the main power transmission system. It is important to consider the dynamics of wind farms in order to accurately determine the transfer capability. The integration of large-scale wind generators into power systems have crucial effects on the existing transfer capability of a transmission line [3].

Power system stability issues and thermal constraints limit transmission capacity. To meet the increasing load demand and satisfy the stability and reliability criteria, either existing transmission and generation facilities must be utilised more efficiently, or new facilities added to the systems. Given the constraints, such as lack of investment and difficulties in getting new transmission line rights-of way, the later is often difficult. The former can be achieved by using flexible AC transmission system (FACTS) controllers, as seen in well-developed power systems throughout the world. The obvious and most effective way to increase a transmission capability (beyond its thermal rating) is to build a new transmission line. However, this is an expensive solution.

The risk of voltage collapse determines the transfer limits in many bulk transmission systems. If voltage limits and voltage stability are the determining factors for the transfer capability, additional sources of reactive power can be installed at critical locations in order to smooth the voltage profile and to increase the reserves against the loss of voltage stability. FACTS devices with suitable controllers allow increased utilisation of existing networks closer to its thermal loading capacity and avoid the need to construct new transmission lines. Among different FACTS devices, static synchronous compensators (STATCOMs) is being increasingly used for enhancing dynamic voltage stability. They have the potential to significantly increase the transient stability margin as well as the voltage stability of a system.

The transfer capabilities of inter-tie transmission lines establish how much power can be exchanged between the areas without compromising system viability, voltage security or dynamic security [4]. In heavily loaded systems, the voltage stability limit dominates, and voltage instability is observed following a large disturbance in a heavily stressed power system with interconnections separated by long distances. Recently, more attention has been paid to maintaining a healthy voltage profile and design controller for voltage stability. Sufficient attention has not been paid to voltage stability in the determination of the available transmission capability (ATC) compared with that to angle stability.

To determine the ATC, methods considering thermal and static power-flow analyses have been widely used. In the literature, approaches for the determination of the steady-state limit, taking into account the system limitations linked to the steady-state conditions such as maximum loadability, bus voltage and transmission current limits, have been presented [5]. A technique based on determining voltage stability limits directly associated with voltage collapse conditions (saddle node bifurcation) is proposed in [6].

An algorithm is presented in [7] to evaluate the first-order effects of network uncertainties, such as load forecast error and simultaneous transfers, on the calculated transfer capability. The dynamics of power system devices are not considered in these papers [6, 7]. Recently, the effects of wind power (WT) integration on the total transfer capability have been investigated using an optimal power-flow technique [8]. This technique does not consider wind farm dynamics and the wind turbine is modelled as a PQ bus. The behaviour of a WT, during and after disturbances, is different from that of a conventional generator. During a fault, a fixed-speed induction generators (FSIG) draws a large amount of reactive power from the system which demonstrates the need to consider the dynamics of wind farms when calculating the dynamic ATC.

Other research directions aim to utilise FACTS devices to enhance the transfer capability of certain lines. If voltage limits and voltage stability are the determining factors for the transfer capability, additional sources of reactive power can be installed at critical location in order to smooth the voltage profile and increase the reserves against the loss of voltage stability. FACTS devices with suitable controllers allow increased utilisation of the existing network, closer to its thermal loading capacity, and avoid the need to construct new transmission lines. Optimal placement of FACTS controllers has been studied to maximise the available transfer limit using second order sensitivity analysis [9]. This approach utilises standard voltage collapse techniques and a variety of static system limits. An optimal power-flow based ATC enhancement model has been formulated to achieve the maximum power transfer for a specified interface with FACTS control, and in which voltage limits and line thermal limits are considered [10].

The impact of FACTS devices on the ATC and its enhancement has been studied using a genetic algorithm to optimise the best location for a static VAr (volt-ampere reactive) compensator (SVC) [11]. In [12], the effects of FACTS devices on the total transfer capability, considering thermal, voltage and transient stability limits of the system, are discussed. Fuzzy control based active and reactive power control of the super-conducting magnetic energy storage (SMES) unit, as well as control of transmission line impedance by the static synchronous series capacitor (SSSC) have been studied to increase the maximum loadability of transmission lines which may be constrained by the transient stability limit [13]. Most of the existing works on the enhancement of the transfer limit use local FACTS controllers which are designed based on linear models. However, conventional generators and wind generators are highly nonlinear and are coupled with each other.

Modern nonlinear large-scale power systems need increasingly sophisticated controllers which require information from the overall network. This implies high transmission costs especially when large distributed systems are used. On the other hand a decentralised controller is effective and cheap since it does not require information exchange between generator units and is based on local measurements. Because the plant structure and parameter uncertainties always exist, it is also very important to design controllers that are robust in terms of modelling uncertainties. An output-feedback robust decentralised switching control has recently been proposed [14]. In it the operating range is divided into several intervals; one controller is designed for each interval and the controllers are switched depending on the operating point. In real power systems, it is difficult to implement the switching controllers as unwanted transients may arise due to the switching.

A decentralised nonlinear controller for a large-scale power system based on the input-output feedback linearisation (FBL) methodology, is proposed in [15]. The controller designed using FBL requires the information about the power system's topology and the states must be measurable. In practice, it is very difficult to measure all states of a power system. In addition, feedback linearisation schemes need exact plant parameters to cancel the inherent system nonlinearities which, for the stability analysis, is an involved task.

Wind generator dynamics have considerable effects on both the voltage and transient stabilities of a power system as well as on its transfer limits. Literature dealing with the impacts of wind generator dynamics on the ATC is scarce. Section 6.5 describes the method used in this chapter to determine the dynamic ATC and also to analyse the important effects of wind power penetration on it. Among different FACTS devices, STATCOM is being increasingly used for enhancing dynamic voltage stability. STATCOMs with a suitable control strategy have the potential to significantly increase the transient stability margin as well as voltage stability of the system.

The main contribution in this chapter is to present a method for designing a decentralized robust STATCOM controller which enhances dynamic voltage stability as well as transient stability and, thereby, increases the ATC. For controller design, we use the decentralised minimax linear quadratic (LQ) output-feedback control design technique. Within the minimax optimal control design framework, robustness is achieved via optimisation of the worst-case quadratic performance of the underlying uncertain system [16]. This method achieves an acceptable trade-off between the system's control performance and robustness. The control design in

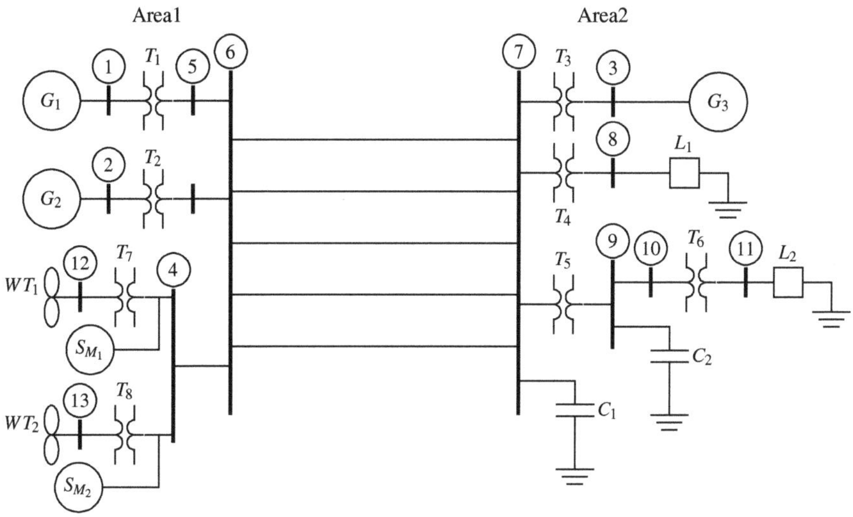

Fig. 6.1 3 machine 2 area test system (G-synchronous generator, WT-wind generator, S-STATCOM and L-load)

this chapter is tested by simulations under various types of disturbances on a test system. For comparison purposes, the performance of a proportional-integral (PI)-based STATCOM controller is also evaluated [17]. The comparison shows the superiority of the designed control method over the conventional PI-based STATCOM controller.

The organisation of the rest of this chapter is: Sect. 6.2 provides the mathematical modelling of the power system devices under consideration and discusses the test system; Sect. 6.3 describes the purposes of determining the ATC and the factors limiting it; Sect. 6.5 presents an algorithm for determining the ATC; Sect. 6.7 summarises the application of the decentralised robust control design technique and presents an algorithm for designing the controller; Sect. 6.6 describes different case studies; Sect. 6.9 discusses the performance of the controller; and Sect. 6.10 provides a summary of this chapter.

6.2 Power System Model

The two area test system shown in Fig. 6.1 consists of 11 buses and 3 generators [18]. It has an area fed by a remote generator G_2 with a nominal capacity of 2,200 MVA through five 500 kV parallel lines. Generator (G_1) models an infinite bus representing a large-inertia interconnected system.

Area 2 contains a 1,600 MVA local synchronous generator (G_3) and two aggregate loads, one industrial served directly via the off-nominal constant ratio transformer

(T_4) and the other a commercial-residential load on bus 11. All the load, P_L = 6,655 MW and Q_L = 2,021 MVAr, for this test system is in area 2 and is connected to the transmission network through two transformers (T_5 and T_6) and a 115 kV transmission line between buses 9 and 10.

The industrial load is represented by two equivalent industrial induction motors (IMs) with different parameters: a large one (IM_1) of 3,375 MVA and a small one (IM_2) of 500 MVA. The 3,000-MW commercial-residential load is half resistive of constant admittance and half motor. The 2,440-MVA single-motor equivalent (IM_3) is an aggregate of motors heavily dominated by the air-conditioning load. The test system parameters are given in Sect. 11.4.

In the simulation, the synchronous generators are modelled through a sixth-order model representing the field winding with one damper winding in the rotor direct-axis and two in the quadrature-axis. The IMs are described through their third-order model, including their rotor mechanical and electrical transients. The motor mechanical load includes both constant and quadratic torque components. Since the time-frame of interest extends to a few seconds, the response of the synchronous generators' prime movers can be neglected. As a result, the mechanical input torque of the generators is considered to be constant during the simulation. The terminal voltage of the local generator is controlled by a fast acting static automatic voltage regulator (AVR) of type IEEE ST1A, equipped with a proportional, summed-type over-excitation limiter (OXL) with a transient field current limitation block. At the initial operation point, the remote generator and the infinite bus deliver about 5,000 MW to the load area. Shunt compensation in area 2 is provided by capacitors C_1 and C_2. For stability analysis we include the transformer and the transmission line in the reduced admittance matrix.

The system described above is modified by adding two wind farms and a STATCOM at each wind farm as shown in Fig. 6.1 and used as a test system. The effect of wind generation on the ATC is analysed by varying the mix of conventional and wind generation. The control design is demonstrated for 5 % of the total generation provided by the wind generators in the two wind farms.

The aim of this chapter is to design STATCOM controllers to minimise variations in the induction generator (IG) slips and thereby improve the ATC. The controllers are designed for STATCOMs, S_{M_1} and S_{M_2}, shown in Fig. 6.1. For each STATCOM controller, the measured variable is its output voltage and the control inputs are the modulation index (k_i) and the firing angle (α_i). In this case k_i is fixed and α_i is used as the control variable. Increases in the dynamic ATC, between the two areas separated by transmission lines between bus 6 and bus 7, due to the designed controller are demonstrated in this chapter.

The ATC is determined using the full nonlinear model of the test system and the controllers are designed using its linearised model. As mentioned previously, owing to the nature of the control problem, decentralised controllers are designed for each STATCOM and an excitation controller is implemented for generator G_3.

6.3 Purpose of Transfer Capability Computations

The term 'transfer capability' refers to the amount of electric power that can be passed through a transmission network from one place to another while maintaining the power system security. This concept is useful for several reasons,

- A system which can accommodate large inter-area transfers is generally more robust and flexible than a system with a limited ability to do so. Thus, the transfer capability can be used as a rough indicator of relative system security.
- The transfer capability is also useful for comparing the relative merits of planned transmission improvements. A transmission expansion that increases the transfer capability between two areas of a grid might be more beneficial for increasing both reliability and economic efficiency than an alternative improvement that provides a lesser increase in the transfer capability.
- Along similar lines the transfer capability can be used as a surrogate for more specific circuit modelling to capture the gross effects of multi-area commerce and provide an indication of the amount of inexpensive power likely to be available to provide generation to deficient and/or high-cost regions.
- Transfer capability computations facilitate energy markets by providing a quantitative basis for assessing transmission reservations.

6.4 Factors Limiting Transfer Capability

The ability of interconnected transmission networks to reliably transfer electrical power may be limited by the physical and electrical characteristics of the systems including any one or more of the following:

6.4.1 Thermal Limit

Thermal limits are due to the thermal capabilities of power system equipment. As the power transfer increases, the current magnitude increases which is a key to thermal damage; for example, in a power plant, the sustained operation of units beyond their maximum operating limits may result in thermal damage to the stator or rotor windings of the unit. Both active and reactive power play roles in the current magnitude. Out in the system, transmission lines and associated equipment must also be operated within thermal limits. A sustained excessive current flow on an overhead line causes the conductors to sag, thus decreasing the ground clearance and reducing safety margins. Extreme levels of current flow will eventually damage the metallic structure of the conductors and cause a permanent sag. Unlike overhead lines, underground cables and transformers must depend on insulation other than air to dissipate their generated heat. These types of equipment are tightly restricted in the amount of current they can safely carry as sustained over-loading will result in a reduction in their services lives due to the resultant damage to their insulation. Although most power system equipment can be safely overloaded, the important aspect is by how much and for how long.

6.4.2 Voltage Limit

Both utility and customer equipment are designed to operate at a certain rated or nominal supply voltage. A large prolonged deviation from this nominal voltage, can adversely affect the performance of, as well as cause serious damage to, system equipment. Current flowing through transmission lines may produce an unacceptably large voltage drop at the receiving end of a system. This is primarily due to the large reactive power loss which occurs as the current flows through the system. If the reactive power produced by generators and other sources is not sufficient to supply a system's demand, the voltage will fall outside the acceptable limits, that is, typically 6 % or more around the nominal value. A system often requires reactive support to help prevent low-voltage problems. The amount of available reactive support often determines the power transfer limits. A system may be restricted to a lower level of active power transfer than desired because the system does not posses the required reactive power reserves to sufficiently support the voltage.

6.4.3 Stability Limit

A transmission network must be capable of surviving disturbances through transient and dynamic time-periods (from milliseconds to several minutes, respectively) following a disturbance. All generators connected to an AC interconnected transmission system operate in synchronism with each other at the same frequency (nominally 50 or 60 Hz). Immediately following a system disturbance, the generators begin to oscillate relative to each other, causing fluctuations in system frequency, line loadings and system voltages. For a system to be stable, the oscillations must diminish as the electrical system attains a new stable operating point, which if not quickly established, will likely cause the generators to lose synchronism with one another and all, or a portion, of the interconnected electrical system may become unstable. Generator instability may damage equipment and cause uncontrolled, widespread interruption of electrical supply to customers.

6.5 Algorithm for Dynamic ATC Assessment

Practical methods to compute the transfer capability are evolving. The computations presently being implemented are usually oversimplified and in many cases do not take sufficient account of effects such as interactions between power transfers, loop flows, nonlinearities, operating policies and voltage collapse blackouts. A number of methods for computing the ATC have been reported in the literature. Continuation power flow (CPF) methods [7] repeat a full-scale AC load-flow solution for each increment of the load above the base case value at the sink bus until a line in the system is overloaded. Although accurate, these methods are not real-time compatible for large systems. As an alternative, there exist DC load flow based methods [19] which are a bit faster than their AC counterparts but they model only real power flow in the lines and assume the network to be without losses.

Methods based on power transfer or outage distribution factors [20] can only cater to the scenarios that are too close to the base case from which these factors are derived. The reported [21] artificial neural network (ANN) method requires a large input vector so that it has to oversimplify the determination of the ATC by limiting it to a special case of power transfer to a single area from all of the remaining areas. This method is unable to track down the bus-to-bus transactions which is the true spirit of deregulation.

In this research, the following procedure is used to analyse the ATC and verify the performance of the designed controller [22]:

1. select the base case and solve the power flow;
2. make step increases in generation and load, and solve the power-flow problem according to the modified system conditions;
3. conduct a stability analysis to check the security limit for large disturbances using the complete nonlinear model;
4. if the security limit is acceptable, go to (2), otherwise go to (5); and
5. the highest feasible increment denotes how much power can additionally be transmitted for the given base scenario.

6.6 Case Studies

The ATC is calculated by increasing the load in Area 2 and the generation in Area 1 in the test system shown in Fig. 6.1. For a security assessment, we apply a three-phase fault at one of the transmission lines between bus 6 and bus 7. The fault is cleared after five cycles by opening the line and the line is restored after a further five cycles. In the load-flow analysis, the acceptable voltage range considered in this case is 0.9–1.1 pu. The dynamic ATC with a conventional generator, evaluated using the method in Sect. 6.5 for the three-phase fault contingency, is 690 MW. Different cases have been conducted in the following section for determining the effects of power system devices on the ATC.

6.6.1 Case I: Generator Reactive Power Limit

Generators have reactive power output limits. After a limit is reached, a generator will not be able to regulate its bus voltage and so it is degraded from a PV bus to a PQ bus. The change in load voltage at bus 11 with increasing transfer from area

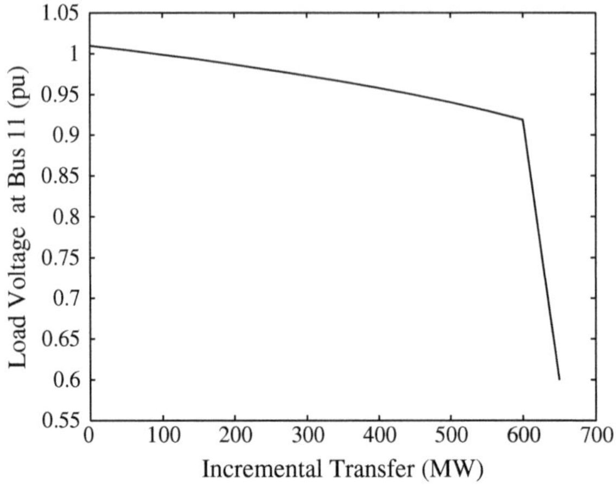

Fig. 6.2 Steady-state transfer capability

1 to area 2 is shown in Fig. 6.2. The ATC is determined for (i) imposing an upper limit of the reactive power supply, and (ii) without forcing the upper limit. The ATC reduces from 690 to 678.5 MW due to the restriction of the reactive power supply from G_3. The voltage collapses when the generator exceeds its maximum reactive power capability, as shown in Fig. 6.2.

6.6.2 Case II: Effects of Dynamic Load

The active power consumption of an induction motor (IM) remains constant even after a voltage drop [23]. Moreover, if the disturbance is such that the electrical torque cannot balance the mechanical load, the motor can decelerate, absorbing higher reactive current, thus inducing a further voltage drop and occasionally a voltage collapse.

Firstly, the ATC is calculated considering only constant impedance load. The stability analysis is carried out with the same load but a 25 % induction motor is added. The power flow through one of the lines with a constant impedance load and a combination of static and dynamic load is shown in Fig. 6.3; from which, it is clear that the dynamic ATC is reduced when the induction motor load is added. The dynamic load reduces the ATC by 5.3 %.

6.6.3 Case III: Effects of Fault Clearing Time

Fault conditions such as fault location and fault duration time are major factors in determining the system stability and thus affects the ATC. A fault condition varies greatly based on the nature of fault and protection device/scheme applied and it

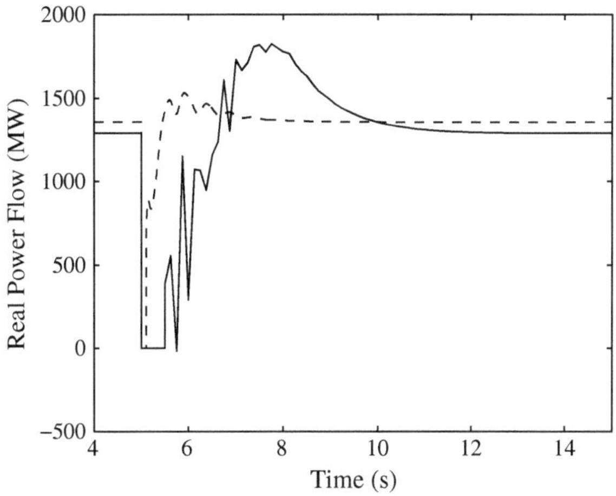

Fig. 6.3 Power flow through one of lines 6–7 for three-phase fault at middle of one of lines 6–7 (*dashed line* static load and *solid line* dynamic load)

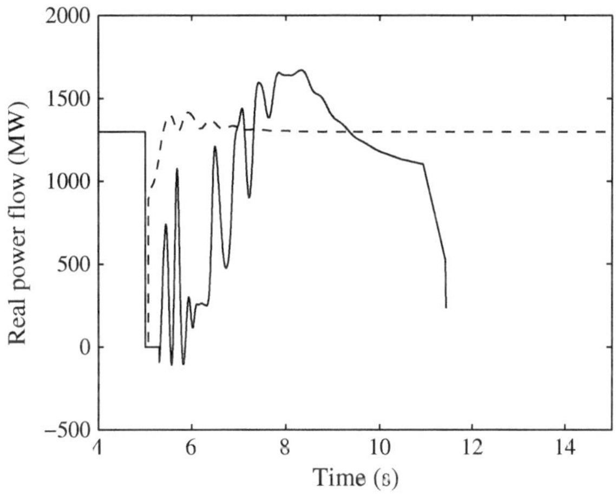

Fig. 6.4 Real power flow through one of lines 6–7 for three-phase fault at middle of one of lines 6–7 (*solid line* FCT − 0.3 s and *dashed line* FCT = 0.15 s)

significantly affects the transfer capability. The dynamic ATC for a fault clearing time (FCT) of 0.15 s is 694 MW. However, the increase in fault clearing time to 0.3 s, reduces the ATC by 0.75 %. A three-phase fault is applied on one of the transmission lines at 5 s and cleared after 0.15 and 0.3 s respectively. Figure 6.4 shows the power flow with same load conditions but for different fault clearing times. From Fig. 6.4, it is clear that for the same ATC, the system becomes stable if fault is cleared after 0.15 s but with fault clearing time 0.3 s, it is unstable.

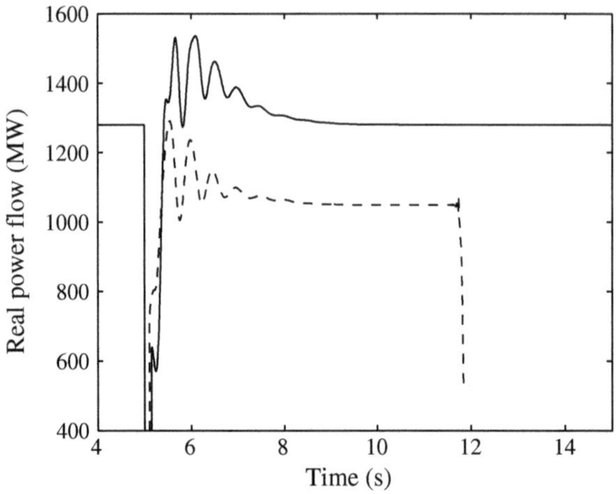

Fig. 6.5 Power flow through one of the lines 6–7 for three-phase fault at middle of one of lines 6–7 (*solid line* STATCOM and *dashed line* shunt capacitance)

6.6.4 Case IV: Effects of Static and Dynamic Compensations

Shunt capacitors have the problem of poor voltage regulation and, beyond a certain level of compensation; a stable operating point is unattainable. Furthermore, the reactive power delivered by a shunt capacitor is proportional to the square of the terminal voltage, during low-voltage conditions the VAr support drops, thus compounding the problem. Figure 6.5 shows the power flows through line 6–7(2) with a shunt capacitor and a STATCOM with a dynamic load. The transfer capability with a static synchronous compensator (STATCOM) is the same for both static and dynamic loads but, with shunt compensation, the ATC reduces by 6.01 % when a dynamic load is added.

6.6.5 Case V: Comparisons of Dynamic Compensating Devices

Since the ATC values are always ultimately limited by heavily loaded circuits and/or nodes with relatively low voltages, the use of the FACTS devices has a potential impact on the ATC. As FACTS devices can control the circuit reactance, voltage magnitude and phase angle, it enables line loading to increase flexibly and, in some cases, all the way up to the thermal limits. In this section, we compare the transfer capability enhancement by using the same capacity (100 MVA) of STATCOM, super-magnetic energy storage (SMES) unit and switched shunt, as shown in Fig. 6.6. The results show that the SMES, which is able to act with active and reactive simultaneous control, is more effective than both the STATCOM and the switched shunt. The

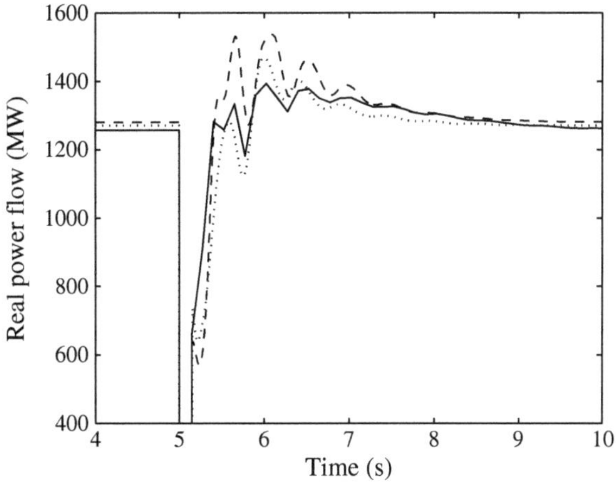

Fig. 6.6 Power flow through one of the lines 6–7 for three-phase fault at middle of one of lines 6–7 (*dotted line* SMES unit, *dashed line* STATCOM and solid line switched capacitance)

Table 6.1 Effects of FSIG on ATC

FSIG integration in %	0	2.5	5	7.5	10
ATC in MW	690	643.5	607	545	90

power flow through one of the lines of 6–7 with switched capacitance is 1,257 MW. The STATCOM increases the power flow by 5.85 % and the SMES unit by 9.38 %.

6.6.6 Effects of Wind Generator Integration on ATC

To show the effects of the wind turbines on the ATC, a portion of the generation provided by G_1 and G_2 is replaced by the wind farms. In the first case, 5% of the conventional power in Area 1 is replaced by fixed-speed wind turbines (FSWTs). The ATC is reduced to 569 MW compared with 690 MW for conventional generation only.

In the second case, we use variable-speed wind turbines (VSWTs) instead of FSWTs. A VSWTs with a doubly-fed induction generator (DFIG) uses constant terminal voltage or unity power factor operation depending on the operating conditions. For the same penetration level and VSWTs with terminal voltage control, the ATC is 689 MW. The ATC is 720 MW with the VSWT operated in the power factor control mode (0.95 lagging). It is clear that, with VSWTs in voltage control mode, the ATC changes only slightly whereas the FSWT reduces it by 17.53 %. The VSWT operating at a lagging power factor of 0.95 increases the ATC by 4.34 %.

The variation of the ATC with the varying amount of penetration of FSWTs is shown in Fig. 6.7. The numerical values are given in Table 6.1. It can be seen in Fig. 6.7 that, initially, the ATC decreases gradually but after a 7.5 % penetration, it

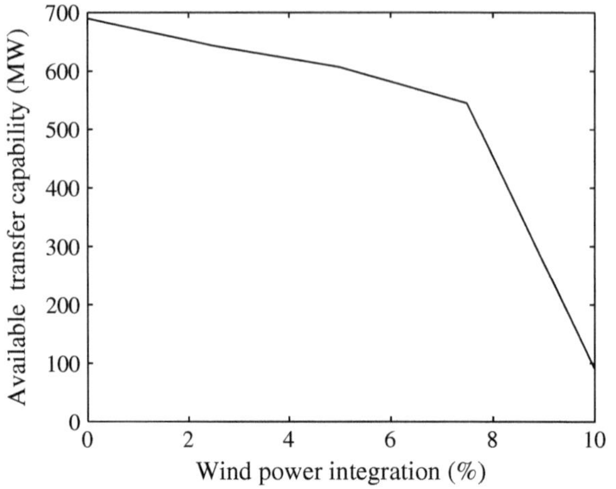

Fig. 6.7 Change in ATC due to FSIG integration

Table 6.2 Compensations to restore ATC

FSIG integration in %	0	2.5	5	7.5	10
STATCOM in MVA	0	12.5	25	44	81
Capacitor in MVAr	0	60	100	160	300

decreases sharply. This sudden drop is due to the non-convergence of power-flow and it underscores the importance of a thorough analysis before replacing conventional generation with wind power.

6.6.7 Compensations to Restore ATC with FSIG

Wind generators are generally connected to power capacitors to improve the power factor. The amount of compensations (both static and dynamic) required to restore the deficiency in ATC due to FSWT integration is shown in Fig. 6.8. Numerical values are shown in Table 6.2.

For a 5 % wind power integration, 2×50 MVAr capacitors are required to restore the ATC to 690 MW. Two 12.5 MVAr capacity STATCOMs can replace the 2×50 MVAr static capacitors and provide a superior dynamic response. The cost of capacitors is \$10–\$20 per kVAr and that of STATCOMs \$55–\$70 per kVAr for systems with a capacity of 100 MVAr or more [24]. The use of STATCOMs does not reduce the cost but they significantly enhance dynamic performance. The reactive power delivered by the shunt capacitor is proportional to the square of the terminal voltage which means that, during low-voltage conditions, VAr support drops, thus compounding the problem. The STATCOM provides extra reactive power sup-

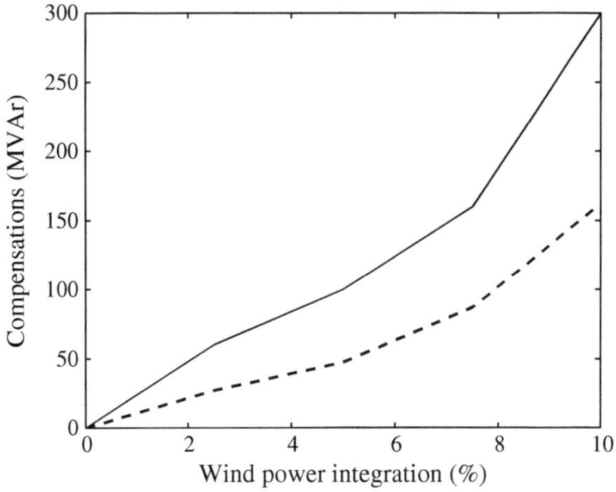

Fig. 6.8 Compensations to restore the ATC (*solid line* capacitor and *dashed line* STATCOM)

port dynamically with a continuous change of output for voltage recovery when the voltage becomes low.

6.7 Decentralised Robust Control

This section presents all the equations required to design decentralized STATCOM controllers. The controller is designed using decentralized output-feedback control synthesis method [14, 25]. The power system model used in this section is described by the following form where a large-scale system (S) is comprised of N subsystems ($S_i, i = 1, 2, \ldots, N$) so:

$$S_i : \dot{x}_i(t) = A_i x_i(t) + B_i u_i(t) + E_i \xi_i(t) + L_i r_i(t), \tag{6.1}$$

$$z_i(t) = C_i x_i(t) + D_i \xi_i(t), \tag{6.2}$$

$$\zeta_i = H_i x_i(t) + G_i u_i(t), \tag{6.3}$$

$$y_i = C_{y_i} x_i(t) + D_{y_i} \xi_i(t), \tag{6.4}$$

where $x_i \in R^{ni}$ is the state vector, $u_i \in R^{mi}$ the control input, $\xi_i \in R^{pi}$ the perturbation, $\zeta_i \in R^{hi}$ the uncertainty output, $z_i \in R^{qi}$ the controlled output, $y_i \in R^{gi}$ the measured output and the input (r_i) describes the effects of the other subsystems $S_1, \ldots, S_{i-1}, S_{i+1}, \ldots, S_N$ on subsystem S_i. The structure of system S is shown in Fig. 6.9.

The system model (6.1)–(6.4) reflects the nature of a generic interconnected uncertain system in which each subsystem is affected by uncertainties that have two sources. Local uncertainties in a large-scale system arise from the presence

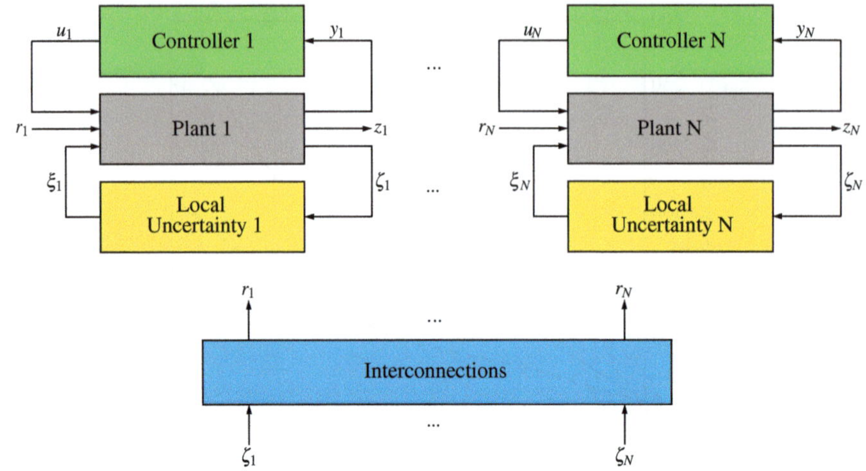

Fig. 6.9 Block diagram of uncertain system

of uncertain dynamics in each subsystem. Such dynamics are only driven by the uncertainty output (ζ_i) of the subsystem (S_i). A second source of uncertainties arises from interactions between the subsystems of the large scale system. Indeed, the partition of a complex uncertain system into a collection of subsystems (S_i) results in the uncertainty in the original system being distributed amongst the subsystems. This provides the motivation for treating the interconnections as uncertain perturbations.

The matrices D_i, G_i and D_{yi} are chosen in such a way that the following assumptions hold [14]:

- **Assumptions 1**: For all $i = 1, \ldots, N$, $D_i^T D_i + G_i^T G_i > 0$, $D_{y_i} D_{y_i}^T > 0$.
- **Assumptions 2**: The pair $(A_i, C_i^T C_i)$, $i = 1, \ldots, N$, is observable.
- **Assumptions 3**: The pair (A_i, B_i), $i = 1, \ldots, N$, is stabilisable.

We also define $\xi_i = \Delta_i \zeta_i$ and $r_i = \sum_{j \neq i} \tilde{\Delta}_{ij} \zeta_j$, where Δ_i and $\tilde{\Delta}_{ij}$ are uncertain gain matrices. The uncertainty and interconnections satisfy the following conditions:

$$\|\xi_i(t)\|^2 \leq \|\zeta_i(t)\|^2 \text{ and } \|r_i(t)\|^2 \leq \sum_{j \neq i} \|\zeta_j(t)\|^2. \tag{6.5}$$

The minimax output-feedback controller designed in this section minimises the following cost subject to the above (6.5) bounds on the local uncertainty and interconnections:

$$\int_0^\infty \sum_{i=1}^N \|z_i(t)\|^2 dt. \tag{6.6}$$

In this case we consider norm bounded constraints, as in (6.5), instead of the more general IQCs (integral quadratic constraints). This means that the designed controllers are suboptimal for norm bounded constraints. As described in [14], the control algorithm finds the infimum of the following function over the set $\mathscr{T}$:

$$\inf_{u_i,\, i=1,\ldots,N} \sup_{\Xi,\Pi} \int_0^\infty \sum_{i=1}^N \|z_i(t)\|^2 dt \leq \inf_{\mathscr{T}} \sum_{i=1}^N x_{i0}^T \left[X_i + \tau_i M_i + \theta_i \bar{M}_i\right] x_{i0}, \quad (6.7)$$

where $[x_{10}, \ldots, x_{N0}]^T$ is the initial condition vector, Ξ a set of all admissible uncertainties, Π a set of admissible interconnection inputs, a set of vectors $\mathcal{T} = \{\{\tau_i\ \theta_i\}_{i=1}^N \in R^{2N}\}$, and $M_i > 0$ and $\bar{M}_i > 0$ are two positive definite symmetrical matrices which satisfy the following conditions:

$$E \int_0^{t_l} \left(\|\zeta_i(t)\|^2 - \|\xi_i(t)\|^2\right) dt > -x_{i0}' M_i x_{i0}, \quad (6.8)$$

$$E \int_0^{t_l} \left(\sum_{n=1,n\neq i}^N \|\zeta_n(t)\|^2 - \|r_i(t)\|^2\right) dt > -x_{i0}' \bar{M}_i x_{i0}, \quad (6.9)$$

where E is the expectation operator, $\{t_l\}_{l=1}^\infty, t_l \to +\infty$ a sequence, and $M_i = M_i' > 0$, $\bar{M}_i = \bar{M}'_i > 0$. Equations (6.8) and (6.9) allow for the effects of the non-zero initial conditions of the uncertain dynamics in the local uncertainty channels and interconnections to be taken into account. The terms on the right hand sides of IQCs (6.8) and (6.9) correspond to the bounds on these uncertainties. These bounds can be written in quadratic forms as: $x_{i0}' M_i x_{i0}$ and $x_{i0}' \bar{M}_i x_{i0}$ [16].

The matrices X_i and Y_i are the solutions to the following pair of parameter-dependent coupled generalised algebraic Riccati equations [25]:

$$A_i^T Y_i + Y_i A_i + Y_i \bar{B}_{2_i} \bar{B}_{2_i}^T Y_i - \left[C_{y_i}^T W_i^{-1} C_{y_i} - \bar{C}_i^T \bar{C}_i\right] = 0, \quad (6.10)$$

$$A_i^T X_i + X_i A_i + \bar{C}^T{}_i \bar{C}_i - X_i \left[B_i R_i^{-1} B_i^T - \bar{B}_{2_i} \bar{B}_{2_i}^T\right] X_i = 0, \quad (6.11)$$

where $R_i = \bar{D}_i^T \bar{D}_i$, $W_i = \bar{D}_{y_i} \bar{D}_{y_i}^T$, $\bar{\theta}_i = \sum_{n=1,n\neq i}^N \theta_n$,

$$\bar{C}_i = \begin{bmatrix} C_i \\ (\tau_i + \bar{\theta}_i)^{1/2} H_i \end{bmatrix}, \quad \bar{D}_i = \begin{bmatrix} D_i \\ (\tau_i + \bar{\theta}_i)^{1/2} G_i \end{bmatrix},$$

$$\bar{B}_{2_i} = \left[\tau_i^{-1/2} E_i\ \theta_i^{-1/2} L_i\right], \quad \bar{D}_{y_i} = \left[\tau_i^{-1/2} D_{y_i}\ 0\right].$$

Then the controller is designed with the equations [14]:

Table 6.3 Participation factors

States	Δs_1	$\Delta E'_{qr_1}$	Δs_2	$\Delta E'_{qr_2}$	$\Delta\omega_3$	$\Delta\delta_3$
Participation factor	0.96	1.0	0.94	0.97	0.89	0.32

$$\dot{x}_{c_i} = \{A_i - \left[B_i R^{-1} B_i^T - \bar{B}_{2_i} \bar{B}_{2_i}^T\right] X_i\} x_{c_i} + [Y_i - X_i]^{-1} C_{y_i}^T W_i^{-1} \left[y_i(t) - C_{y_i} x_{c_i}(t)\right], \tag{6.12}$$

$$u_i = -R_i^{-1} B_i^T X_i x_{c_i}. \tag{6.13}$$

The solutions are required to satisfy the following conditions: $\tau_i > 0, \theta_i > 0, X_i \geq 0$, $Y_i \geq 0$ and $Y_i > X_i$.

6.8 Controller Design for Test System

The problem considered here is the design of a robust LQ output-feedback decentralised STATCOM controller which works in the presence of interconnection effects. To demonstrate the control design process the controllers are designed for two 12.5 MVAr STATCOMs connected to the two wind farms and an excitation controller for generator G_3. The STATCOMs (S_{M_1}, S_{M_2}) and the wind generators (WT_1, WT_2) are shown in Fig. 6.1. The two wind farms are equipped with FSIGs which supply 5 % of the total load and each is represented by an aggregated wind generator model [26].

Modal analysis is performed on the interconnected system to obtain an idea of the dominant modes which need to be controlled [18]. The dominant mode for the test system is $-0.098 \pm j3.463$ with a damping ratio 0.028. The normalised significantly contributing participation vector for the dominant mode is shown in Table 6.3. From the participation vector, it is clear that, as both the wind farms contribute significantly to the dominant mode, controllers should be designed for both the wind generators. This emphasises the need to design decentralised controllers.

The test system considered in this chapter is divided into three subsystems: (1) Wind Farm 1 and STATCOM 1; (2) Wind Farm 2 and STATCOM 2; and (3) G_1, G_2, and G_3. STATCOM controllers are designed for subsystems 1 and 2, and an excitation controller for subsystem 3 which is implemented on generator G_3. All the generators and exciters are represented by an aggregated equivalent seventh-order model [27].

The first step in designing controllers for the STATCOMs is to determine the matrices and define the variables in problem formulation (6.1)–(6.4). To obtain the subsystem matrices in (6.1)–(6.4), the complete system is first linearised about the desired equilibrium point. For each subsystem the state variables are divided into two parts. One part consists of the states of the devices in the subsystem, called x_i and the other part consists of the rest of the states, called r_i. The matrices A_i and L_i are appropriately chosen from the complete linearised model equations.

6.8.1 Subsystems 1 and 2

The uncertainty output (ζ_i) and the perturbation input (ξ_i) are chosen such that

$$\zeta_i = \left[\Delta s_i, \Delta E_{dr_i}, \Delta E_{qr_i}, \Delta v_{dc_i}\right]^T \text{ and } \xi_i = \zeta_i.$$

Owing to this choice of uncertainty output and perturbation input, the inequalities in (6.5) are satisfied. The state vector for the wind farm subsystems is ($i = 1, 2$):

$$x_i = \left[\Delta\omega_{wt_i}, \Delta\omega_{m_i}, \Delta\gamma_i, \Delta s_i, \Delta E_{dr_i}, \Delta E_{qr_i}, \Delta v_{dc_i}, \Delta v_{tm_i}\right]^T.$$

The uncertainty term, represented by $E_i\xi_i$, is obtained by increasing the load by 10 %, finding the new equilibrium point, linearising the system about that point, and taking the difference between the subsystem A-matrices for the nominal load and the increased load [28]. This difference in the A-matrices is E_i.

For the subsystems with wind generators ($i = 1, 2$):

$$C_i = \left[0\,0\,0\,1\,0\,0\,0\,0\right], \quad C_{yi} = \left[0\,0\,0\,0\,0\,0\,0\,1\right], \quad H_i = \begin{bmatrix} 0\,0\,0\,1\,0\,0\,0\,0 \\ 0\,0\,0\,0\,1\,0\,0\,0 \\ 0\,0\,0\,0\,0\,1\,0\,0 \\ 0\,0\,0\,0\,0\,0\,1\,0 \end{bmatrix}. \tag{6.14}$$

The above choice of matrices means that the controlled output is the variation in the induction generator slip and the measured output is the change in the STATCOM terminal voltage. The control input is the firing angle α_i of the STATCOMs.

Matrices D_i, G_i, and D_{yi} for the subsystems with wind generators are chosen as:

$$D_i = 10^{-4}\left[1\,1\right], \; G_i = 10^{-6}\begin{bmatrix} 1\,1 \\ 1\,1 \\ 1\,1 \\ 1\,1 \end{bmatrix}, \; D_{yi} = 10^{-4}\left[0\,0\,0\,1\,1\,1\,1\,0\right],$$

$$x_{i0} = \left[0.1 \ldots 0.1\right]^T. \tag{6.15}$$

6.8.2 Subsystem 3

The uncertainty output (ζ_3) and the perturbation input (ξ_3) are chosen such that

$$\zeta_3 = \left[\Delta E'_q, \Delta E'_d, \Delta\psi_{1d}, \Delta\psi_{2q}\right]^T \text{ and } \xi_3 = \zeta_3.$$

The matrices for subsystem 3 with all the generators are:

$$C_3 = \begin{bmatrix}1\,0\,0\,0\,0\,0\,0\end{bmatrix}, \quad C_{y3} = \begin{bmatrix}0\,1\,0\,0\,0\,0\,0\end{bmatrix}, \quad H_3 = \begin{bmatrix}0\,0\,1\,0\,0\,0\,0\\0\,0\,0\,1\,0\,0\,0\\0\,0\,0\,0\,1\,0\,0\\0\,0\,0\,0\,0\,1\,0\end{bmatrix}. \tag{6.16}$$

This choice means that the controlled variable is the generator angle deviation and the measured variable is the speed deviation from the synchronous speed. The other matrices in this subsystem model representation are:

$$D_3 = 10^{-6}, \quad G_3 = 10^{-6}\begin{bmatrix}0\,0\,0\,1\end{bmatrix}^T, \quad D_{y3} = 10^{-4}\begin{bmatrix}0\,0\,1\,1\,1\,1\,0\end{bmatrix}, \tag{6.17}$$
$$x_{30} = [0.1, \ldots, 0.1]^T.$$

The matrices M_i and $\bar{M}_i$ can be chosen to be arbitrary positive definite matrices; we select them as identity matrices.

The algorithm for designing the proposed controller can be summarised as follows,

- **Step 1**: For a given equilibrium point, obtain the matrices in (6.1)–(6.4) according to the procedure outlined in Sect. 6.7;
- **Step 2**: Solve the optimisation problem (6.7). This is done by using a line search technique for positive values of τ_i and θ_i. The Matlab function *fmincon* can be used to conduct the line search with a proper initialisation. In the design presented in this chapter the line search is initialised with $\tau_i = 0.0015$ and $\theta_i = 0.000015$. For this case, the function *fmincon* converges without any perceptible delay. To be certain that the solution is not numerically ill-conditioned, it is ensured that the solutions to the Eqs. (6.10)–(6.11) give positive definite X_i and Y_i for the values of τ_i and θ_i, $i = 1, \ldots, N$ at which the infimum is achieved in (6.7);
- **Step 3**: Substitute the optimising values of τ_i and θ_i into the Riccati Equations (6.10), (6.11) and obtain X_i and Y_i;
- **Step 4**: The designed controllers are given by Eqs. (6.12)–(6.13).

For the test system considered in this chapter, the optimum value of the objective function is obtained for $\tau_1 = 0.0645$, $\tau_2 = 0.0468$, $\tau_3 = 0.0167$, $\theta_1 = 0.0005$, $\theta_2 = 0.0045$, and $\theta_3 = 0.0001$. The optimal minimax value of the performance cost for the test system is 0.2156 with a 5 % wind generator integration and 2×12.5 MVA STATCOMs.

6.9 Controller Performance Evaluation

For a 5 % wind energy penetration and 2×12.5 MVA STATCOM controllers designed in this section, the dominant mode for the closed-loop system is $-0.424 \pm j0.47831$ with a damping ratio of 0.66291. From this it is clear that the closed-loop sys-

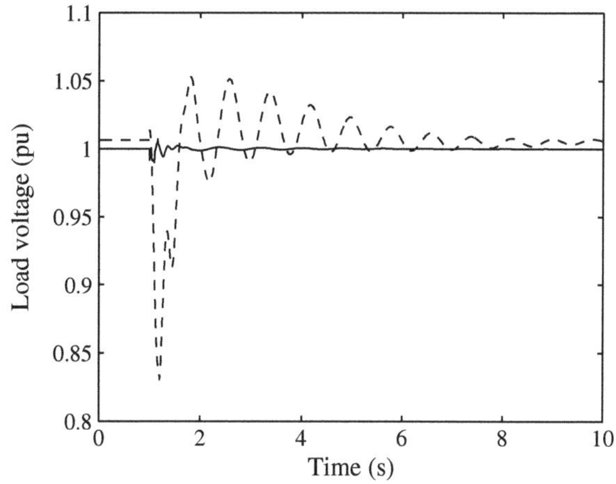

Fig. 6.10 Load voltage for outage of one of lines 6–7 (*solid line* designed STATCOM control and *dashed line* capacitor bank)

tem is well-damped. In contrast, the critical mode for the open-loop system with 2×50 MVAr capacitors is $-0.1578 \pm j2.6$ and the damping ratio is only 0.061.

The performance of the designed robust decentralised STATCOM controller is tested by simulating responses to two contingencies on the test system: (a) the outage of one transmission line; and (b) a three-phase short-circuit at the middle of one of the transmission lines between bus 6 and bus 7.

6.9.1 Outage of One Transmission Line

The simulation is performed with the line opened at 1 s and subsequently reclosed after 0.15 s. Figure 6.10 shows the load voltage at bus 11, due to the outage of one of the transmission lines between buses 6 and 7, with (i) 2×50 MVAr static capacitor banks, and (ii) the designed controllers for 2×12.5 MVA STATCOMs. The steady-state values for load voltage at bus 11 for conditions (i) and (ii) are 1.066 and 1.00 pu, respectively. From Fig. 6.10, it can be seen that STATCOM controllers provide better dynamic performances as compared with those of the 2×50 MVAr capacitor banks in terms of both damping and overshoot and settling times. This behaviour is explained by the low damping (0.061) of the open-loop system with fixed capacitor compensation. For a total of 90 MVAr and less fixed capacitor compensation, the system is not able to recover to its pre-fault voltage.

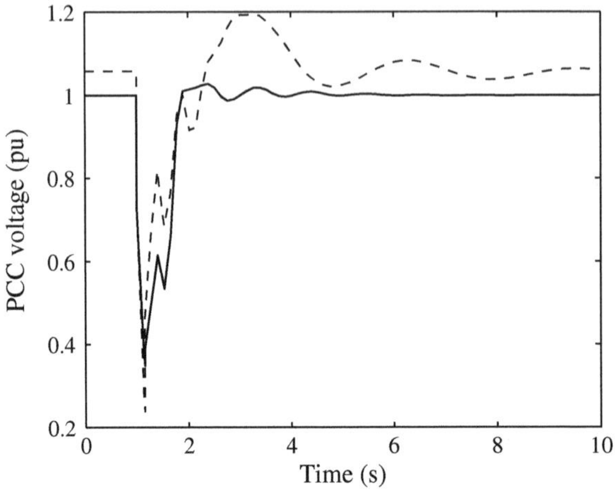

Fig. 6.11 PCC voltage for three-phase fault at middle of one of lines 6–7 (*solid line* designed STATCOM control and *dashed line* capacitor bank)

6.9.2 Three-Phase Short-Circuit

In this simulation a symmetrical three-phase fault is applied at the middle of one of the transmission lines between buses 6 and 7. The fault is cleared after five cycles. Figure 6.11 depicts variations of the PCC (point of common coupling) voltage at bus 4. The power transferred through lines 4–6 is shown in Figs. 6.12 and 6.13 from which, it can be concluded that more power can be transmitted through the transmission line with the designed STATCOM controllers during a transient. The reactive power drawn by wind generators and supplied by a STATCOM is shown in Fig. 6.14. It can be seen that the total reactive power output of STATCOMs is in phase of the reactive power consumed by the wind generators.

6.9.3 Comparison of Designed and PI-Based STATCOM Controllers

To evaluate the designed controller performance the dynamic ATC is calculated with 5 % of FSIGs using a PI-based STATCOM controller and the proposed robust STATCOM controller. The maximum ATC with the PI-based STATCOM is 687 MW whereas, with the proposed robust STATCOM controller, it is 698 MW, i.e., the ATC is increased by 1.6 % using this robust control algorithm.

To test the dynamic performance, a simulation is performed with the increased ATC (698 MW) by applying the same three-phase fault as in the previous simulation. Figures 6.15 and 6.16 show the load voltage at bus 11 and the speed of a wind

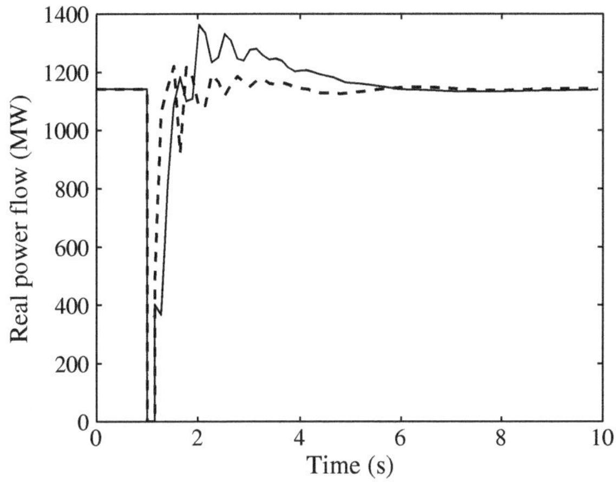

Fig. 6.12 Real power flow through one of lines 4–6 for three-phase fault at middle of one of lines 6–7 (*solid line* designed STATCOM control and *dashed line* capacitor bank)

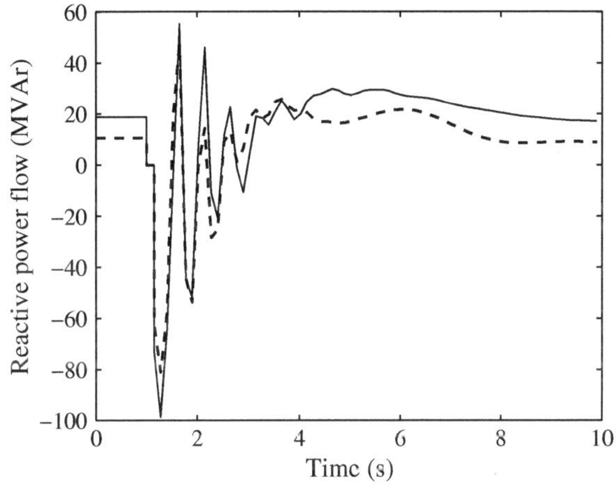

Fig. 6.13 Reactive power flow through one of lines 4–6 for three-phase fault at middle of one of lines 6–7 (*solid line* designed STATCOM control and *dashed line* capacitor bank)

generator (WF_1) with the proposed STATCOM controller and a PI-based STATCOM controller. With the PI-based controller, the speed continues to increase due to the imbalance between the mechanical power extracted from the wind and the electrical power delivered to the grid even after the fault is cleared. When a disturbance or fault occurs, the voltages at the terminals of the WT drop significantly causing the electromagnetic torque and electrical power output of the generator to be greatly reduced. However, given that the mechanical input torque is almost constant when typical non-permanent faults occur in a wind farm, this leads to an acceleration of the

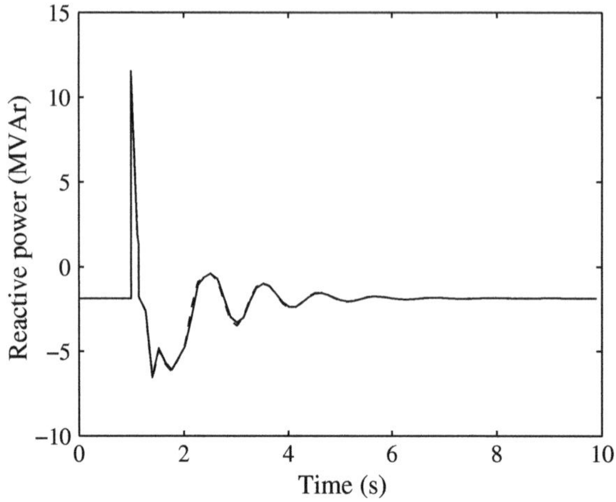

Fig. 6.14 Reactive power for three-phase fault on middle of one of lines 6–7 (*solid line* reactive power output by the designed STATCOM (S_{M_1}) controller and *dashed line* reactive power drawn by wind generator (WF_1))

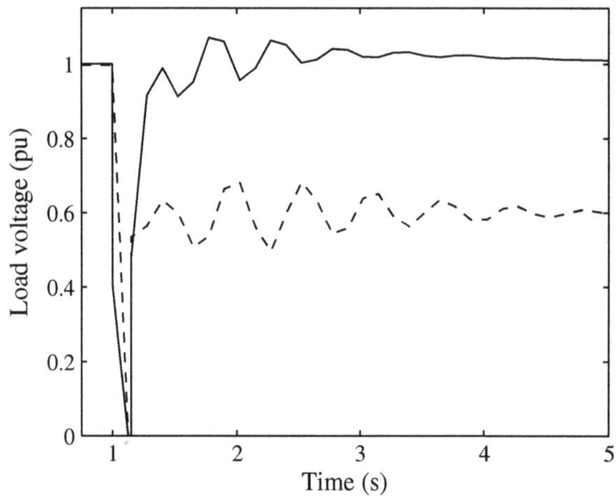

Fig. 6.15 Load voltage at bus 11 for three-phase fault at middle of one of lines 6–7 (*solid line* designed STATCOM control and *dashed line* PI-based STATCOM)

machine rotor. Furthermore, the voltages gradually decrease and the wind generators have to be disconnected from the grid to protect them and avoid voltage collapse.

The reference input for each PI-based STATCOM controller in this case has a reactive power (Q_{ref} =12.5 MVAr). For this reference reactive power, the load-flow converges for two different values of the load voltage at bus 11, 1.0 and 0.6 pu. In this simulation, the post fault voltage with the PI-based STATCOM settles to the lower 0.6 pu voltage equilibrium point. The designed controller provides a satisfactory

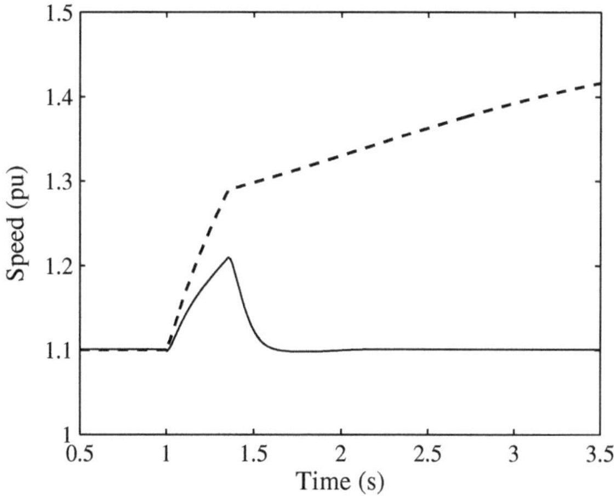

Fig. 6.16 Speed of wind generator (WF_1) for three-phase fault at middle of one of lines 6–7 (*solid line* designed STATCOM control response and *dashed line* PI-based STATCOM)

dynamic response for this contingency. From this, we conclude that the proposed controller performs better than the conventional PI-based controller in terms of both the transfer capability and dynamic performance of the system.

6.10 Chapter Summary

In this chapter, the impacts of wind farms dynamic on the ATC of a heavily stressed transmission line is investigated. As the penetration level of FSIGs increases, the ATC substantially decreases. The amounts of both static and dynamic compensation required for different levels of wind power integration are determined in order to obtain the same ATC as that of conventional generators. The STATCOM with reduced capacitors provides better performance in terms of the ATC than does a capacitor bank alone. The controller is robust in the presence of parameter interconnection effects and uncertainty. The performance of the proposed STATCOM controller is compared with a conventional PI controller based STATCOM. The dynamic voltage stability, as well as transient stability, is improved and, thereby, the ATC increases significantly when the designed robust STATCOM controller is applied instead of a conventional PI-based STATCOM.

Today, the focus regarding interactions between an electrical network and wind energy installation has shifted since the loss of such a considerable part of power production (as wind energy constitute in some regions) due to network disturbances can no longer be accepted. Accordingly, the important issue is to avoid a disconnection of a wind energy installation during a network disturbance. The fault ride-through capabilities of FSIGs with STATCOMs will be investigated and a suitable control technique for enhancing them will be outlined in the next chapter.

References

1. Rao KN, Amarnath J, Kumar KA (2007) Voltage constrained available transfer capability enhancement with FACTS devices. ARPN J Eng Appl Sci 2(2):1–9
2. Ackermann T, Soder L (2002) An overview of wind energy-status 2002. Renew Sustain Energ Rev 6(1–2):67–128
3. Hossain MJ, Pota HR, Kumble C (2009) Transfer limit enhancement using decentralized robust STATCOM control for wind farm. In: Asia-Pacific power and energy engineering conference. Wuhan, China, pp 1446–1450
4. North American Electric Reliability Council (NERC) (2000) Available transfer capability definitions and determination
5. Ayasum CA, Fischl R (1997) On the region-wise analysis of available transfer capability (ATC). In: Proceedings of North American power symposium (NAPS), Wyoming, pp 464–469
6. Klump RP, Overbye TJ (1996) A transmission-based voltage stability measure for available transfer capability (ATC) calculations. In: Proceedings of North American power symposium (NAPS), MIT, pp 351–357
7. Gravener MH, Nwankpa C (1999) Available transfer capability and first order sensitivity. IEEE Trans Power Syst 14(2):512–518
8. Ramezani M, Haghifam MR (2007) Modelling and evaluation of wind turbines on total transfer capability. In: IEEE Power Engineering Society General Meeting. Tampa, FL, pp 1–6
9. Canizares CA, Berizzi A, Marannino P (1998) Using FACTS controllers to maximize available transfer capability. In: Proceedings of bulk power system dynamics and control IV—restructuring. Santorini, Greece, pp 633–641
10. Xiao Y, Song YH, Liu CC, Sun YZ (2003) Available transfer capability enhancement using FACTS devices. IEEE Trans Power Syst 18(1):305–312
11. Farahmand H, Roshidi-Nejad M, Fotirhi-Firoozabad M (2004) Implementation of FACTS devices for ATC enhancement using RPF technique. In: Proceedings of large engineering systems conference on power engineering, pp 30–35
12. Yu X, Singh C, Jakovljevic S, Ristanovic D, Huang G (2003) Total transfer capability considering FACTS and security constraints. IEEE PES Trans Distrib Conf Exposition 1:73–78
13. Sadeghzadeh SM, Ehsan M, Said NH, Feuillet R (1998) Improvement of transient stability limit in power system transmission lines using fuzzy control of FACTS devices. IEEE Trans Power Syst 13(3):917–922
14. Li L, Ugrinovskii VA, Orsi R (2007) Decentralized robust control of uncertain markov jump parameter systems via output feedback. Automatica 43(11):1932–1944
15. Tuglie ED, Iannone SM, Torelli F (2008) Feedback-linearization and feedback-feedforward decentralized control for multimachine power system. Electr Power Syst Res 78(3):382–391
16. Petersen IR, Ugrinovskii VA, Savkin AV (2000) Robust control design using H_∞ methods. Springer, London
17. Rao P, Crow ML, Yang Z (2000) STATCOM control for power system voltage control applications. IEEE Trans Power Delivery 15(4):1311–1317
18. Kundur P (1994) Power system stability and control. Mcgraw-Hill, New York
19. Hamoud G (2000) Assessment of available transfer capability of transmission systems. IEEE Trans Power Syst 15(1):27–32
20. Kumar A, Srivastava SC (2002) AC power transfer distribution factors for allocating power transactions in a deregulated market. IEEE Power Eng Rev 22(7):42–43
21. Luo X, Patton AD, Singh C (2000) Real power transfer capability calculations using multi-layer feed-forward neural networks. IEEE Trans Power Syst 15(2): 903-908
22. Cheng Y, Chung TS, Chung CY, Yu CW (2006) Dynamic voltage stability constrained ATC calculation by a QSS approach. Int J Electr Power Energ Syst 28(6):408–412
23. Cutsem TV, Vournas C (1998) Voltage stability of electric power systems. Kluwer Academic, Norwell, MA
24. Kueck J, Kirby B, Rizy T, Li F, Fall N (2006) Reactive power from distributed energy. Electr J 19(10):27–38

25. Athanasius GX, Pota HR, Subramanyam PB, Ugrinovskii V (2007) Robust power system stabiliser design using minimax control approach: validation using real-time digital simulation. In: 46th IEEE conference on decision and control, New Orleans, Louisiana, pp 2427–2432
26. Fernandez L, Garcia C, Saenz JR, Jurado F (2009) Equivalent models of wind farms by using aggregated wind turbines and equivalent winds. Energ Convers Manag 50(3):691–704
27. Germond AJ, Podmore R (1978) Dynamic aggregation of generating unit models. IEEE Trans Power Apparatus Syst PAS-97(4):1060–1069
28. Yoon MG, Ugrinovskii VA, Pszczel M (2007) Gain-scheduling of minimax optimal state feedback controllers for uncertain lpv systems. IEEE Trans Automatic Control 52(2):311–317

Chapter 7
Control for Fault Ride-Through Capability Augmentation

Abstract This chapter presents the design and implementation of a new control scheme for reactive power compensation, voltage regulation and transient stability enhancement for wind turbines equipped with fixed-speed induction generators in large interconnected power systems. Grid codes and the low-voltage-ride-through (LVRT) requirements for the grid are discussed in details. A systematic procedure is proposed to design decentralized multi-variable controllers for large interconnected power systems using minimax output-feedback control design method and the controller design procedure is formulated as an optimization problem involving rank-constrained linear matrix inequalities (LMIs). In this chapter it is shown that STATCOM with energy storage system (STATCOM/ESS), controlled via robust control technique, is an effective device for improving the LVRT capability of fixed-speed wind turbines.

7.1 Introduction

Traditionally, wind power generation has been treated as a distributed small generation or negative load [1]. Wind turbines (WTs) have been allowed to be disconnected when a fault is encountered. Such a perspective does not require wind turbines to participate in voltage and frequency control and their disconnection is considered insignificant in terms of loss of production. However, the penetration of wind power in some countries such as Denmark (18.5 %), Spain (7.8 %) and Germany (4.3 %) is high [2]. These figures are for equivalent annual production of wind power over total electricity demand of above, the maximum penetration during peak hours can be 4–5 times these figures [3].

Globally, wind power development is experiencing dramatic growth. According to the global wind energy council (GWEC) 15,197 MW WTs were installed in 2006, an increase of 32 % over 2005 [4]. The total global wind energy capacity had increased to 74,223 MW by the end of 2006, from 59,091 MW in 2005. The European wind energy association (EWEA) has set a target of satisfying 23 % of Europe's electricity

J. Hossain and H. R. Pota, *Robust Control for Grid Voltage Stability: High Penetration of Renewable Energy*, Power Systems,
DOI: 10.1007/978-981-287-116-9_7, © Springer Science+Business Media Singapore 2014

need using wind by 2030 [4]. The exponential growth of the wind industry reflects the increasing demand for clean, safe and domestic energy, and can be attributed to government policies associated with environmental concerns, and research and development of innovative cost-reducing technologies.

Wind energy has emerged as the fastest growing source of renewable energy and is expected to see continued strong growth in the immediate future. As the total base of installed wind capacity grows with the installation of additional WTs and new wind farms, compliance with interconnection criteria becomes increasingly important. Wind power generation is required to provide a certain reliability of supply and a certain level of stability. Motivated by the above issues, many grid operators have started to introduce new grid-codes which treat wind power generation in a special manner. Most interconnection standards today require wind farms to have the ability to withstand severe faults, usually called the fault ride-through (FRT) capability or, in some cases, the low-voltage ride-through (LVRT) capability.

WTs and generators are complex systems, with large flexible structures working under very turbulent and unpredictable environmental conditions, and connected to an electrical grid with a highly variable power demand. The efficiency and reliability of a wind generator depends heavily on the applied control strategy. A number of problems, e.g., highly nonlinear behaviour, large model uncertainty due to the interactions of aerodynamics, mechanical and electrical subsystems, stability problems, the need to maximise wind energy conversion, load reduction strategies, mechanical-fatigue minimisation problems, reliability issues, availability aspects, and costs per kWh reduction strategies, make the design of high-performance control systems mandatory.

Presently, thirty percent of the installed wind power is produced by squirrel-cage induction generators (SCIGs) which are directly connected to the grid and operate at an almost fixed-speed [5]. They are preferred as wind generators for their low cost, low maintenance, and due to their rugged brushless construction and asynchronous operation. A directly connected induction generator (IG) is not able to contribute to power system regulation and control in the same way as can a conventional field-excited synchronous generator as it needs reactive power support to be connected to stiff grids. However, WTs are usually connected at weak nodes or at distribution levels at which the network was not originally designed to transfer power into the grid [6]. This increases the need for dynamic reactive power support to ride-through severe faults [7].

When a disturbance or fault occurs, the voltage at the terminals of a WT drops significantly, causing the electromagnetic torque and electrical power output of the generator to be greatly reduced [8]. However, the mechanical input torque is almost constant during typical non-permanent faults which causes the machine to accelerate. As the slip of the IG increases, the reactive power absorbed from the connecting power system increases. Therefore, unless the WT is prevented from over-speeding, the post-fault voltage on the network is not likely to return to its pre-fault value. Normally, after the fault is cleared, a large amount of reactive-power is drawn by the generators. If this is not available, the machine will speed out of control and become disconnected from the power system. While the loss of power from a small-capacity wind farm may be acceptable, large wind farms are subject to grid-code requirements and must be able to ride-through these types of transient disturbances.

If the post-fault voltage level is restored, the magnetic field of the generators is re-established and the electromagnetic torque is restored. Therefore, a quick recovery of voltage and the re-establishment of the electromagnetic torque are crucial [9]. The over-speeding of a generator may also be limited by controlling the input mechanical torque. Turbines equipped with a pitching system have the advantage of actively controlling the input mechanical torque by pitch control. For fixed-speed induction generators (FSIGs) with short-circuited rotors, there are no active control methods available to control the reactive power in order to help voltage recovery. Therefore, a fast reactive power control device, such as a static synchronous compensator (STATCOM), can help in both voltage recovery and the re-establishment of the magnetic field and torque of the machine [10].

Most of the work on STATCOM control for enhancing the LVRT capability of FSIGs concentrates on the control of the STATCOM output current and DC bus voltage regulation for a given reference reactive current using a modelling strategy similar to that used for the field-oriented control of three-phase AC machines [11–14]. In most cases, there are two main control objectives for the converters. One is to regulate the DC terms (DC voltage for a voltage-source converter and DC current for a current-source converter) to constant values [15]. The other objective is to control the AC-side reactive power (or power factor). Two loops are designed separately, one for each control objective, and the interaction between them is not usually considered [15]. Decoupled control of the AC- and DC-side voltages of a STATCOM is difficult to achieve in practice due to the inherent coupling between the d- and q-axis variables through the load.

The conventional converter model of a STATCOM is a multi-input multi-output nonlinear model, and the difficulty of controlling it is mainly due to its nonlinear behaviour [16]. There are several ways of dealing with nonlinearities. A simple way is to use two separate proportional-integral (PI) controllers [17] to control the DC-term and the reactive power [18, 19]. However, in these cases, the response time is usually slow, and it is difficult to find appropriate PI parameters in a systematic way [20]. Another method is to linearise the system around an operating point and then design a linear controller [21]. The two main problems with this method are: (a) the controller is not effective for large disturbances; and (b) the design is dependent on the operating point. This motivates the use of advanced control techniques that consider nonlinear interactions and ensure stability for large disturbances, thus keeping the wind farm connected to the main grid under both fault and post-fault conditions.

A robust linear control algorithm is proposed in [22] to handle the nonlinearity of a WT over a range of wind velocities. This range is divided into several intervals, one controller is designed for each range and the appropriate controller is switched depending on the operating point. In practical power systems, it is difficult to implement switching controllers as unwanted transients may arise due to the switching. Only the nonlinearity due to the input wind velocity is considered in this paper [22].

The authors in [23] propose a proportional-integral-derivative (PID) pitch angle controller for a fixed-speed active-stall wind turbine. The controller is designed using the root-locus method and only the nonlinearities of the system are taken into account to determine its second-order transfer function (using step response test) which represent the system more accurately than a linear representation. The actual transfer function of the wind turbines is of higher order than considered in [23], also this method cannot capture the nonlinearity accurately. Due to the nonlinear nature of the wind turbine, step response tests have to be conducted in different operating regions, leading to different transfer functions around their respective operating points. The design and implementation of a single linear controller for different transfer functions is a complex task. In this method, it is easy to deal with nonlinearities, but only those due to small disturbances can be captured. To fully capture the nonlinearity, a method using mean-value theorem is proposed in this book and an excitation controller is designed in which the unstructured uncertainty representation is presented [24]. This representation is simple but conservative.

Since STATCOMs are only able to provide reactive power control, their application is limited to reactive power support. To overcome this problem, STATCOMs with battery energy storage systems (STATCOM/BESS) have emerged as more promising devices for power system applications given that they are able to provide both real and reactive power control [25–27]. However, as the BESS is based on chemical processes, it has some limitations such as, a slow response time and short service life. Another alternative proposed in this book is to simultaneously control the STATCOM (reactive power) and the pitch angle of the wind turbine (real power) to enhance the LVRT capability of induction generators in wind farms.

Several control methods for controlling the pitch angle have been reported, e.g., the classical PID control [28, 29], gain-scheduling control [30], robust control [31] and other nonlinear controllers [32]. However, most strategies are based on a linear WT model around a specific operating point [33]. An LQG method for designing pitch control is discussed in [34]. It is known that an LQG controller provides good robustness in terms of gain margins and phase margins. However, it is unable to provide robustness against uncertainties in the operating conditions [35].

Robust control in power systems deals with the application of new techniques in linear system theory to enhance their voltage and transient stabilities. The authors in [36] propose an H_∞ pitch angle control design using a linear matrix inequality (LMI) approach to reduce the fluctuating power of wind generators. In [37], a robust coordinated control method is proposed to smooth fluctuations in the generated power via pitch angle and battery charge-discharge control. An H_∞ controller is used with an LMI approach to achieve system robustness [37]; however, the nonlinearities of WTs have not been considered for the design of these linear controllers. To overcome the limitations of linear controllers, a nonlinear control technique for STATCOMs has been proposed to improve the power quality and LVRT capability of wind turbines [38]. Nonlinear controllers usually have more complicated structures and are harder to implement in practice than linear controllers. From an industrial point of view, it is preferable to use simple linear robust controllers in WTs; however, for robust performance, the nonlinearities need to be taken into account when the controllers are designed.

Although simultaneous control of a STATCOM and the pitch angle of the wind turbine can provide both active and reactive-power controlling abilities, the response of pitch angle controller is slow. To overcome this problem, STATCOM with energy storage system (STATCOM/ESS) can be used to supply the reactive power, increase the capability to damp electromechanical oscillations, and enhance the low voltage ride-through (LVRT) capability of fixed-speed wind turbines. Although a STATCOM/ESS has great potential to fulfill the requirements of grid-code to connect wind turbines, considerable advances in the control of this system are still needed for its practical implementation.

The design of linear optimal control (LOC) for large interconnected systems requires information transfer from the remote areas which are inherently dependent on centralised processing and long-distance wide-band communications. Although many valuable contributions have been made, a centralised solution to the LOC problem will lead to a very complicated actuator structure which is too expensive and difficult to implement.

As a result, during the last decade, interest has shifted towards the implementation of decentralised controllers [39]. The aim is to use only locally measurable output variables instead of full-state feedback and try to approach the performance of LOC with global state feedback [40]. To design decentralised controllers for large systems, reducing the dimension of the system models to a manageable size is necessary in order to reduce the controller bandwidth. Many researchers have worked on the decomposition of interconnected power systems into several subsystems, as well as dynamic equivalencing methods [41]. Some of these decomposition methods are mathematical in nature and may not have a physical meaning; however, in most cases, couplings between the subsystems in an interconnected power system are omitted or extremely simplified [41].

In recent years, the design of decentralized controllers for interconnected large power systems has been widely investigated and intensively studied with most attention being on guaranteeing the connective stability of the overall system despite the interconnection terms [42–44]. Results concerning the robust decentralised stabilisation of interconnected power systems based on approaches which explicitly take into account the interaction terms have been reported in [39, 45]. In the work of [46], an interesting decentralised turbine/governor controller scheme for power systems is presented. However, the local state feedback controllers designed through this approach need the complete state information which may not be feasible. Furthermore, the nonlinear terms are not explicitly included in the controller design.

Recently a control method for limiting the torque and enhancing the LVRT capability of grid-connected cage induction machines during the recovery process after grid faults by using a STATCOM is proposed in [47]. The authors in [48] propose a novel damping control algorithm for a STATCOM in a series compensated wind park for mitigating sub-synchronous resonance (SSR) and damping power system oscillations. An efficient control strategy to improve the LVRT capability in doubly-fed induction generators is proposed in [49]. These control techniques are mainly aimed to maximise the output power, increase the reactive current during low-voltage and reduce the peak rotor fault current. However, the nonlinearity and their interactions among wind farms are not considered in these papers. However, it is essential to consider the nonlinearities and interconnection effects in order to design controllers for multi-machine power systems, and also quantify the deviation of the operating point from the equilibrium point for which the system maintains closed-loop stability.

During faults and in post-fault operation the system state can be significantly far from the desired equilibrium point. In most situations the post-fault uncontrolled system has unstable post-fault trajectory. The difficulty in providing the LVRT capability is due to the nonlinearities in the power system model [50]. Linear controllers have a limited range of operation which normally does not include post-fault voltage conditions [50]. A solution to the LVRT problem is to design a globally stabilising controllers [51]. Unfortunately these controllers often need a full-state feedback and are not robust to modelling uncertainties. The next option is to use a linear controller which is robust to the change in the linear model with changing operating conditions—a necessary outcome of the underlying nonlinear model. As mentioned above, robust controllers do exist for power systems but few of them have been able to systematically provide robustness against such large deviations as is required for LVRT.

In this research a method is presented which can be used to design a linear controller that is robust to accommodate post-fault low-voltage conditions [52]. Based on this design method, the synthesis for the following robust controllers are presented to enhance the FRT capability of FSIGs in this chapter:

(a) STATCOM controller with unstructured uncertainty representation;
(b) Simultaneous STATCOM and pitch angle control;
(c) STATCOM controller with structured uncertainty representation; and
(d) Decentralised STATCOM/ESS controller.

In addition, the effects of STATCOM rating and wind farm integration on the FRT capability of FSIGs are studied analytically using the power-voltage and torque-slip relationships and also using detailed simulations.

The organisation of this chapter is: Section 7.2 discusses the current grid-codes requirements for wind farms; Section 7.3 provides the fault ride-through scheme for induction generators; a basic idea about critical clearing time and critical speed is given in Sect. 7.4; Section 7.5 describes the robust STATCOM control design technique and also contains simulation studies for validating its performance; design and performance analysis of simultaneous STATCOM and pitch angle controls are provided in Sect. 7.6; Section 7.7 contains STATCOM control technique with structured uncertainty representation; a systematic procedure to design decentralised STATCOM/ESS controller is included in Sect. 7.8. Section 7.9 presents the summary of this chapter.

7.2 Grid-Code Requirements for Wind Farm Connections

Large-capacity wind farms have created several challenges for transmission network operators. The intermittent nature of the wind causes power quality and stability problems. Unpredictable power penetration affects the reliability and stability of a power grid. Therefore, the grid-code rules being imposed on large wind farms stipulate that they are operated as conventional power plants.

The operating procedures and principles governing the relationship between the transmission system operator and the users of the transmission system, (e.g., generators, suppliers or non-embedded customers) are set out in the grid-code. With reference to the grid connection of wind farms, there are a number of issues specified by this proposed grid-codes, i.e., FRT, frequency range, frequency control, reactive power range capability and voltage control.

7.2.1 Fault Ride-Through

Under the grid-code proposals of some countries, a wind farm, whether connected directly to the transmission system or to the distribution system, is required to remain connected and feed power into the system for any solid single- or multi-phase short-circuit fault occurring on the high-voltage (HV) transmission system. The period of zero voltage on the HV system for this requirement is limited to 140 ms. In addition to continuing to operate in a stable manner, the mechanical power to a WT should not be deliberately reduced during this period.

7.2.2 Frequency–Power Variations

Grid-codes require that wind farms should be able to operate continuously at any system frequency between 47.5 and 52 Hz for at least 20 s. Additionally, wind farms should meet the following power frequency characteristics:

- from 49.5 to 50.4 Hz: power output should not vary with changes in system frequency;
- from 49.5 to 47.0 Hz: as the frequency reduces the power output should not reduce more than proportionately; and
- from 50.4 to 52.0 Hz: power output should be reduced by at least 2 % for each 0.1 Hz increase in system frequencies.

7.2.3 Frequency Control

The grid-code proposals require that wind farms should have the capabilities to provide frequency response and respond to a 0.5 Hz change in system frequency by changing the output by 10 % of the wind farm capability within 10 s.

7.2.4 Reactive Power Range Capability

The latest grid-code proposals require that over the normal active power operating range of the wind farm, a reactive power capability of 0.95 power factor lagging to 0.95 power factor leading (based on full output power) should be available at the connection point. These limits reduce the impact of fluctuating wind power on the grid voltage. A wider range of power factor would be required for a remote area when low power is generated by the wind farm.

7.2.5 Voltage Control

The grid-code proposals will require each wind farm to be capable of controlling voltage at the point of its connection to the public electricity system. During a transmission system voltage dip: (i) a wind farm shall be required to provide active power in proportion to its retained voltage and maximise the reactive current to the transmission system without exceeding the WT generator limit and (ii) it will have to provide at least 90 % of its maximum available active power as quickly as the technology allows and, in any event, within 1 second of the transmission system voltage recovering to its normal operating range.

7.3 Fault Ride-Through Schemes for Wind Turbines

Each type of WT has specific vulnerabilities when subjected to grid faults. In order to fulfil the requirements imposed by grid utilities, it must be equipped with the ability to ride-through in the case of faults in the grid. The phenomena that occur in an IG without a FRT capability during a grid fault are described first. Later, how a STATCOM connected with a wind turbine can help to meet grid-codes is presented. Typical FRT requirements demand that a wind farm remains connected to the grid for voltage levels as low as 5 % of the nominal voltage (for up to 250 ms) [53] as shown in Fig. 7.1.

The sequence of events during grid faults can be described as follows: when a fault occurs in the grid, the terminal voltage drops rapidly; the magnitude of the voltage drop is dependent on the fault distance from the generator; and due to a reduced terminal voltage, the machine loses electric torque, which leads to a rotor acceleration. Meanwhile, the mechanical torque applied to the rotor can be considered constant during this event. If the electrical torque at this point is higher than the mechanical torque, the generator will eventually be back to its normal operating point. However, when the electrical torque is lower than the mechanical torque, the speed will continuously increase which could result in electrical torque reduction leading to an unstable situation.

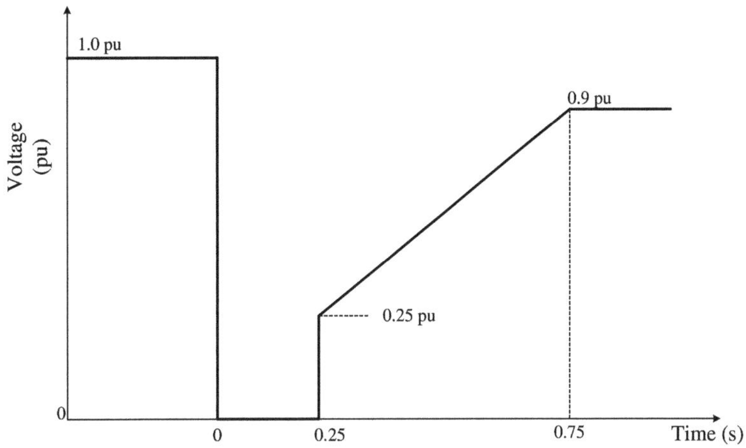

Fig. 7.1 Voltage limit criteria according to grid-code

In the case of the two-mass model of a drive train, the situation is even worse. This is because, in the instance of voltage recovery, although the generator speed is able to recover to the normal operating point, the turbine speed does not decrease instantaneously because some of the energy in the mechanical shaft is stored in the twisted shaft instead of being directly utilised to decelerate the turbine speed. This allows the turbine speed to continuously increase for a short time.

During the next period, the energy is released from the twisted shaft and creates the opposite effect. This charge and discharge of twisted shaft energy creates oscillations in the generator and the turbine speed. In reality, the electrical torque also suffers oscillations due to terminal voltage variation caused by active and reactive power fluctuations. At the same time, the turbine torque also fluctuates slightly due to a small variation in the power coefficient. All these factors interacting with each other create composite oscillations. This clearly indicates that instability might not occur only due to the fault but that there is also a considerably high mechanical stress in the shaft.

For a wind park with induction generators, the main requirements for compensation devices are to compensate for any reactive power demand of the induction generators in steady-state to control reactive power exchange with the system, and to help the recovery of wind parks after system disturbances to improve ride-through capability. The former can be fulfilled by a simple solution with conventional switched shunt capacitors while the latter requires dynamic compensation devices with more advanced control algorithms.

The STATCOM has to be equipped with a set of functions in order to help wind parks to fulfill the grid-code requirements. These functions are listed below:

- Steady-state reactive power supply or absorption. This function can be fulfilled by following a reactive power set-point, a set-point for a power factor at the connection point of the wind park or by operating according to a linear reactive power versus voltage characteristic (Q/V characteristic).

- The implementation of the latter case also fulfils the voltage control requirement often asked for in the grid-codes. The grid companies often require certain flexibility to change the basic behaviour of the voltage control scheme. A reduced set of changeable parameters has to be available, especially the target voltage and the slope of the linear characteristic.
- Smoothly follow a set-point ramp. No steps occur, e.g., with solutions based on switched passive components.
- The dynamic requirements of the grid-codes are met, e.g., a step in the set-point is followed within less than 1 s without notable overshoots or oscillations.
- During voltage dips (balanced or unbalanced), the STATCOM injects reactive current in the order of nominal STATCOM current and therefore helps to support the grid voltage.

7.4 Critical Clearing Time and Critical Voltage

The stability analysis of a power system may consider the determination of its critical clearing time (CCT), for a given fault, in order to find the maximum value of the fault clearing time for which the post-fault system is stable. If the fault is cleared within this time, the system will remain stable. However, if the fault is cleared after this time the power system will lose its stability. The calculation of the CCT is very important from the protection point of view.

Mathematically, a CCT is a complex function of pre-fault system conditions (operating point, topology, system parameters), fault structure (type and location) and post fault conditions that themselves depend on the protective relaying plan employed. In this research, the CCT is first estimated by using the following equations and then the exact value is determined from simulations.

An approximate value of the CCT can be calculated from the following equation:

$$\dot{s} = \frac{1}{2H}\left[T_m - T_e\right]. \tag{7.1}$$

During a solid three-phase short-circuit at the generator terminals, $T_e = 0$ and then (7.1) can be written as:

$$\dot{s} = \frac{1}{2H}T_m. \tag{7.2}$$

Integrating both sides

$$s = \int_0^t \frac{1}{2H}T_m + s_0. \tag{7.3}$$

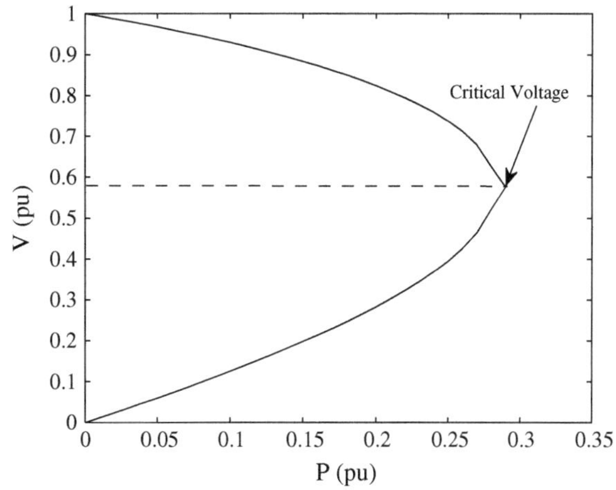

Fig. 7.2 P–V relationship

If s_c is the critical speed (CS) of a machine, then the CCT can be given as:

$$t_c = \frac{1}{T_m} 2H(s_c - s_0). \tag{7.4}$$

The critical speed is given by the intersection between the torque-speed curve for the specified system and the mechanical torque [54].

The critical voltage can be obtained from the P–V curve [55], which determines the maximum power that may be transferred between two parts of the system before voltage collapse. The information it provides can help the analyst or transmission planner determine how to strengthen the power system against the risk of voltage collapse. The P–V curve is formed by varying system load or transfer and plotting it against voltage. This curve can provide real power and voltage margins using the knee of the curve as reference point. Figure 7.2 is an example of a P–V curve where the critical point is caused by a reactive power limit. On the lower voltage side of the P–V curve (below critical voltage):

- increased load admittance reduces load power;
- adding a shunt capacitor reduces bus voltage;
- tap changers reduces voltage;
- stable for impedance load; and
- unstable for constant power load.

From the power-flow equations of an infinite-bus versus a single load connected through a lossless transmission line, the relationship between voltage and power is given by [55]:

$$V = \sqrt{\frac{E^2}{2} - QX \pm \sqrt{\frac{E^4}{4} - X^2P^2 - XE^2Q}}, \tag{7.5}$$

where the symbols carry their usual meanings. Equations (7.4) and (7.5) are solved to estimate the CCT and critical voltage of IGs.

7.5 Robust STATCOM Control with Unstructured Uncertainty

In this section, a novel robust controller for a STATCOM is presented to enhance the FRT capability of fixed-speed induction generators. The wind generator is a highly nonlinear system, which is modelled in this work as a linear term plus a nonlinear term, which is the Cauchy remainder term in the Taylor series expansion of the equations used to model the wind farm. Bounds derived for this Cauchy remainder term are used to define an uncertain linear model for which a robust control design is performed. The controller resulting from this robust design provides an acceptable performance over a wide range of conditions needed to operate the wind farm during severe faults. The performance of the designed controller is demonstrated by large disturbance simulations on a test system.

7.5.1 Test System

The test system shown in Fig. 7.3 consists of two main buses connected via two long parallel transmission lines. WTs are connected to the first bus via transformers and the other bus is directly connected to the grid. Each IG works at the rated operating point and supplies 2 MW of active power. The data for this system is given in Appendix-V. We use an aggregated model of a wind farm to design the proposed controller. The assumption is reasonable when the power system under consideration is large and the purpose is to observe the effect of penetration on the external network rather than within the wind farm. The load is modelled as a constant impedance load. The wind farm, rated at 50 MW, is normally not allowed to operate under severe fault conditions and the addition of the STATCOM with appropriate control is expected to increase the stability margin as well as the FRT capability of the wind farm.

The dynamics of the test system is represented by the following equations:

$$\dot{\omega}_m = (1/2H_m)\left[T_{ae_i} - K_s\gamma - D_m\omega_m\right], \tag{7.6}$$

$$\dot{\omega}_G = (1/2H_G)\left[K_s\gamma - T_e - D_G\omega_G\right], \tag{7.7}$$

$$\dot{\gamma} = 2\pi f(\omega_m - (1/N_g)\omega_G) \tag{7.8}$$

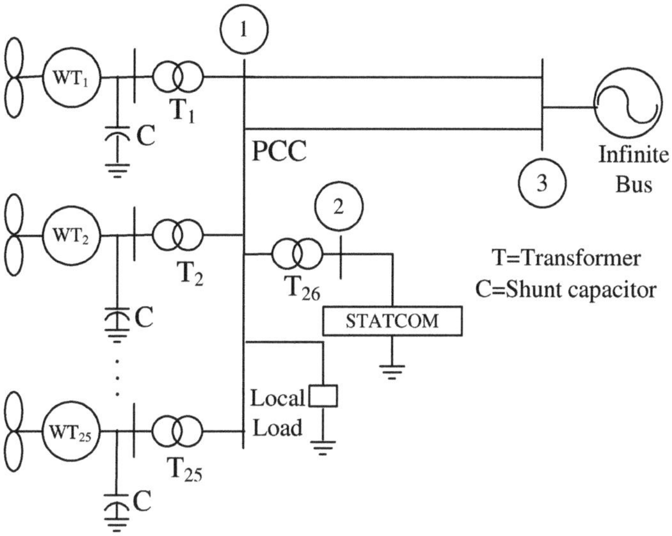

Fig. 7.3 Single wind farm infinite bus study system

$$\dot{s} = (1/2H_G)\,[T_m - T_e]\,, \tag{7.9}$$

$$\dot{E}'_{qr} = -(1/T'_o)\left[E'_{qr} - (X - X')i_{ds}\right] - s\omega_s E'_{dr}, \tag{7.10}$$

$$\dot{E}'_{dr} = -(1/T'_o)\left[E'_{dr} + (X - X')i_{qs}\right] + s\omega_s E'_{qr}, \tag{7.11}$$

$$\dot{v}_{dc} = -P_s/(Cv_{dc}) - v_{dc}/(R_C C), \tag{7.12}$$

$$\dot{v}_{tm} = -\frac{v_{tm}}{T_m} + K_m V_t, \tag{7.13}$$

where P_s is given by

$$\begin{aligned} P_s = {} & |E|^2 G_{22} + |E||V_\infty|\,[B_{23}\sin\alpha + G_{23}\cos\alpha] \\ & + |E||E'_{dr}|\,[B_{21}\sin(\delta - \alpha) + G_{21}\cos(\delta - \alpha)] \\ & + |E||E'_{qr}|\,[B_{21}\cos(\delta - \alpha) - G_{21}\sin(\delta - \alpha)] \end{aligned} \tag{7.14}$$

where G_{23} and B_{23} are the real and imaginary parts of the equivalent transfer impedances between the terminal buses of STATCOM and infinite bus, and G_{21} and B_{21} between terminal buses of STATCOM and wind farm, δ the rotor angle of the wind generator, $E = kv_{dc}\angle\alpha$, the STATCOM AC terminal voltage, α the bus angle of the STATCOM in the reduced network, and $k = \sqrt{(3/8)}m$ the constant and m the modulation index. The other symbols carry their usual meanings given in Chap. 2.

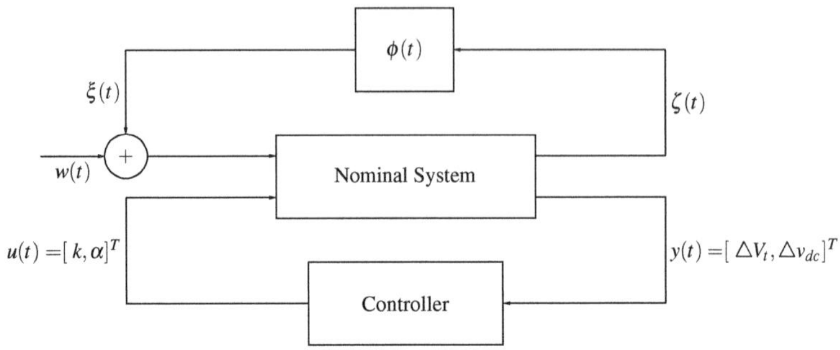

Fig. 7.4 Block diagram of uncertain system

7.5.2 Linearisation and Uncertainty Modelling

Conventionally, a linear controller is designed based on the Taylor series around an equilibrium point by retaining the linear term and neglecting the higher order terms. In this work, to quantify the neglected higher order terms, we propose the use of a linearisation scheme which retains the contributions of the higher order-terms in the form of the Cauchy remainder.

The nonlinear system (7.6)–(7.13) is reformulated using the technique proposed in this book as follows:

$$\Delta\dot{x} = A\Delta x + (L - A)\,\Delta x + (M - B_1)\,\Delta u + B_1\Delta u, \tag{7.15}$$

where $A = \frac{\partial f}{\partial x}|_{\substack{x=x_0\\u=u_0}}$, $B_1 = \frac{\partial f}{\partial u}|_{\substack{x=x_0\\u=u_0}}$, $u = [k, \alpha]^T$ and
$\Delta x = \left[\Delta\omega_m, \Delta\omega_G, \Delta\gamma, \Delta s, \Delta E_{dr}, \Delta E_{qr}, \Delta v_{dc}, \Delta v_{tm}\right]^T$.

The system (7.15) is of the form which allows for an application of the minimax control design technique [56]. To apply this technique, we rewrite system (7.15) in terms of the block diagram shown in Fig. 7.4. In this figure, we introduce a fictitious signal ξ such that:

$$(L - A)\,\Delta x + (M - B_1)\,\Delta u = B_2\xi(t), \tag{7.16}$$

where

$$B_2 = \text{diag}\left(0, \frac{1}{2H_G}, 0, \frac{1}{2H_G}, \frac{X_s - X_s'}{T_0'}, \frac{X_s - X_s'}{T_0'}, \frac{1}{C}, \frac{1}{T_m}\right), \text{ and}$$
$$\zeta = \sqrt{\beta}\left(\tilde{C}_1\Delta x + \tilde{D}_1\Delta u\right), \ \phi = (1/\sqrt{\beta})\left[\tilde{\phi} \ \ \tilde{\psi}\right].$$

The nonlinearities considered in this test system are only due to s, E'_{dr}, E'_{qr} and v_{dc}, with this matrices $\tilde{C}_1$ and $\tilde{D}_1$ are chosen as

$$\tilde{C}_1 = \begin{bmatrix} 0\,0\,0\,1\,0\,0\,0\,0 \\ 0\,1\,0\,0\,1\,0\,0\,0 \\ 0\,0\,1\,0\,0\,1\,0\,0 \\ 0\,0\,0\,0\,0\,0\,1\,0 \end{bmatrix}, \quad \tilde{D}_1 = \begin{bmatrix} 1\,1 \\ 1\,1 \\ 1\,1 \\ 1\,1 \end{bmatrix},$$

The expressions for obtaining $\tilde{\phi}$ and $\tilde{\psi}$ are given in Appendix-VI. To apply this reformulation, the following inequality needs to be satisfied

$$\|\xi(t)\|^2 \leq \|\zeta(t)\|^2. \tag{7.17}$$

To facilitate control design, the power system model is finally summarised as:

$$\Delta\dot{x}(t) = A\Delta x(t) + B_1\Delta u(t) + B_2\xi(t), \tag{7.18}$$
$$y(t) = C_2\Delta x(t) + D_2\xi(t)), \tag{7.19}$$
$$\zeta(t) = C_1\Delta x(t) + D_1 u(t), \tag{7.20}$$

Condition (7.17) will enable us to apply the minimax LQG control design methodology to obtain a controller for the underlying nonlinear system. Robustness properties of the minimax LQG controller ensure that this controller stabilises the nonlinear system (7.18)–(7.20) for all instances of linearisation errors.

The output matrix $C_2 = \left[0, 0, 0, 0, 0, 0, 0, 1; 0, 0, 0, 0, 0, 0, 1, 0\right]$. Equations (7.18)–(7.20) provide a new representation of the power system model which contains both the linear and higher-order terms. The new formulation presented in this section is used with the minimax linear quadratic Gaussian (LQG) controller synthesis approach to design a STATCOM controller for nonlinear wind generators.

7.5.3 Minimax LQG STATCOM Controller

This section presents the minimax LQG control algorithm. Minimax LQG methodology can be considered as a robust version of standard LQG controller design and it combines the advantages of both LQG and H_∞ control [57]. Within the minimax optimal control design framework, robustness is achieved via optimisation of the worst-case quadratic performance of the underlying uncertain system. This helps to achieve an acceptable trade-off between control performance and robustness of the system. The minimax LQG method, described in [56, 57], for uncertain systems of the form shown in Fig. 7.4, is applied to the following stochastic version of the power system model (7.18)–(7.20):

$$\Delta\dot{x}(t) = A\Delta x(t) + B_1\Delta u(t) + B_2\xi(t) + B_2 w(t), \tag{7.21}$$
$$y(t) = C_2\Delta x(t) + D_2\xi(t)), \tag{7.22}$$
$$\zeta(t) = C_1\Delta x(t) + D_1\Delta u(t), \tag{7.23}$$

where $w(t)$ is a unity Gaussian white noise. The underlying physical system (power system) does not include noise-like inputs. The white noise term is a technical addition to enable the design of a robust output feedback controller which computes control inputs to drive the system to its equilibrium point in the presence of uncertain disturbances in the system such as those due to the effect of nonlinearities. It is suggested in [57, p. 342] that the optimal minimax LQG controller for the above system (7.21)–(7.23) is also a quadratically stabilising robust controller for the deterministic system (7.18)–(7.20) with norm bounded uncertainty subject to (7.17). This motivates using the stochastic minimax LQG control design methodology to design a robust controller for the problem in this research. As compared to the standard LQG control this minimax LQG controller provides robustness due to uncertainties which is important for the control design of wind generators.

In the minimax LQG problem for the stochastic system (7.21)–(7.23) the following quadratic cost functional is considered

$$J = \lim_{T\to\infty} \frac{1}{2T} E \int_0^T \left(\Delta x(t)^T R \Delta x(t) + \Delta u(t)^T G \Delta u(t) \right) dt, \tag{7.24}$$

where $R \geq 0$ and $G > 0$, $R \in R^{n\times n}$, $G \in R^{m\times m}$ and E is the expectation operator. The work in [56, 57] considers the minimisation of the maximum value of the cost over all uncertainties, such as ϕ in Fig. 7.4, satisfying an integral quadratic constraints (IQC) [57]. Uncertainties satisfy an IQC if they belong to the set ϕ_{IQC} which consists of all ϕ such that for input signals $\zeta(t)$ and the output $\xi(t) = \phi(\zeta(t))$ there exists a constant $d > 0$ and a sequence of times $\{t_l\}$, $l = 1, 2, \ldots, t_l \to \infty$, such that the following inequality holds for all l

$$E \int_0^{t_l} \|\xi(t)\|^2 dt \leq E \int_0^{t_l} \|\zeta(t)\|^2 dt + d. \tag{7.25}$$

The uncertainty class considered in this chapter is a subset of the class of IQC uncertainties. This means that the optimum cost ($\inf_{\tau>0} V_\tau$) is an upper bound on the optimum cost for the norm-bounded uncertainty considered in this chapter.

The optimum cost is the infimum of function V_τ obtained over all $\tau > 0$. The function V_τ is

$$\begin{aligned} V_\tau = &\frac{1}{2}\text{tr}\Big[Y_\infty R_\tau + \left(Y_\infty C_2^T + B_2 D_2^T \right) \Gamma^{-1} \\ &\times \left(C_2 Y_\infty + D_2 B_2^T \right) X_\infty \left(I - \frac{1}{\tau} Y_\infty X_\infty \right)^{-1} \Big], \end{aligned} \tag{7.26}$$

where τ is a free parameter and the matrices X_∞ and Y_∞ are the solution to the following pair of parameter-dependent algebraic Riccati equations [56]:

$$\left(A - B_2 D_2^T \Gamma^{-1} C_2\right) Y_\infty + Y_\infty \left(A - B_2 D_2^T \Gamma^{-1} C_2\right)^T$$
$$- Y_\infty \left(C_2^T \Gamma^{-1} C_2 - \frac{1}{\tau} R_\tau\right) Y_\infty + B_2 \left(I - D_2^T \Gamma^{-1} D_2\right) B_2^T = 0, \quad (7.27)$$

and

$$X_\infty \left(A - B_1 G_\tau^{-1} \gamma_\tau^T + \left(A - B_1 G_\tau^{-1} \gamma_\tau^T\right) X_\infty\right.$$
$$+ \left(R_\tau - \gamma_\tau G_\tau^{-1} \gamma_\tau^T\right) - X_\infty \left(B_1 G_\tau^{-1} B_1^T - \frac{1}{\tau} B_2 B_2^T\right) X_\infty = 0. \quad (7.28)$$

Solutions of the above two Riccati Equations (7.27) and (7.28) are required to satisfy the following conditions: $Y_\infty > 0$, $X_\infty > 0$, the spectral radius of the matrix $X_\infty Y_\infty$ is $\rho(X_\infty Y_\infty) < \tau$, $R_\tau - \gamma_\tau^T G_\tau^{-1} \gamma_\tau \geq 0$, $R_\tau = R + \tau C_1^T C_1$, $G_\tau = G + \tau D_1^T D_1$, $\gamma_\tau = \tau C_1^T D_1$.

The minimax LQG optimal controller u^*, with the τ^* at which the infimum of V_τ is reached, is given by [56]:

$$\dot{\hat{x}}_c = \left(A - B_1 G_{\tau^*}^{-1} \gamma_{\tau^*}^T\right) \hat{x}_c - \left(\left(B_1 G_{\tau^*}^{-1} B_1^T - \frac{1}{\tau^*} B_2 B_2^T\right) X_\infty\right) \hat{x}_c + \left(I - \frac{1}{\tau^*} Y_\infty X_\infty\right)^{-1}$$
$$\times \left(Y_\infty C_2^T + B_2 D_2^T\right) \Gamma^{-1} \left(y - \left(C_2 + \frac{1}{\tau^*} D_2 B_2^T X_\infty\right) \hat{x}_c\right), \quad (7.29)$$
$$u^* = - G_{\tau^*}^{-1} \left(B_1^T X_\infty + \gamma_{\tau^*}^T\right) \hat{x}_c. \quad (7.30)$$

The controller u^* guarantees the following minimax property

$$\sup_{\zeta \in \phi_{\mathrm{IQC}}} J(u^*, \zeta) = \inf_u \sup_{\zeta \in \phi_{\mathrm{IQC}}} J(u, \zeta) = \inf_\tau V_\tau. \quad (7.31)$$

To obtain the controller for the system considered in this section, the parameter, τ, is chosen to minimise the quantity V_τ. A line search is carried out to find the value of $\tau > 0$ which attains the minimum value of the cost function V_τ. This line search involves solving the Riccati Equations (7.27) and (7.28) for different values of τ and finding that value which gives the smallest V_τ in (7.26). This allows us to construct a controller of the form in (7.30). In this section, this suboptimal controller for the norm-bounded uncertainty is designed and implemented on the test system.

7.5.4 Case Studies

The test system shown in Fig. 7.3 is used to analyse the effect of wind power integration and STATCOM rating on the FRT capability of WTs. The FRT capability of a wind generator is expressed in this book as voltage and transient stability margins.

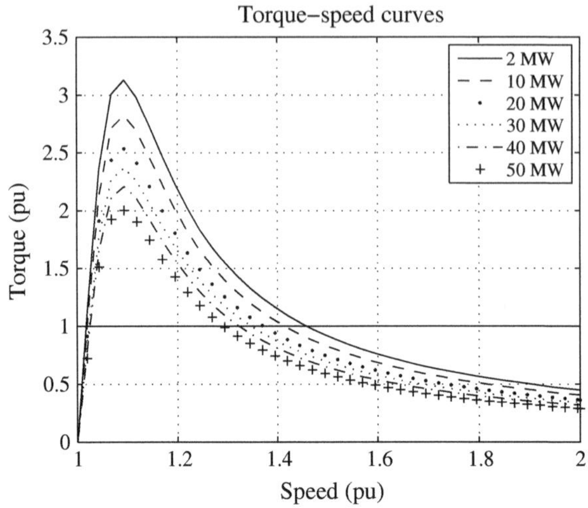

Fig. 7.5 Torque versus speed curve of growing wind power

The voltage stability margin is defined as the difference between the operating voltage and the critical voltage. The transient stability margin is given as the difference between the speed after a specified fault duration and the critical speed of the generator. The torque-speed characteristics for increasing wind generation are given in Fig .7.5.

The CS and CCT for increasing wind generation, without compensation and control, are given in Table 7.1. The power-voltage relationship for different wind generation is shown in Fig. 7.6. It can be seen that as the number of wind generators increases, the corresponding CS and CCT and terminal voltage decrease. The maximum difference between the estimated value obtained from (7.4) and that obtained using the detailed simulation is 3.74 % for the CS and 7.8 % for the CCT. The estimated speed is greater than the values obtained from the detailed simulations. This error is caused by the transients at the time of reclosing, since some time is needed to re-magnetise the IG before it is able to output the electrical torque given by the steady-state torque-speed characteristics.

The CS for the different STATCOM MVA ratings for a 50 MW wind farm, with PI controllers and terminal voltage feedback, are given in Table 7.2. It is observed that the STATCOM increases significantly the CS and, thereby, the stability limit as well as the FRT capability of the IG which result in a corresponding increase of the CCT during a three-phase fault. From the data in Table 7.2, it can be concluded that the system with a higher rating STATCOM can have longer ride-through times for short-circuit faults.

The critical speed from simulations with 50 MW generation, a 25 MVAr capacitor and a 10 MVA STATCOM is 1.35 pu. Figure 7.7 shows the speed of the IG for a

Table 7.1 Effect of growing wind power

Wind power (MW)	Parameters	Critical speed and CCT from calculation	Critical speed and CCT from simulation
2	ω_{critical}	1.47	1.415
	CCT	0.45	0.485
10	ω_{critical}	1.42	0.137
	CCT	0.40	0.423
20	ω_{critical}	1.37	1.298
	CCT	0.35	0.382
30	ω_{critical}	1.345	1.298
	CCT	0.325	0.344
40	ω_{critical}	1.31	1.295
	CCT	0.29	0.318
50	ω_{critical}	1.285	1.274
	CCT	0.265	0.281

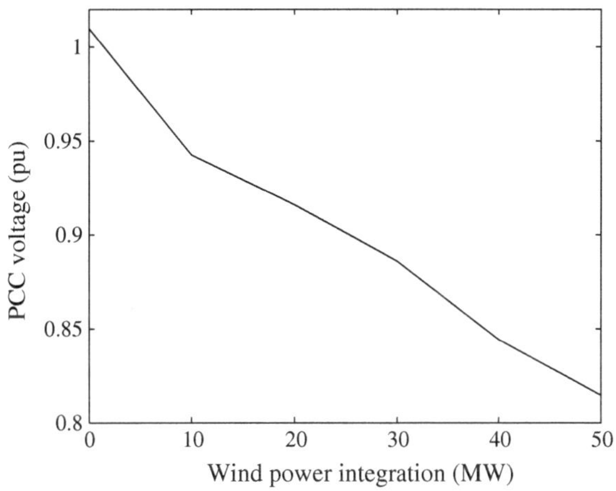

Fig. 7.6 P–V relationship at bus PCC

three-phase fault at bus 2 with (a) a 50 MVAr shunt capacitor only and (b) combination of a 10 MVA STATCOM and 25 MVAr STATCOM from which it is clear that with the addition of a STATCOM in the system, instability of the IG can be avoided because the reactive power delivered by the shunt capacitor is proportional to the square of the terminal voltage which means that during low voltage conditions, VAr (Volt-Ampere Reactive) support drops, thus compounding the problem. The STATCOM is expected to provide extra reactive power support dynamically, with a continuous and fast change of output for voltage recovery when the voltage becomes low.

Table 7.2 Effect of STATCOM ratings

STATCOM (MVA)	Error (s)	Critical speed and CCT from calculation	Critical speed and CCT from simulation
0	ω_{critical}	1.285	1.174
	CCT	0.265	0.281
10	ω_{critical}	1.35	1.28
	CCT	0.315	0.33
25	ω_{critical}	1.40	1.31
	CCT	0.38	0.396
40	ω_{critical}	1.44	1.38
	CCT	0.42	0.437

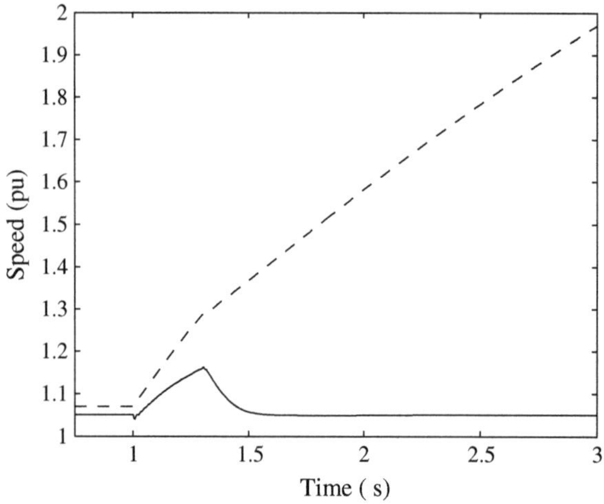

Fig. 7.7 Effect of STATCOM on FRT capability (*Solid line* with STATCOM and *dashed line* without STATCOM)

7.5.5 Control Design Algorithm and Performance Evaluation

First, we carry out several simulations to obtain an idea of the operating range during transients by applying large disturbances. The controller is then designed for stable operation in the region of interest. The basic steps are:

Step 1: From simulations of the faulted system, obtain the range of the variation of all state variables and form a volume, Ω, with corner points given by $(x_{0_p} - x_{f_p})$ and $(x_{0_p} + x_{f_p})$, $p = 1, \ldots, 7$, where $2x_{f_p}$ is the largest variation of the pth state variable about its equilibrium value, x_{0_p};

Step 2: Obtain

$$\beta^* = \max_{x^{*p} \in \Omega} \left\{ \beta : ||\phi(t)||^2 < 1 \right\};$$

Step 3: Check if a feasible controller with $\beta = \beta^*$, exists, i.e., if there is a scalar τ such that there is a feasible solution to the coupled Riccati Equations (7.27)–(7.28).

Step 4: If we obtain a feasible controller in the above step, either enlarge the volume, Ω, i.e., increase the volume of the region Ω, or if we have arrived at the largest possible volume, perform an optimal search over the scalar parameter, τ, to get the infimum of V_τ. If there is no feasible solution with the chosen $\beta = \beta^*$, reduce the volume, Ω, and go to Step 2.

This process enables the selection of the largest range for which a feasible controller is obtained. For the given power system model, we are able to obtain a feasible controller with the value of $\beta = 0.96$. The controller is stabilising for all variation of states in the polytope region Ω formed by end points $\left[\bar{s}, \bar{E}'_{dr}, \bar{E}'_{qr}, \bar{\omega}_m, \bar{\omega}_G, \bar{\gamma}, \bar{v}_{dc}, \bar{v}_{tm}\right]^T$ and $\left[\underline{s}, \underline{E}'_{dr}, \underline{E}'_{qr}, \underline{\omega}_m, \underline{\omega}_G, \underline{\gamma}, \underline{v}_{dc}, \underline{v}_{tm}\right]^T$ with the values $\bar{s} = s_0 + 0.43$ pu, $\underline{s} = s_0 - 0.43$ pu, $\bar{E}'_{dr} = E'_{dr0} + 0.29$ pu, $\underline{E}'_{dr} = E'_{dr0} - 0.29$ pu, $\bar{E}'_{qr} = E'_{qr0} + 0.29$ pu, $\underline{E}'_{qr} = E'_{qr0} - 0.29$ pu, $\bar{\omega}_m = \omega_{m0} + 0.32$ pu, $\underline{\omega}_m = \omega_{m0} - 0.32$ pu, $\bar{\omega}_G = \omega_{G0} + 0.43$ pu, $\underline{\omega}_G = \omega_{G0} - 0.43$ pu, $\bar{\gamma} = \gamma_0 + 28°$, $\underline{\gamma} = \gamma_0 - 28°$, $\bar{v}_{dc} = v_{dc0} + 0.35$ pu, $\underline{v}_{dc} = v_{dc0} - 0.35$ pu, $\bar{v}_{tm} = v_{tm0} + 0.225$ pu, $\underline{v}_{tm} = v_{tm0} - 0.225$ pu, $\bar{k} = k_0 + 0.27$ pu, $\underline{k} = k_0 - 0.27$ pu, $\bar{\alpha} = \alpha_0 + 35°$ and $\underline{\alpha} = \alpha_0 - 35°$.

The performance of the proposed controller for a 10 MVA STATCOM is evaluated for the sudden outage of one of the lines serving the wind farm. One of the lines connecting buses 1 and 3 is taken out of service at 1 s and restored after 300 ms. The generator speed and terminal voltage with the PI and proposed STATCOM controller are shown in Figs. 7.8 and 7.9, respectively. It is clear that the proposed controller can stabilise the voltage as well as the IG and performs better than the PI control.

Simulations are performed for a severe symmetrical three-phase short-circuit at one of the long transmission lines. Figures 7.10 and 7.11 show the speed and terminal voltage of the IG with the PI and the proposed controllers.

The three-phase fault is applied at 1 s and cleared at 1.35 s. The speed of 1.32 pu at the fault clearing is greater than the critical speed of 1.28 pu obtained for the PI controller with numerical simulations. Thus, with PI controller, the speed continues to increase even after the fault is cleared. Furthermore, the voltage gradually decreases and the wind generators have to be disconnected from the grid to protect them and avoid voltage collapse. From Figs. 7.10 and 7.11, it can be seen that the proposed controller performs better than the PI controller and results in a higher CCT.

7.6 Simultaneous STATCOM and Pitch Angle Control

In this section, a procedure to design robust multi-variable controllers for a STATCOM and for the pitch angle of a fixed-speed induction generator is presented with the objective of enhancing the LVRT capability of wind farms. The control problem in

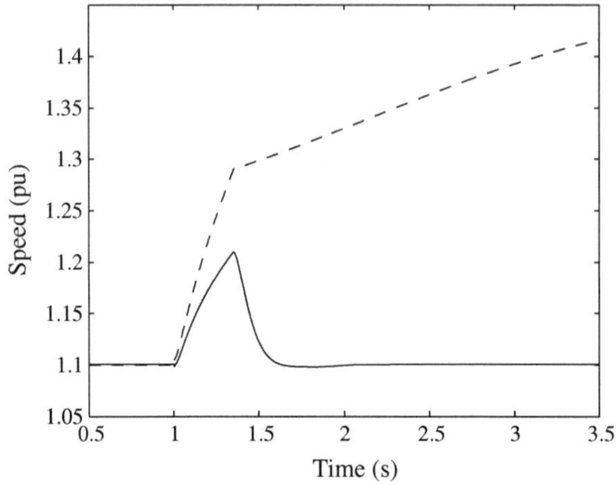

Fig. 7.8 Speed for outage of one of lines 1–3 (*Solid line* designed and *dashed line* PI controller)

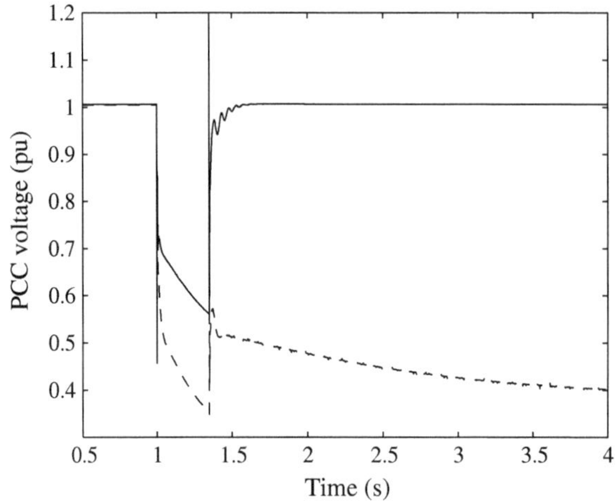

Fig. 7.9 PCC voltage for outage of one of lines 1–3 (*Solid line* designed and *dashed line* PI controller)

this case prompts us to choose the minimax LQG design method for simultaneous STATCOM and pitch angle controls to augment the LVRT capability of wind turbines. The control design in this section is tested by simulations under various types of disturbances on a test system. The performances of two other controllers are also tested: (a) a robust STATCOM controller designed according to previous section; and (b) a PI-based STATCOM controller proposed in [12] and compared with that of the controller proposed in this section.

The test system, shown in Fig. 7.3, is used to design the controller. Pitch control performs power reduction by rotating each blade about its axis in the direction of

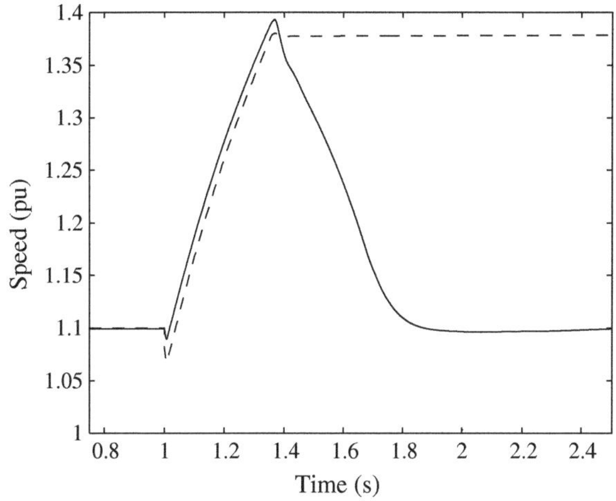

Fig. 7.10 Speed for outage of one of lines 1–3 (*Solid line* designed and *dashed line* PI controller)

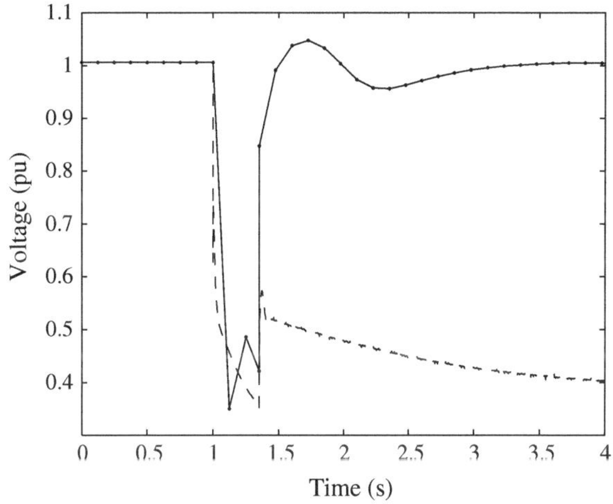

Fig. 7.11 PCC voltage for outage of one of lines 1–3 (*Solid line* designed and *dashed line* PI controller)

reducing the angle of attack. In comparison with the passive stall, the pitch control provides an increased energy capture at rated wind speed and above. Constant-speed wind turbines can be equipped with pitch drives which quickly increase the pitch angle when an acceleration of the rotor is detected. This reduces the mechanical power and consequently limits the rotor speed and the reactive power consumption after the fault. Figure 7.12 depicts the pitch angle controller. In this work the pitch rate limit is set to 8 (deg/s), $\theta_{\min} = -5°$, $\theta_{\max} = 45°$, $\dot{\theta}_{\min} = -10$ (deg/s), $\dot{\theta}_{\max} = 10$ (deg/s) and time constant is 0.2 s.

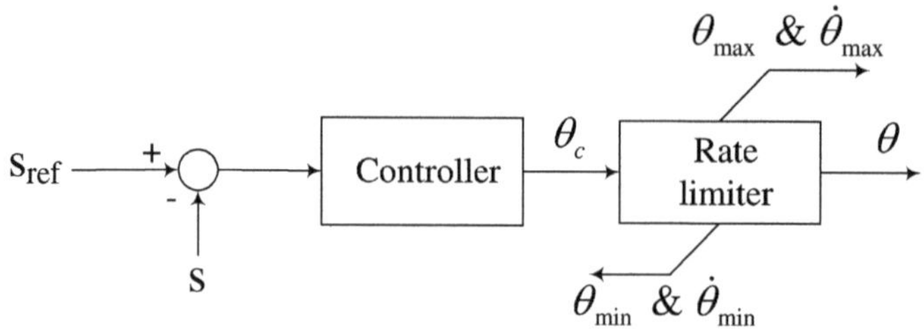

Fig. 7.12 Pitch angle control strategy

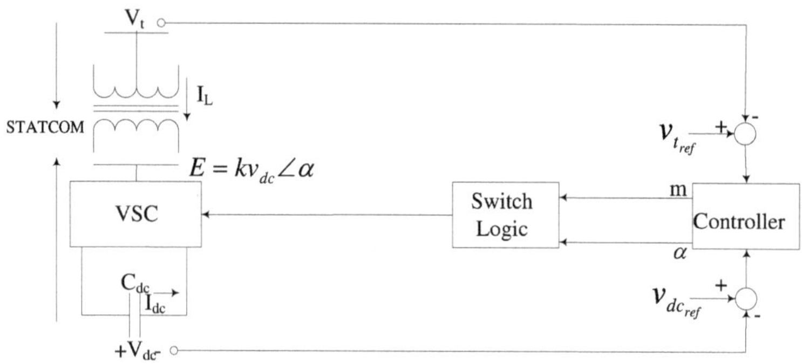

Fig. 7.13 STATCOM control strategy

The STATCOM control strategy used in this section is shown in Fig. 7.13. The pulse-width modulation (PWM) control technique is used because a PMW based STATCOM offers faster response and capability for harmonic distribution [58]. In this method, the compensation is achieved by measuring the rms voltage at PCC and the DC capacitor voltage. The output voltage magnitude of the VSC relates to the DC side voltage and is also a function of the control phase angle and the modulation ratio of the PWM.

Figures 7.14 and 7.15 show the frequency response of the linearised open-loop system for the following operating conditions: (i) nominal system; (ii) one transmission line outage; (iii) 10 % load increase; and and (iv) 10 % load decrease. Bode plots in Figs. 7.14 and 7.15 show the change in output terminal voltage and slip with respect to the firing angle input, from the plots it is clear that the transfer function does not change much over a wide range of operating conditions. The same is true for other inputs (modulation index and pitch angle) to the system. Bode plots 7.14 and 7.15 also indicate that controllers designed based only on the linearised system will have similar dynamics. It is only by considering the often neglected nonlinear terms that we can guarantee controller performance over the desired operating region.

Figure 7.16 shows PCC (point of common coupling) voltage response for one transmission line outage. From the simulation result, it is clear that the nonlinearities in the system affect the dynamic performance significantly and that the trajectory widely deviates from the equilibrium point. The system nonlinearities can be captured

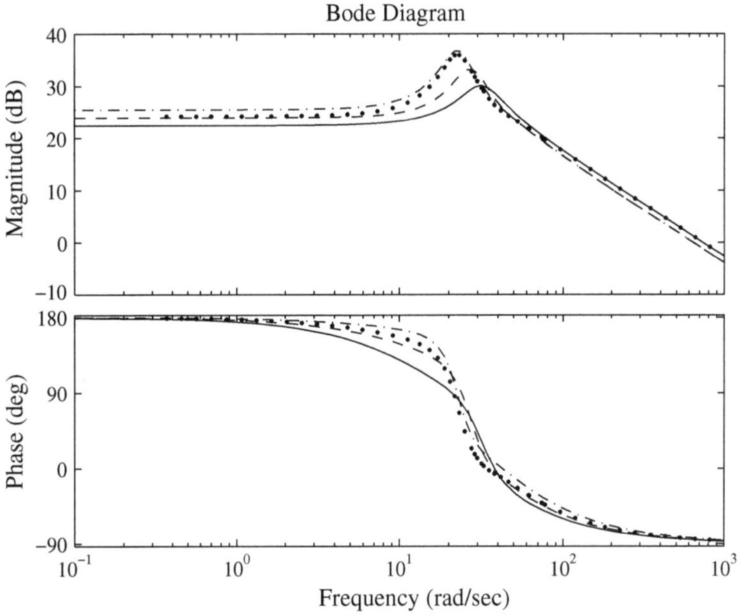

Fig. 7.14 Bode plot (open loop linear system) of the transfer function from input (firing angle) to terminal voltage (*Solid line* (-) nominal system, the *dashed* (–) *line* one line outage, the *dotted line* (.) 10 % increase in load and the *dashdot line* (-.) and 10 % decrease in load)

by using the formulation presented in this research. A linear controller with the proposed linearisation algorithm, which ensures stability for a pre-defined region in the state-space is designed in this section.

Several simulations are carried out to gain an idea about the operating range during transients by applying large disturbances. From simulations we estimated the operating region Ω for severe faults with the corner points $\left[\bar{s}, \bar{E}'_{dr}, \bar{E}'_{qr}, \bar{\omega}_m, \bar{\omega}_G, \bar{\gamma}, \bar{v}_{dc}, \bar{v}_{tm}\right]^T$ and $\left[\underline{s}, \underline{E}'_{dr}, \underline{E}'_{qr}, \underline{\omega}_m, \underline{\omega}_G, \underline{\gamma}, \underline{v}_{dc}, \underline{v}_{tm}\right]^T$ given by: $\bar{s} = s_0 + 0.115$ pu, $\underline{s} - s_0 - 0.115$ pu, $\bar{E}'_{dr} = E'_{dr0} + 0.398$ pu, $\underline{E}'_{dr} = E'_{dr0} - 0.398$ pu, $\bar{E}'_{qr} = E'_{qr0} + 0.374$ pu, $\underline{E}'_{qr} - E'_{qr0} - 0.374$ pu, $\bar{\omega}_m - \omega_{m0} + 0.336$ pu, $\underline{\omega}_m = \omega_{m0} - 0.336$ pu, $\bar{\omega}_G = \omega_{G0} + 0.483$ pu, $\underline{\omega}_G - \omega_{G0} - 0.483$ pu, $\bar{\gamma} - \gamma_0 + 25°$, $\underline{\gamma} - \gamma_0 - 25°$, $\bar{v}_{dc} = v_{dc0} + 0.216$ pu, $\underline{v}_{dc} = v_{dc0} - 0.216$ pu, $\bar{v}_{tm} = v_{tm0} + 0.225$ pu, $\underline{v}_{tm} = v_{tm0} - 0.225$ pu, $\bar{\theta} = \theta_0 + 17°$, $\underline{\theta} = \theta_0 - 17°$, $\bar{m} = m_0 + 0.475$, $\underline{m} = m_0 - 0.475$, $\bar{\alpha} = \alpha_0 + 28°$ and $\underline{\alpha} = \alpha_0 - 28°$.

The controller in Fig. 7.17 is designed in the same way as described in the previous section. For the given power system model, we are able to obtain a feasible controller with the value of $\beta = 0.975$ for the region Ω with the corner points of $\left[\bar{s}, \bar{E}'_{dr}, \bar{E}'_{qr}, \bar{\omega}_m, \bar{\omega}_G, \bar{\gamma}, \bar{v}_{dc}, \bar{v}_{tm}\right]^T$ and $\left[\underline{s}, \underline{E}'_{dr}, \underline{E}'_{qr}, \underline{\omega}_m, \underline{\omega}_G, \underline{\gamma}, \underline{v}_{dc}, \underline{v}_{tm}\right]^T$ given by the values $\bar{s} = s_0 + 0.265$ pu, $\underline{s} = s_0 - 0.265$ pu, $\bar{E}'_{dr} = E'_{dr0} + 0.42$ pu, $\underline{E}'_{dr} = E'_{dr0} - 0.42$ pu, $\bar{E}'_{qr} = E'_{qr0} + 0.405$ pu, $\underline{E}'_{qr} = E'_{qr0} - 0.405$ pu, $\bar{\omega}_m = \omega_{m0} + 0.436$ pu,

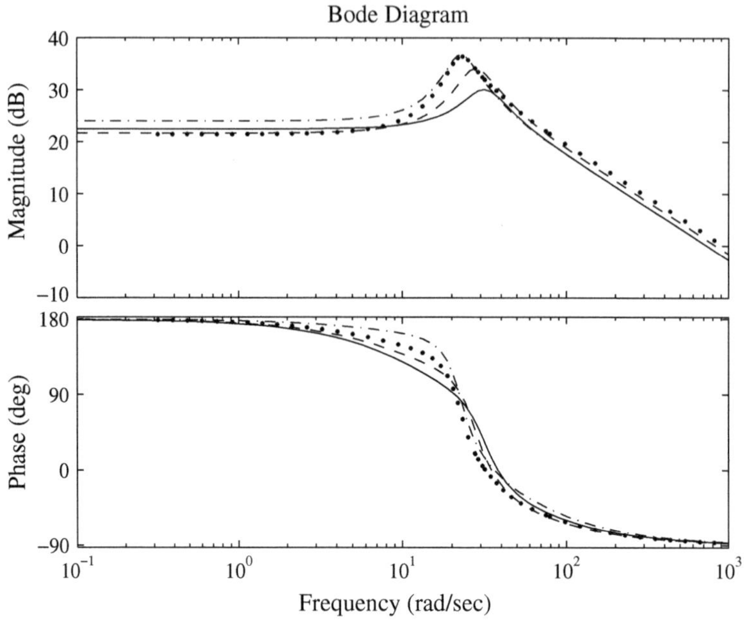

Fig. 7.15 Bode plot (open loop linear system) of the transfer function from input (firing angle) to slip (*Solid line* (-) nominal system, the *dashed* (–) *line* one line outage, the *dotted line* (.) 10 % increase in load and the *dashdot line* (-.) 10 % decrease in load)

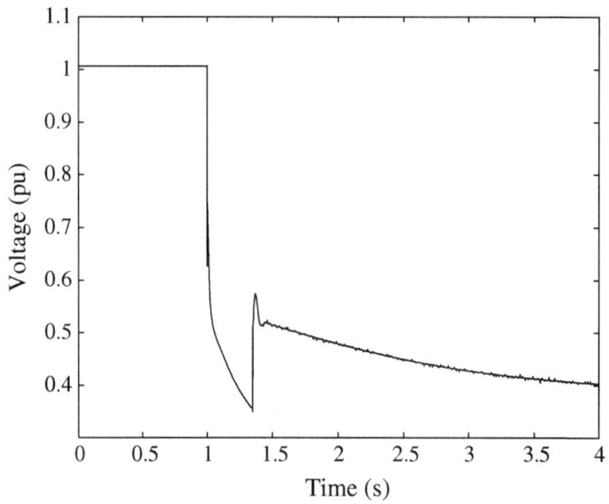

Fig. 7.16 PCC voltage for outage of one of lines 1–3

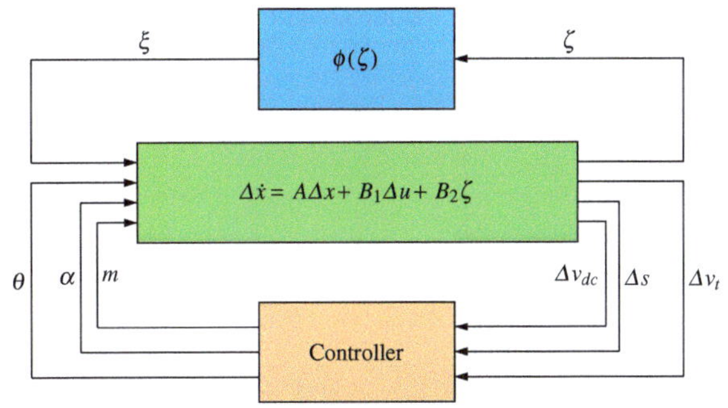

Fig. 7.17 Control block diagram

Table 7.3 Closed-loop eigenvalues

$-6.8 \pm i27.2$	$-5.4 \pm i21.9$	-5.3	-15.2	-8.9	-5.0

$\underline{\omega}_m = \omega_{m0} - 0.436$ pu, $\bar{\omega}_G = \omega_{G0} + 0.523$ pu, $\underline{\omega}_G = \omega_{G0} - 0.523$ pu, $\bar{\gamma} = \gamma_0 + 40^\circ$, $\underline{\gamma} = \gamma_0 - 40^\circ$, $\bar{v}_{dc} = v_{dc0} + 0.365$ pu, $\underline{v}_{dc} = v_{dc0} - 0.365$ pu, $\bar{v}_{tm} = v_{tm0} + 0.345$ pu, $\underline{v}_{tm} = v_{tm0} + 0.345$ pu, $\bar{\theta} = \theta_0 + 35^0$, $\underline{\theta} = \theta_0 - 35^0$, $\bar{m} = m_0 + 0.475$, $\underline{m} = m_0 - 0.475$ $\bar{\alpha} = \alpha_0 + 25^\circ$ and $\underline{\alpha} = \alpha_0 - 25^\circ$. This range in variations of the state variables is larger than that of noted earlier in this subsection for several large disturbances. Although the designed controller is not global-stabilising, we can be confident that it will stabilise the system for most contingencies.

7.6.1 Controller Performance Evaluation

The closed-loop eigenvalues of the test system are given in Table 7.3. The dominant mode for the closed loop is $-5.34 \pm i21.86$ and the damping is 0.244. From the eigenvalues, it can be seen that the closed-loop system is well-damped. Feasible low-frequency gains and suitable cut-off frequency of the designed controllers can be seen from the Bode plots of the designed controllers shown in Figs. 7.18, 7.19, 7.20 and 7.21.

The performances of the following three controllers is compared with respect to the CCT and CS: (a) a PI-based STATCOM controller ($\Delta\theta = 0$); (b) a single-input robust STATCOM controller ($\Delta\theta = 0$); and (c) two-input two-output simultaneous STATCOM and pitch controllers. The results of the comparisons are shown in Table 7.4. The CCT and CS for the three-phase fault with a 10 MVA STATCOM are 0.35 s and 1.33 pu using the simultaneous STATCOM and pitch controllers, compared with 0.33 s and 1.305 pu using the single input robust STATCOM controller and 0.315 s and 1.28 pu using the PI-based STATCOM. It can be concluded that an appropriate combination of the voltage control by reactive power and the speed

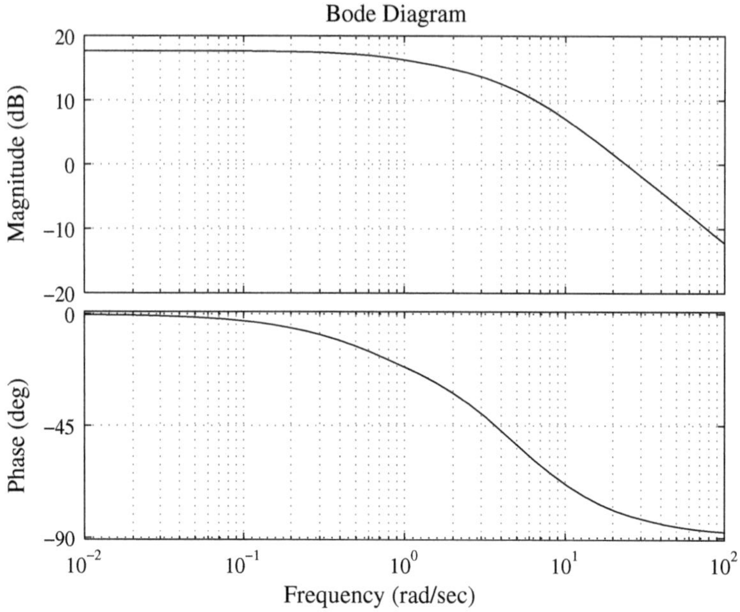

Fig. 7.18 Bode plot of STATCOM controller—firing angle versus terminal voltage

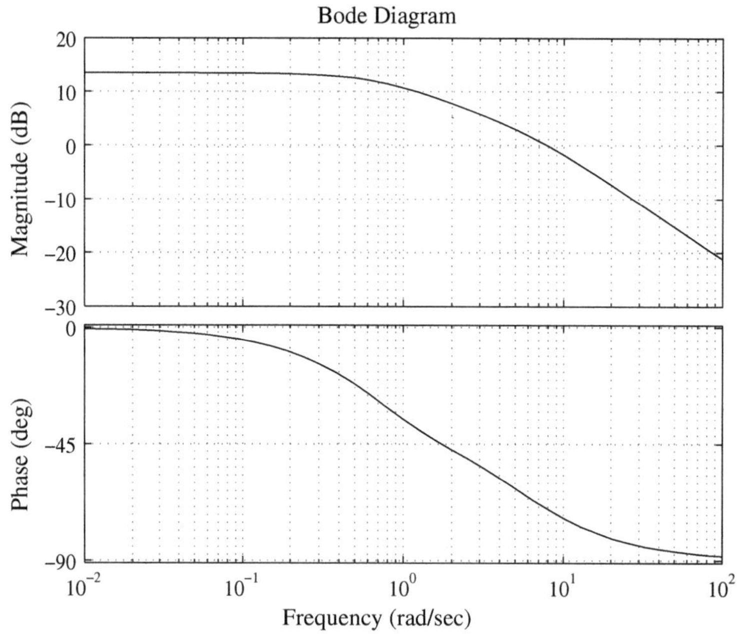

Fig. 7.19 Bode plot of Pitch controller—pitch angle versus terminal voltage

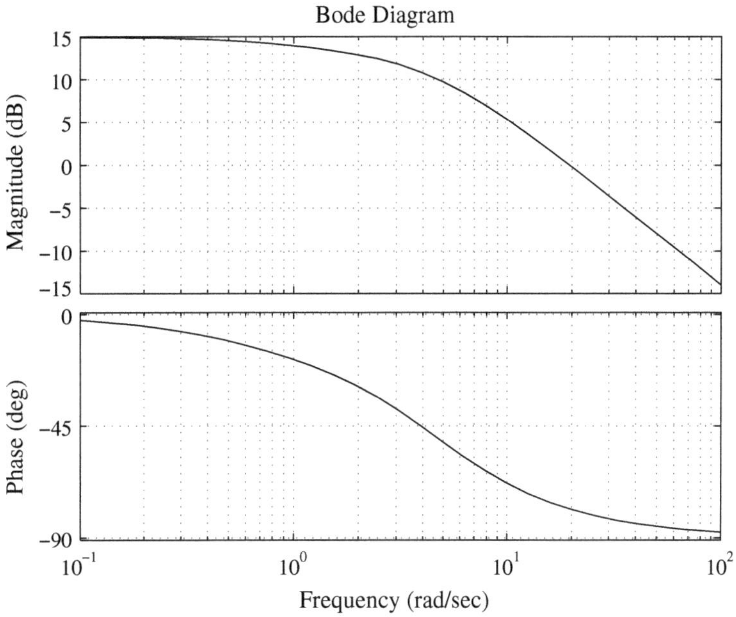

Fig. 7.20 Bode plot of STATCOM controller—slip versus firing angle

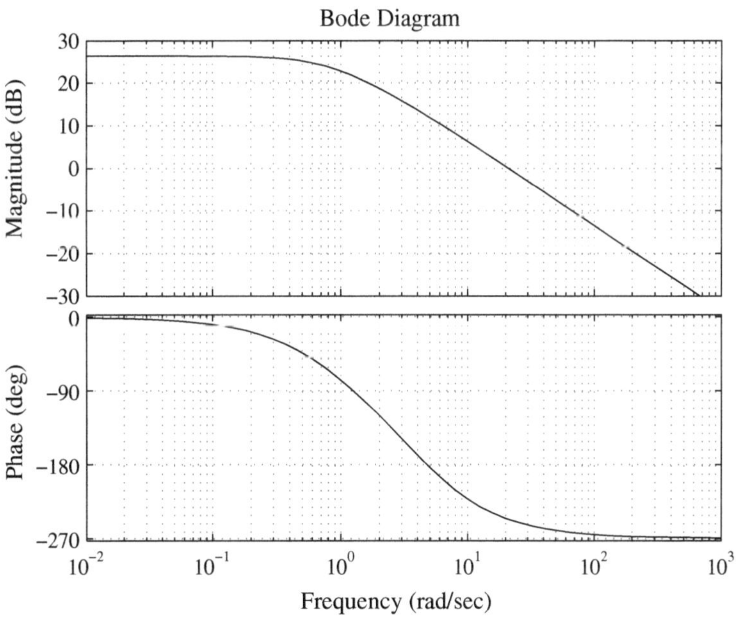

Fig. 7.21 Bode plot of pitch controller—slip versus pitch angle

Table 7.4 Performance comparison: (*a*) proposed controller; (*b*) robust STATCOM; and (*c*) PI-based STATCOM

STATCOM (MVA)	(a) Proposed controller		(b) Robust SISO		(c) PI control	
	CS (pu)	CCT (s)	CS (pu)	CCT (s)	CS (pu)	CCT (s)
10	1.33	0.35	1.305	0.33	1.28	0.315
25	1.43	0.418	1.35	0.405	1.31	0.396
40	1.47	0.458	1.45	0.43	1.38	0.415

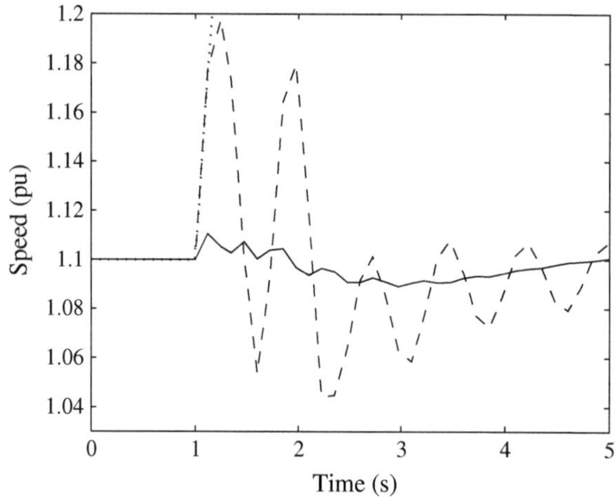

Fig. 7.22 Speed for outage of one of lines 1–3 (*Solid line* simultaneous STATCOM and pitch angle controllers, *dashed line* robust STATCOM and *dashdot line* PI based STATCOM controller)

control by torque is an effective way of improving the stability and enhancing the FRT capability of the relevant IG-based wind turbines.

The performance of the proposed controller for a 10 MVA STATCOM is evaluated for: (i) a sudden outage of one of the lines serving the wind farm; and (ii) a three-phase fault at one of the parallel lines. The generator speeds and terminal voltages with the PI-based STATCOM, the robust STATCOM controller and the proposed STATCOM and pitch controllers, due to the outage of one of the transmission lines for 300 ms, are shown in Figs. 7.22 and 7.23. It is clear that the proposed controller is able to stabilise the voltage as well as the IG speed with a fault clearing time of 0.35 s. The tuned PI controller is not able to stabilise voltage and the wind generator. In this case, the gain of the tuned (trial and error method) PI controller is obtained as $K_P = 0.3$ and $K_I = 10.25$. The robust STATCOM controller is almost as effective as the STATCOM and pitch controllers when restoring terminal voltage but is less effective in preventing the wind generators from over-speeding.

Simulations are performed for a severe three-phase short-circuit at one of the long transmission lines. The three-phase fault is applied at 1 s and cleared at 1.33 s. The CCT for the three-phase fault is 0.35 s with the STATCOM and pitch con-

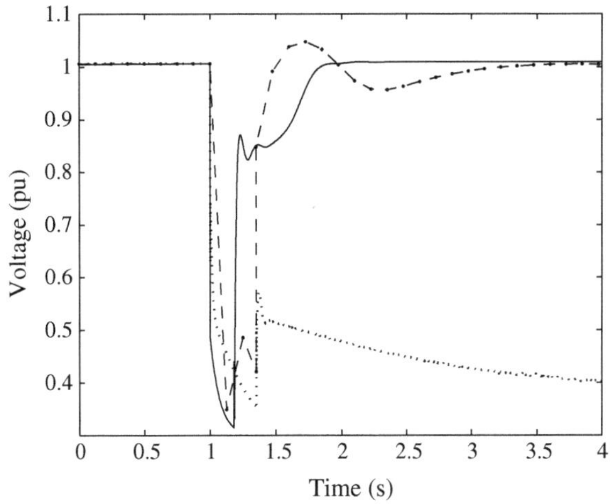

Fig. 7.23 PCC voltage for outage of one of lines 1–3 (*Solid line* simultaneous STATCOM and pitch angle controllers, *dashed line* robust STATCOM and *dashdot line* PI based STATCOM controller)

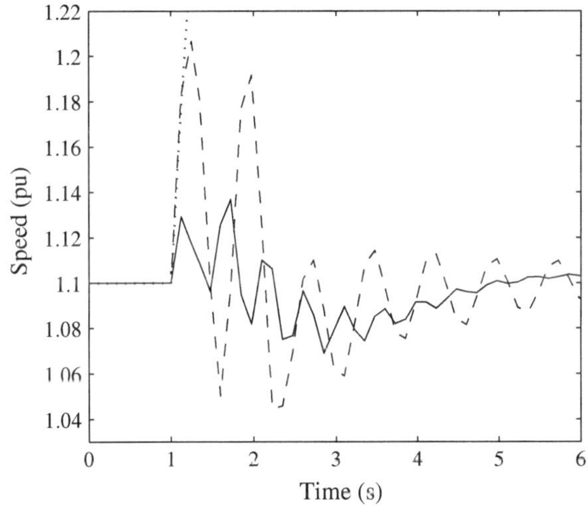

Fig. 7.24 Speed for three-phase fault at one of lines 1–3 (*Solid line* simultaneous STATCOM and pitch angle controllers, *dashed line* robust STATCOM and *dashdot line* PI based STATCOM controller)

trollers, compared to 0.33 s with the robust STATCOM controller and 0.315 s with the PI based STATCOM. Figures 7.24 and 7.25 show the speeds and terminal voltages, respectively, of the IG with the PI-based STATCOM, the robust STATCOM controller and the proposed controller from which it can be seen that the proposed

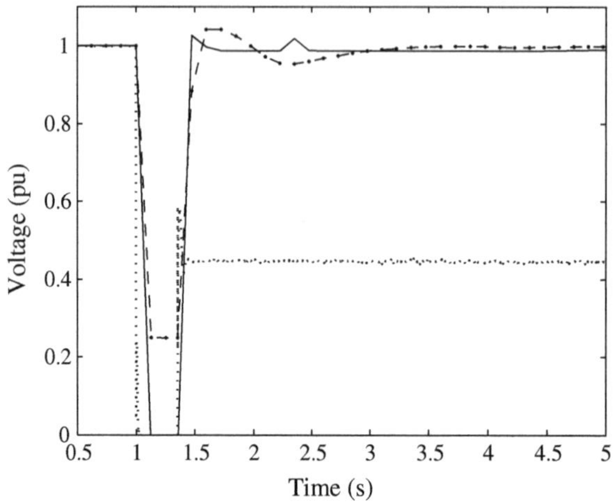

Fig. 7.25 PCC voltage for the three-phase fault at one of lines 1–3 (*Solid line* simultaneous STATCOM and pitch angle controllers, *dashed line* robust STATCOM and *dashdot line* PI based STATCOM controller)

simultaneous STATCOM and pitch angle controllers perform better than the PI-based and single-input robust STATCOM controller and results in a higher CCT.

As the power system stabilising pitch angle controller controls the active power of the wind farm, the oscillations in the speed become damped which is especially visible between the simulation times of 1.0 s and 6.0 s. It is clear that controlling the voltage by only the robust STATCOM controller might not be enough to keep the system stable. Sometimes, in order to stabilise the system quickly, it could be necessary to use the pitch angle control. The combined strategy of robust STATCOM and pitch angle control is more effective in recovering system operation and such combined control makes the system ride-through the fault without having to disconnect the generators from the system.

7.7 STATCOM Controller for Wind Farm with Structured Uncertainty

The main feature of synthesis methods for unstructured uncertainties is that they allow one to find closed-form solutions to control synthesis problems [59]. The inherent limit of these approaches is that they may be quite conservative in practical applications. Research on techniques and algorithms for problems with structured uncertainties is quite recent [59]. A robust STATCOM controller with structured uncertainties is designed and implemented on a test system in this section.

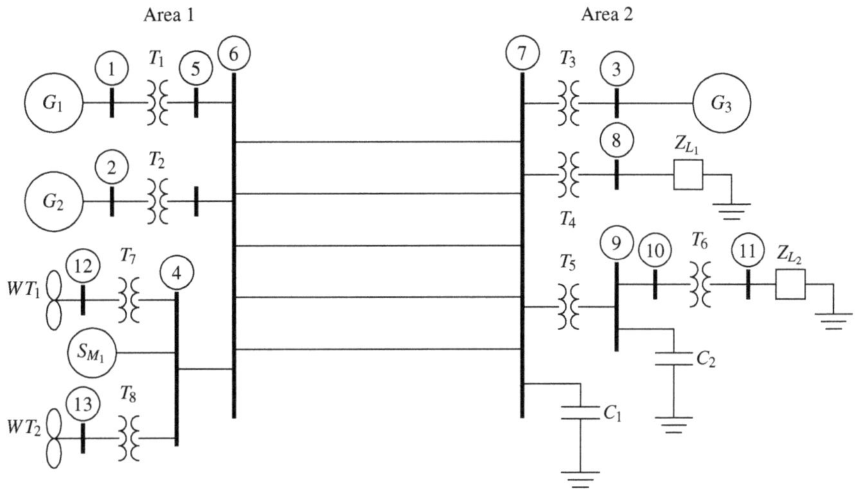

Fig. 7.26 3 machine 2 area study system with wind farms (*G* synchronous generator, *WT* wind generator, *S* STATCOM and *L* load)

7.7.1 Test System and Control Task

The test system shown in Fig. 7.26 consists of eleven buses and three generators. The parameters for the system are given in Appendix-IV. The total load for this system is $P_L = 6,655$ MW and $Q_L = 2,021$ MVAr and the generation is $P_G = 6,871$ MW, $Q_G = 1,738$ MVAr. The generation in a remote area (generators $G_1 = 3,981$ MW, and $G_2 = 1,736$ MW) is connected to the main load through five transmission lines. The remaining load ($P = 1,154$ MW) is supplied by the local generator, G_3. The load at bus 11 is modelled as 50 % constant impedance and 50 % constant current for both active and reactive power and the load at bus 8 is modelled as constant MVA for both active and reactive power.

We design a robust STATCOM controller for the modified test system where the generators $G_2 = 0$ MW, $G_3 = 0$ MW, $WT_1 = 1,736$ MW, and $WT_2 = 1,154$ MW. Generators G_2 and G_3 are disconnected from the system. The remaining power is supplied from G_1 which is considered in this case as an infinite bus. A STATCOM is connected at bus 4 to meet the connection requirements for power system grids. The wind generators are arranged in two parallel lines and we represent each of them by an aggregated wind generator model [60]. To appreciate the nature of the control task, we carry out the modal analysis for the open loop system. The dominant mode for the test system is $-0.105 \pm j0.71$. The participation vector for the dominant mode is shown in Table 7.5. The participation vector indicates that the states E'_{qr1}, E'_{qr2}, s_1 and s_2 have the most significant contribution to the dominant mode. The dominant mode is related to both reactive and active power mismatch. The reactive power can be controlled by the designed STATCOM controller and a conventional

Table 7.5 Participation factors

States	Δs_1	$\Delta E'_{dr1}$	$\Delta E'_{qr1}$	Δs_2	$\Delta E'_{dr2}$	$\Delta E'_{qr2}$
Participation factor	0.96	0.048	1.0	0.94	0.04	0.97

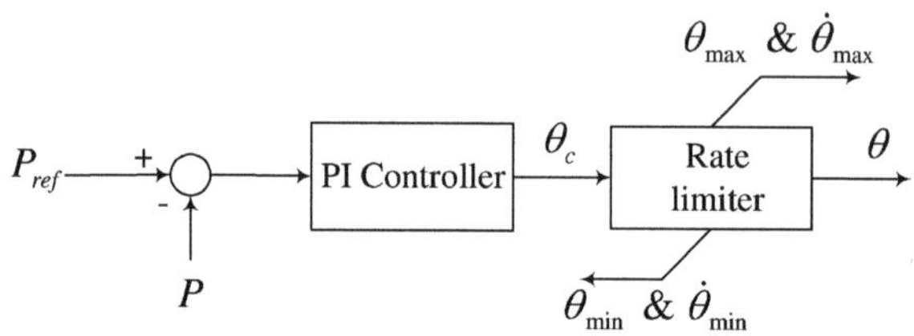

Fig. 7.27 Pitch angle control strategy

pitch controller is used to control real power. For the test system, the state vector is $x = \left[s_1, E'_{dr1}, E'_{qr1}, s_2, E'_{dr2}, E'_{qr2}, V_{dc}, V_{tm}\right]^T$.

Figure 7.27 depicts the pitch angle controller. In this work the pitch rate limit is set to 8 (deg/s), $\theta_{\min} = -5°$, $\theta_{\max} = 45°$, $\dot{\theta}_{\min} = -10$ (deg/s), $\dot{\theta}_{\max} = 10$ (deg/s) and time constant is 0.2 s. In this case, the gain of the tuned (trial and error method) PI controller is obtained as $K_P = 5$ and $K_I = 25$.

7.7.2 STATCOM Control Strategy

STATCOM employment has to counteract the well known instability problems related to the induction generator operation conditions, by providing a controlled reactive power, with the consequent stability margin improvement. The rationale of the proposed control strategy is based upon the ability of the inverter to distribute electrical power between the induction generator and load, guaranteeing the load required voltage profile. More specifically, at steady state conditions, the inverter does not exchange active power with the load, so that only the induction generator supplies the active power amount required by the load and inverter losses.

The STATCOM control strategy used in this chapter is shown in Fig. 7.28. In this method, the compensation is achieved by measuring the rms voltage at PCC. The output voltage (E) magnitude of the VSC relates to the DC side voltage and is also a function of the phase angle and the modulation ratio of the PWM. In this case k is fixed and α is used as the control variable. The inverter control consists of regulating the voltage amplitude and the phase delay angle (α) between the emf E and the inverter output voltage (v_t).

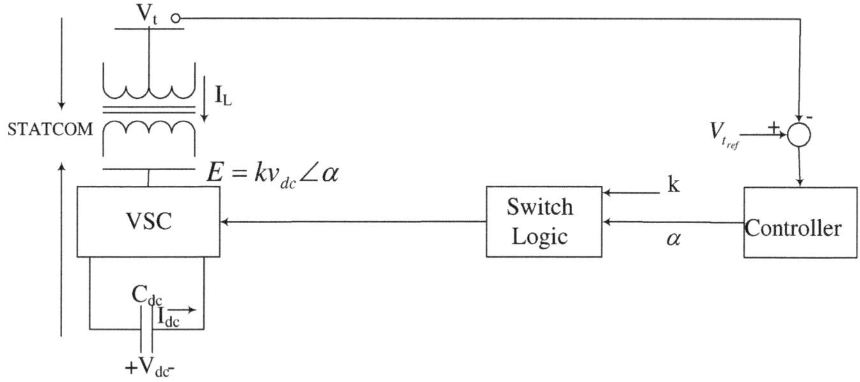

Fig. 7.28 STATCOM control strategy

7.7.3 Linearisation and Uncertainty Modelling

Linear controllers are designed based on the Taylor series approximation around an equilibrium point. This linearisation technique limits the applicability of the linear model to small deviations from the equilibrium point. In general, the range of these small deviations is difficult to quantify. To quantify the neglected higher order terms, we propose the use of a linearisation scheme which retains the contributions of the higher order terms in the form of the Cauchy remainder. In the design of the linear controller, a bound on the Cauchy remainder is incorporated as an uncertain term thus quantifying the deviations permitted in the linear model.

The test system dynamic is written in the compact form as:

$$\Delta\dot{x} = A\Delta x + (L - A)\Delta x + B_1\Delta u + (M - B_1)\Delta u, \tag{7.32}$$

where $A = \left.\frac{\partial f}{\partial x}\right|_{\substack{x=x_0\\u=u_0}}$ and $B_1 = \left.\frac{\partial f}{\partial u}\right|_{\substack{x=x_0\\u=u_0}}$.

We rewrite system (7.32) in terms of the block diagram shown in Fig. 7.29, where

$$(L - A)\,\Delta x + (M - B_1)\,\Delta u = \sum_{k=0}^{7} B_{2k}\xi_k(t), \tag{7.33}$$

where $\xi_1(t), \ldots, \xi_k(t)$ are known as the uncertainty inputs. The matrices $[B_{20}, \ldots, B_{27}]$, $\left[\tilde{C}_{10}, \cdots, \tilde{C}_{27}\right]$ are calculated such that

$$(L - A)\,\Delta x + (M - B_1)\,\Delta u = \sum_{k=0}^{7} B_{2k}\tilde{\phi}_k(\tilde{C}_{1k}\Delta x) + \sum_{k=0}^{7} B_{2k}\tilde{\psi}_k(\tilde{D}_{1k}\Delta u) \tag{7.34}$$

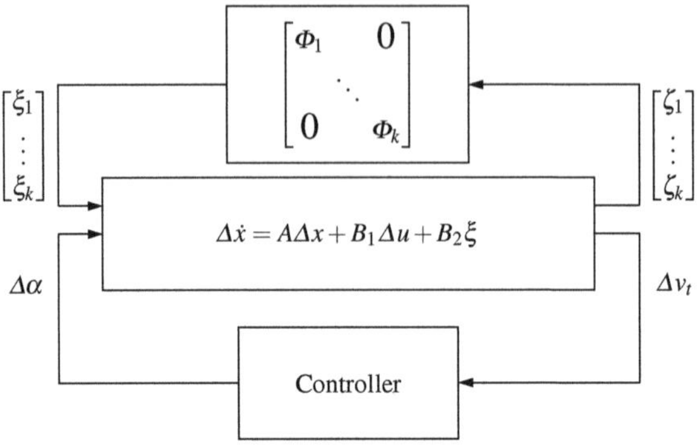

Fig. 7.29 Control strategy of structured uncertain system

where $\xi_k = \tilde{\phi}_k \tilde{C}_{1k} \Delta x + \tilde{\psi}_k \tilde{D}_{1k} \Delta u$, $k = 0, \ldots, 7$, and

$$
\begin{aligned}
B_{20} &= \left[\tfrac{1}{2H_{m1}}\ 0\,0\,0\,0\,0\,0\,0\right]^T, \quad B_{21} = \left[0\ \tfrac{T'_{01}}{X_1 - X'_1}\ 0\,0\,0\,0\,0\,0\right]^T, \\
B_{22} &= \left[0\,0\ \tfrac{T'_{01}}{X_1 - X'_1}\ 0\,0\,0\,0\,0\right]^T, \quad B_{23} = \left[0\,0\,0\ \tfrac{1}{2H_{m2}}\ 0\,0\,0\,0\right]^T, \\
B_{24} &= \left[0\,0\,0\,0\ \tfrac{T'_{02}}{X_2 - X'_2}\ 0\,0\,0\right]^T, \quad B_{25} = \left[0\,0\,0\,0\,0\ \tfrac{T'_{02}}{X_2 - X'_2}\ 0\,0\right]^T, \\
B_{26} &= \left[0\,0\,0\,0\,0\,0\ \tfrac{1}{To1}\ 0\right]^T, \quad B_{27} = \left[0\,0\,0\,0\,0\,0\,0\ K_m X_{s1}\right]^T,
\end{aligned}
\tag{7.35}
$$

The nonlinearities in this system are due to s_i, E_{dr_i}, E_{qr_i}, and v_{dc}, $i = 1,\ 2$, with this the matrices $\tilde{C}_{1k}$ and $\tilde{D}_{1k}$ are chosen such that

$$
\tilde{C}_{1k} = \begin{bmatrix}
1\,0\,0\,0\,0\,0\,0\,0 \\
0\,1\,0\,0\,0\,0\,0\,0 \\
0\,0\,1\,0\,0\,0\,0\,0 \\
0\,0\,0\,1\,0\,0\,0\,0 \\
0\,0\,0\,0\,1\,0\,0\,0 \\
0\,0\,0\,0\,0\,1\,0\,0 \\
0\,0\,0\,0\,0\,0\,1\,0
\end{bmatrix}, \quad \tilde{D}_{1k} = [1\ 1\ 1\ 1\ 1\ 1\ 1]^T,
$$

$$
(L - A)\,\Delta x + (M - B_1)\,\Delta u = \sum_{k=0}^{7} B_{2k}\xi_k(t), \quad k = 0, \ldots, 7 \tag{7.36}
$$

where $\xi_k = \tilde{\phi}_k \tilde{C}_{1k} \Delta x + \tilde{\psi}_k \tilde{D}_{1k} \Delta u$, and $\phi_k(t) = \frac{1}{\sqrt{\beta_k}} \left[\tilde{\phi}_k(t)\ \ \tilde{\psi}_k(t)\right]$, where β_k are scaling factors which affect the magnitude of the uncertain outputs ζ_k, $k = 0, \ldots, 7$.

In general, $x^{*p}, p = 1, \ldots, 8$, are not known beforehand, it is difficult to obtain the exact value of $(L - A)$ and $(M - B_1)$, but it is possible to obtain a bound on $\tilde{\phi}_k$ and $\tilde{\psi}_k$ over the operating range and parameter β_k is chosen to ensure

$$\|\phi_k(t)\|^2 \leq 1, \quad k = 0, \ldots, 7. \tag{7.37}$$

From this, we have

$$\|\xi_k(t)\|^2 \leq \beta_k \|\left(\tilde{C}_{1k}\Delta x + \tilde{D}_{1k}\Delta u\right)\|^2. \tag{7.38}$$

and we recover the IQC (integral quadratic constraint) [57],

$$\|\xi_k(t)\|^2 \leq \|\zeta_k(t)\|^2, \quad k = 0, \ldots, 7 \tag{7.39}$$

The expressions for obtaining $\tilde{\phi}_k(t)$ and $\tilde{\psi}_k(t)$ are given in Appendix-VIII. The system (7.32) can now be written as

$$\Delta\dot{x} = A\Delta x + B_1\Delta u + \sum_{k=0}^{7} B_{2k}\xi_k(t). \tag{7.40}$$

To facilitate control design, the power system model is finally summarised as

$$\Delta\dot{x}(t) = A\Delta x(t) + B_1\Delta u(t) + \sum_{k=0}^{7} B_{2k}\xi_k(t), \tag{7.41}$$

$$y(t) = C_2\Delta x(t) + \sum_{k=0}^{7} D_{2k}\xi_k(t)), \tag{7.42}$$

$$\zeta_k(t) = C_{1,k}\Delta x(t) + D_{1,k}u(t), \quad k = 0, \ldots, 7 \tag{7.43}$$

where $\zeta_k,\ k = 0, \ldots, 7$, are known as the uncertainty outputs and $y(t)$ is the measured output.

The output matrix is defined as $C_2 = \left[0, 0, 0, 0, 0, 0, 0, 1\right]$. We choose $D_{20} = 0.01$, $D_{21} = 0.01$, $D_{22} = 0.01$, $D_{23} = 0.1$, $D_{24} = 0.1$, $D_{25} = 0.01$, $D_{26} = 0.1$ and $D_{27} = 0.005$. Equations (7.41)–(7.43) provide a new representation of the power system model which contains both the linear and higher-order terms. The new formulation presented in this section is used to design a robust output feedback STATCOM controller for a nonlinear power system.

7.7.4 STATCOM Controller Design

The control design problem considered in this section is of providing a stabilising robust output-feedback control algorithm for a system containing structured uncertainty described by a certain IQC [57, 61]. The output feedback control method is applied to the uncertain systems of the form shown in Fig. 7.29.

It is shown in [61] that the linear robust control theory can be applied to (7.41)–(7.43) subject to the following constraint:

$$\int_0^{t_i} \|\xi_k(t)\|^2 dt \le \int_0^{t_i} \|\zeta_k(t)\|^2 dt, \quad \forall i \text{ and } \forall k = 0, \ldots, 7. \tag{7.44}$$

The necessary and sufficient condition for the absolute stabilisability of the uncertain system (7.41)–(7.43) is given in terms of the existence of solutions to a pair of parameter dependent algebraic Riccati equations [57]. The Riccati equations under consideration are defined as follows for the given constants $\tau_1 > 0, \ldots, \tau_7 > 0$:

$$\left(A - \tilde{B}_2 \tilde{D}_2^T \Gamma_\tau^{-1} C_2\right) Y + Y \left(A - \tilde{B}_2 \tilde{D}_2^T \Gamma_\tau^{-1} C_2\right)^T + Y \left(C_\tau^T C_\tau - C_2^T \Gamma_\tau^{-1} C_2\right) Y$$
$$+ \tilde{B}_2 \left(I - \tilde{D}_2^T \Gamma_\tau^{-1} \tilde{D}_2\right) \tilde{B}_2^T = 0, \tag{7.45}$$

$$X \left(A - B_1 G_\tau^{-1} D_\tau^T C_\tau\right) + \left(A - B_1 G_\tau^{-1} D_\tau^T C_\tau\right)^T X + C_\tau^T \left(I - D_\tau G_\tau^{-1} D_\tau^T\right) C_\tau$$
$$+ X \left(\tilde{B}_2 \tilde{B}_2^T - B_1 G_\tau^{-1} B_1^T\right) X = 0, \tag{7.46}$$

where

$$C_\tau = \begin{bmatrix} C_{10} \\ \sqrt{\tau_1} C_{11} \\ \vdots \\ \sqrt{\tau_7} C_{17} \end{bmatrix}; \quad D_\tau = \begin{bmatrix} D_{10} \\ \sqrt{\tau_1} D_{11} \\ \vdots \\ \sqrt{\tau_7} D_{17} \end{bmatrix}; \; \tilde{B}_2 = \left[B_{20} \; \frac{1}{\sqrt{\tau_1}} B_{21} \; \cdots \; \sqrt{\tau_7} B_{27} \right];$$

$$G_\tau = D_\tau^T D_\tau; \tilde{D}_2 = \left[D_{20} \; \frac{1}{\sqrt{\tau_1}} D_{21} \; \cdots \; \sqrt{\tau_7} D_{27} \right]; \quad \Gamma_\tau = \tilde{D}_2 \tilde{D^T}_2.$$

The original control problem is to stabilise the uncertain system via robust control. However, by introducing $\tau_1, \ldots, \tau_k$, the problem of absolutely stabilising an uncertain system becomes equivalent to an output-feedback H_∞ control problem, the solution to which is well-known [62]. The solutions to the above Riccati equations should satisfy the following conditions to guarantee the closed-loop stability, $X > 0$, $Y > 0$, and the spectral radius of the matrix, XY is $\rho(XY) < 1$.

The uncertain system (7.41)–(7.43) is required to satisfy the following assumptions. Let the matrices B_2, C_1, D_1, D_2, G and Γ be defined by

$$B_2 = \left[B_{20} \cdots B_{27} \right]; \quad D_2 = \left[D_{20} \cdots D_{27} \right];$$

$$C_1 = \begin{bmatrix} C_{10} \\ \cdots \\ C_{17} \end{bmatrix} \quad D_1 = \begin{bmatrix} D_{10} \\ \cdots \\ D_{17} \end{bmatrix}; \quad G = \sum_{k=0}^{7} D'_{1k} D_{1k};$$

and $\Gamma = \sum_{k=0}^{7} D'_{2k} D_{2k}$. With the above choice, the pair (A, B_1) is stabilizable, $G > 0, \Gamma > 0$, the pair (A, C_2) is detectable, the pair $(A - B_1 G^{-1} D'_1 C_1, (I - D_1 G^{-1} D'_1) C_1)$ is observable, and the pair $(A - B_2 D'_2 \Gamma^{-1} C_2, B_2 (I - D_2 \Gamma^{-1} D'_2))$ is controllable. The output-feedback controller is [61]:

$$\dot{x}_c = A_c x_c(t) + B_c y(t), \quad u(t) = C_c x_c(t), \tag{7.47}$$

$$\text{where } A_c = A + B_1 C_c - B_c C_2 + \left(\tilde{B}_2 - B_c \tilde{D}_2\right) \tilde{B}'_2 X, \tag{7.48}$$

$$B_c = (I - YX)^{-1} \left(Y\tilde{C}_2 + \tilde{B}_2 \tilde{D}'_2\right) \Gamma_\tau^{-1}, \tag{7.49}$$

$$C_c = -G_\tau^{-1} \left(B'_1 X + D'_\tau C_\tau\right). \tag{7.50}$$

7.7.5 Controller Design Algorithm

Firstly, we carry out several simulations by applying large disturbances in order to obtain an estimate of the operating region during LVRT transients. The controller is designed in the following way to ensure stability in the operating range of interest:

(i) For a given equilibrium point, obtain the matrices for the system representation (7.41)–(7.43) according to the procedure outlined in Sect. 7.7.3.
(ii) Choose an operating range $(x^{*p} - x^p)$ $p = 1, \ldots, 8$;
(iii) Determine the maximum value of β_k, $k = 0, \ldots, 7$, over all values of L and M in this range;
(iv) Design a robust controller given by (7.47)–(7.50);
(v) If the controller is feasible, go to step (vi), otherwise stop; and
(vi) Increase the range $(x^{*p} - x^p)$ and go to step (ii).

The process described above enables the selection of the largest range for which a feasible controller is obtained. The equilibrium point for this system is ($s_{i0} = 0.013$, $E'_{dri0} = 0.2186$, $E'_{qri0} = 0.9176$, $v_{dc0} = 1.3$, $v_{tm0} = 1$) pu, $i = 1, 2$. For the given power system model, we obtain the values of β_k given in Table 7.6 for the region Ω with the corner points of $\left[\bar{s}_1, \bar{E}'_{dr_1}, \bar{E}'_{qr_1}, \bar{s}_2, \bar{E}'_{dr_2}, \bar{E}'_{qr_2}, , \bar{v}_{dc}, \bar{v}_{tm}\right]^T$ and $\left[\underline{s}_1, \underline{E}'_{dr_1}, \underline{E}'_{qr_1}, \underline{s}_2, \underline{E}'_{dr_2}, \underline{E}'_{qr_2}, \underline{v}_{dc}, \bar{v}_{tm}\right]^T$ given by the values $\bar{s}_i = s_{i0} + 0.45$ pu, $\underline{s}_i = s_{i0} - 0.45$ pu, $\bar{E}'_{dr_i} = E'_{dri0} + 0.27$ pu, $\underline{E}'_{dr_i} = E'_{dri0} - 0.27$ pu, $\bar{E}'_{qr_i} = E'_{qri0} + 0.28$ pu, $\underline{E}'_{qr_i} = E'_{qri0} - 0.28$ pu, $\bar{v}_{dc} = v_{dc0} + 0.35$ pu, $\underline{v}_{dc} = v_{dc0} - 0.35$ pu, $\bar{v}_{tm} = v_{tm0} +$

Table 7.6 Values of $\beta_k,\ k = 0, \ldots, 7$

β_0	β_1	β_2	β_3	β_4	β_5	β_6	β_7
0.85	0.95	0.45	0.98	0.68	0.65	0.79	0.94

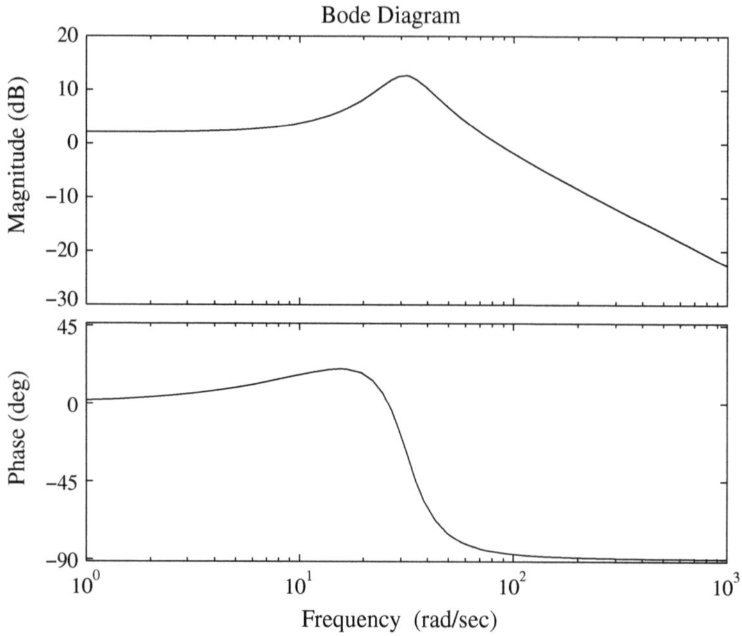

Fig. 7.30 Bode plot of open-loop system

0.45 pu, $\underline{v}_{tm} = v_{tm_0} - 0.45$ pu, $\bar{\alpha} = \alpha_0 + 25°$ and $\underline{\alpha} = \alpha_0 - 25°$, $i = 1, 2$. For this problem, $\tau_1 = 0.0005$, $\tau_2 = 0.0106$, $\tau_3 = 0.0346$, $\tau_4 = \tau_5 = \tau_6 = \tau_7 = 0.0045$.

7.7.6 Controller Performance Evaluation

Figures 7.30 and 7.31 show the open- and closed-loop frequency responses of the test system. It can be seen from Fig. 7.30 that there is a resonance peak in the magnitude response in the open-loop system and also a sharp drop in the phase angle. The shown in Fig. 7.31 has a higher damping ratio and smaller overshoot.

7.7.7 Stability During Low Voltages

The performance of the proposed controller for a 150 MVA STATCOM is evaluated for a three-phase fault at one of the parallel lines between bus 6 and bus 7. The CCT and CS with the proposed control are 0.18 s and 0.215 pu, respectively. To compare

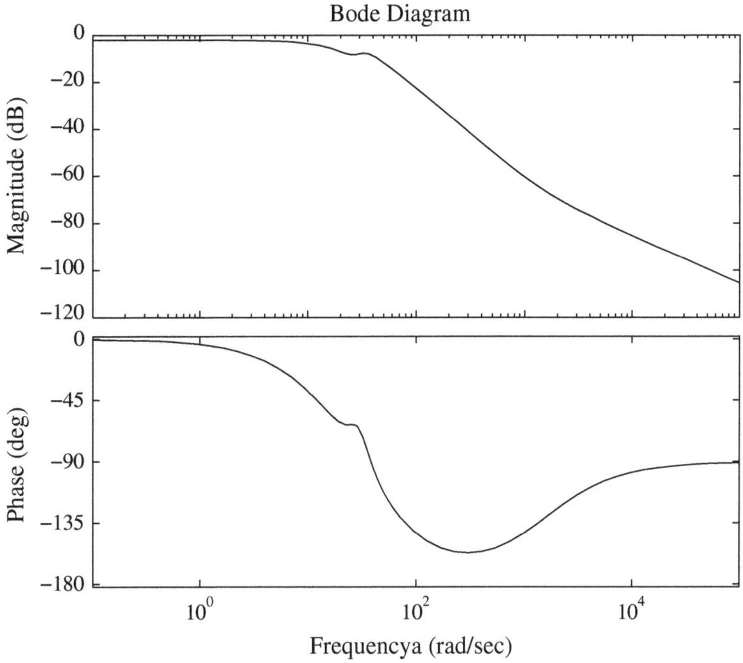

Fig. 7.31 Bode plot of closed-loop system

performances, we also determine CCT and CS with PI based STATCOM which are 0.165 s and 0.19 pu.

Figures 7.32 and 7.33 show the speeds and terminal voltages of the IG with the PI-based and proposed controllers, respectively. The fault is applied at 1 s and cleared at 1.18 s. From Figs. 7.32 and 7.33, it is clear that the proposed controller can stabilise the voltage and speed of the IG with a fault clearing time of 0.18 s. The slip of 0.195 pu at the fault clearing is greater than the CS of 0.19 pu obtained for the PI-based controller using numerical simulations. As the slip at post-fault is higher than the CS with the PI-based STATCOM, the speed continues to increase and voltage gradually decreases even after the fault is cleared. As a consequence, the system becomes unstable.

The designed controller guarantees stability if the system operating point, after the fault is cleared, falls within the region for which the controller is designed. We can conclude that the proposed controller performs better than the PI-based controller and results in a higher CCT.

7.7.8 Turbine Response to a Change in Wind Speed

Initially the wind speed applied to each turbine is 8 m/s, then starting at 5 s for wind speed is rammed to 9 m/s in 1 seconds and then returns to original value. Then, at 10 s

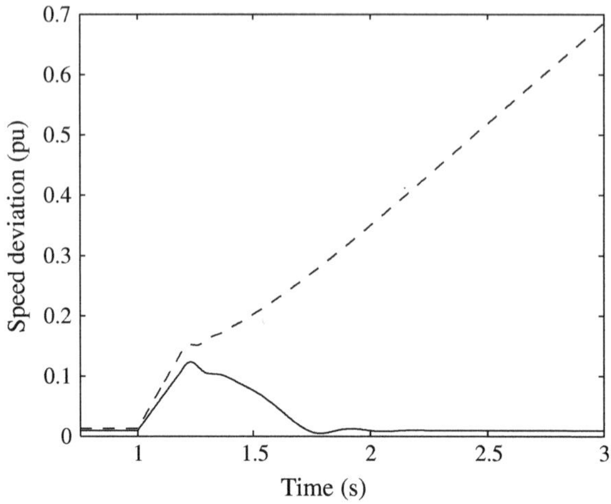

Fig. 7.32 Generator speed for three-phase fault at one of lines 6–7 (*Solid line* designed and *dashed line* PI controller)

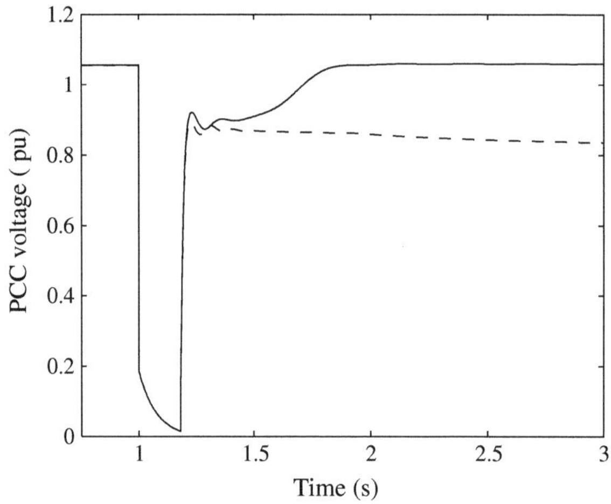

Fig. 7.33 Voltage at bus 4 for three-phase fault at one of lines 6–7 (*Solid line* designed and *dashed line* PI controller)

a temporary fault is applied at wind farm 1. As the asynchronous machine operates in generator mode, its speed is slightly above the synchronous speed (1.011 pu).

Figure 7.34 show the response of wind speed. real power output of wind generator, speed of wind generator, pitch angle output of pitch controller and reactive power output of the designed STATCOM controller. The generator speed shown in Fig. 7.36 increases quickly due to the increased aerodynamic torque. The electrical power increases too, until the pitch controller reacts by modifying the pitch angle as shown in Fig. 7.35. Over that time frame the turbine speed will have increased from

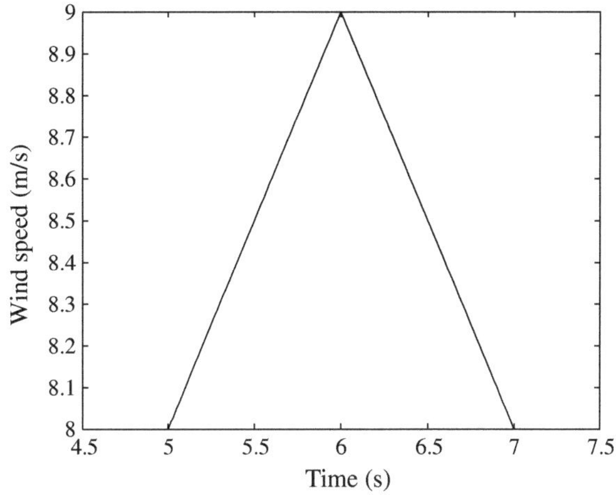

Fig. 7.34 Wind speed

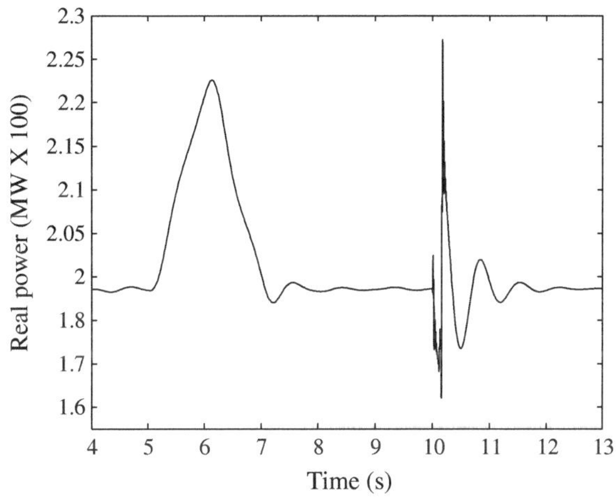

Fig. 7.35 Real power output of wind generator

1.0028 to 1.0047 pu. Initially, the pitch angle of the turbine blades is 7.95 degree. When the output power exceed rated power, the pitch angle shown in Fig. 7.37 is increased from 7.95 to 12 deg in order to bring output power back to its nominal value. The designed STATCOM controller increases the reactive power supply to keep the voltage constant which is visible from Fig. 7.38, while the speed controller prevents the generator speed becoming too high. It can be concluded that the proposed controller performs well in the case of change of input wind speed and severe three-phase fault.

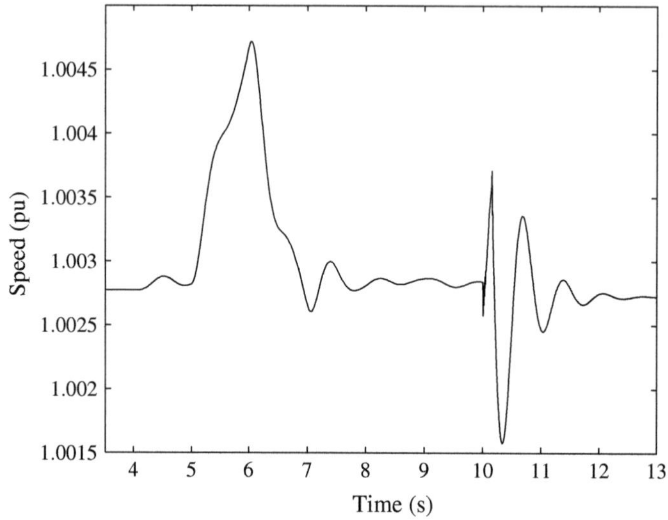

Fig. 7.36 Speed of wind generator

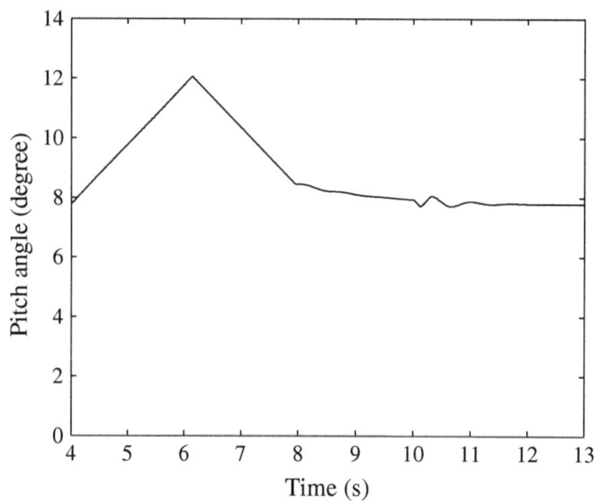

Fig. 7.37 Pitch angle

7.8 Decentralised STATCOM/ESS Controller

The design and implementation of a new control scheme for reactive power compensation, voltage regulation, and transient stability enhancement for wind turbines equipped with fixed-speed induction generators in large interconnected power systems is presented in this section. A systematic procedure is proposed to design decentralised multi-variable controllers for large interconnected power systems using

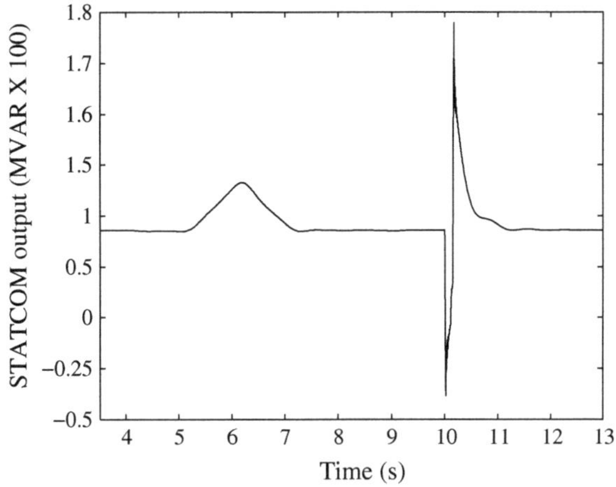

Fig. 7.38 Reactive power output of designed STATCOM controller

minimax output-feedback control design method and the controller design procedure is formulated as an optimisation problem involving rank-constrained linear matrix inequalities (LMIs). In this section, it is shown that STATCOM with energy storage system (STATCOM/ESS), controlled via robust control technique, is an effective device for improving the LVRT capability of fixed-speed wind turbines.

7.8.1 Test System and Control Task

A one-line diagram of the New England system [63] is shown in Fig. 7.39 and its parameters are given in Appendix-XI. This system is modified by replacing the four conventional generators at buses 31–34 by four wind farms and used as the test system in this section. The modified system network consists of six thermal power plants and four wind farms. The conventional generation, the wind generation and the total load in this system are 3760.48, 2432.93 and 6150.5 MW, respectively. We use an aggregated wind generator [60] and synchronous generator models [64] for controller design.

Two 150 kVA 2-level VSC based STATCOM/ESS are connected to the system at buses 32 and 34 through shunt coupling transformers at 110 kV to regulate their respective terminal bus voltages at the point of common coupling (PCC). The most effective locations and best feedback-signals for the STATCOM/ESS and PSSs are found by the method of comparing the residues [65] which are the products of modal controllability and observability gramians. The modal controllability indicates the degree of influence of the given input to the mode in question. The modal observability is a measure of the modal information contained in a feedback signal. They are independent and hence can be computed separately. Buses 36 and 39 are the best

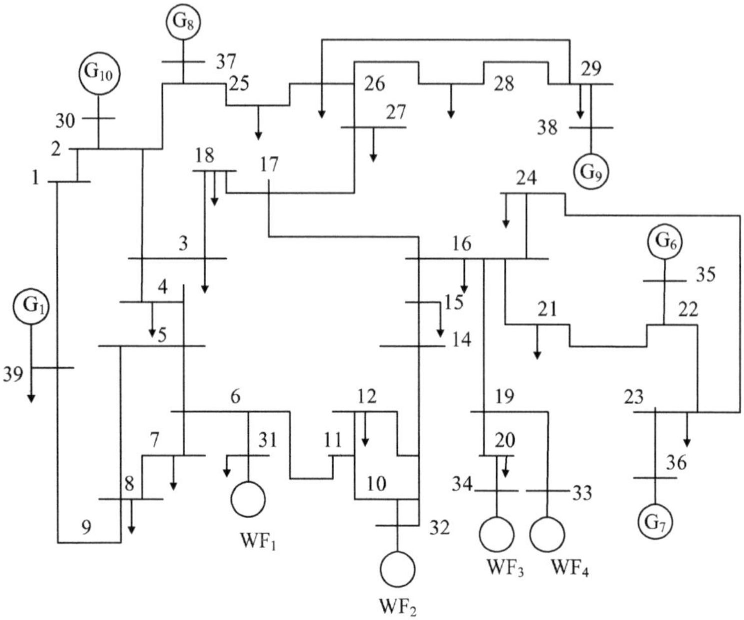

Fig. 7.39 10 machine 39 bus study system

locations for the PSSs. The selection of the STATCOM/ESSs and PSSs locations and the best feedback signal are made by comparison of the residues at all locations.

The eigenvalue analysis of the original system (before replacing the conventional generation by wind farms) shows that it has a dominant mode at $-0.0131 \pm j0.711$ with a damping ratio of 0.019. The modified system has three critical modes and their values with most significant normalized participation vectors are shown in Table 7.7. The mode $-0.062 \pm j2.21$ is an electromechanical mode with a damping ratio of 0.028. The other two modes, with eigenvalues of 0.018 and 0.14, are monotonic modes associated with both the rotor electrical dynamics of IGs. These two monotonic modes are introduced due to the replacement of synchronous generation with IGs. In this case, attention is directed to the design of robust control for these unstable modes.

The test system considered in this case is divided into four subsystems based on the coherent groups (generators swing together) of generating units: (i) wind farms 1 and 3, (ii) wind farms 2 and 4, (iii) G_6, G_9, and G_{10} and (iv) G_7, G_8, and G_1. One STATCOM/ESS controller is a part of each subsystem 1 and 2 and PSSs are parts of subsystems 3 and 4. The PSSs are designed using the standard process, as given in [66]. For subsystems 1 and 2, the state vector is $x_i = \left[\Delta\omega_{m_i}, \Delta\omega_{G_i}\Delta\gamma_i, \Delta s_i, \Delta E'_{dr_i}, \Delta E_{qr_i}, \Delta v_{dc_i}, \Delta v_{tm_i}, \Delta v_{sc_i}\right], i = 1, 2$. For STATCOM/ESS controllers, the control input $u_i = [\Delta m_i \ \Delta\alpha_i]^T, y_i = \left[\Delta v_{t_i} \ \Delta v_{dc_i}\right]^T$, and for (PSSs), $u_i = \Delta V_{s_i}$ and $y_i = \Delta\omega_i$, where V_{s_i} is the PSS output signal and ω_i is the rotor speed of the synchronous generator. In STATCOM/ESS control design,

Table 7.7 Critical modes and participation factors

Modes	Participation factors		
$-0.062 \pm j2.21$	$\Delta\delta_6 = 1$	$\Delta\delta_1 = 0.7738$	$\Delta E_{dr1} = 0.74$
0.018	$\Delta V_{dc1} = 1$	$\Delta E_{dr1} = 0.52$	$\Delta s_1 = 0.49$
0.14	$\Delta V_{dc2} = 1$	$\Delta E_{dr2} = 0.93$	$\Delta s_2 = 0.29$

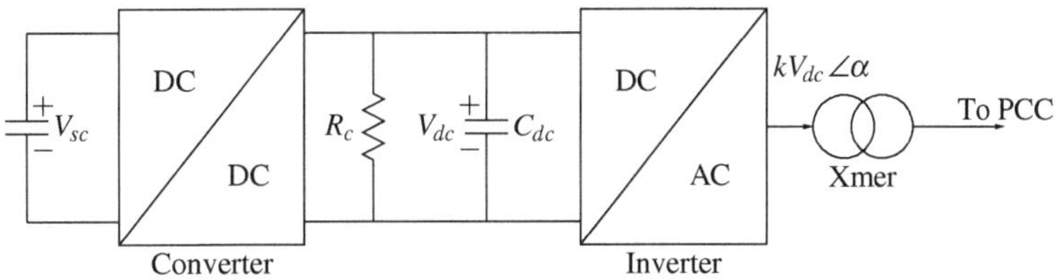

Fig. 7.40 Power circuit of STATCOM/ESS

slip is the controlled variable because the sensitivity of the monotonic modes to the slip frequency is high and a little increment of the generator speed can decrease the damping of these modes.

For the stability analysis, the transformer and the transmission line is included in the reduced admittance matrix. The nonlinear model of a WT used in this study includes a static model of the aerodynamics, a two mass model of the drive train, an actuator model and a third order model of the induction generator [67]. Also, a single-axis third-order generator model and an IEEE-ST1A type excitation system are used [68].

The main components of a STATCOM/ESS, as shown in Fig. 7.40, are a normal STATCOM and a supercapacitor-based energy storage system which is comprised of a supercapacitor and a bi-directional DC–DC buck-boost converter to control the charge and discharge of the supercapacitor modules. The switching losses are represented by a resistance in parallel with the DC capacitor. The aim of this module is to store energy in the supercapacitor and then deliver it to the grid via the DC link when required. The DC–DC converter operates in the buck mode to recharge the supercapacitor, whereas the boost mode transfers the stored energy to the DC link [69]. A conventional controller is used to control the buck-boost converter [69].

The STATCOM/ESS controller is depicted in Fig. 7.41. The controller converts the commanded signals into PWM switching commands for the STATCOM/ESS to regulate the modulation gain and firing angle. The firing angle α_i mainly affects the variation of the active power exchanged between the system and the STATCOM, and the duty cycle ratio m_i mainly regulates the magnitude of the STATCOM's output voltage and therefore the system voltage.

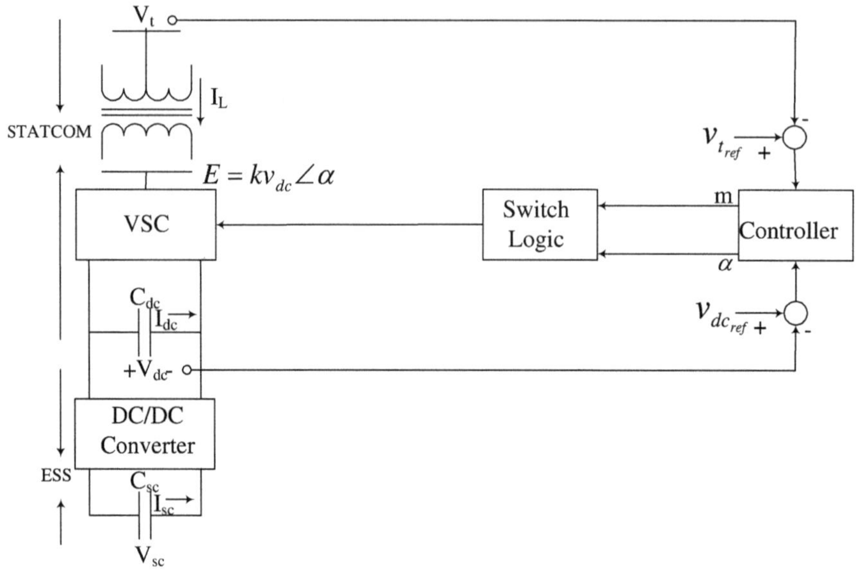

Fig. 7.41 STATCOM/ESS control strategy

7.8.2 Problem Formulation

The reformulation proposed in this research using Cauchy remainder allows us to represent the nonlinear large scale power system S comprising n subsystems S_i of the following form:

$$S_i : \Delta\dot{x}_i = A_i\Delta x_i + B_i\Delta u_i + E_i\xi_i + L_i r_i, \tag{7.51}$$

$$z_i = C_i\Delta x_i + D_i\Delta u_i, \tag{7.52}$$

$$\zeta_i = H_i\Delta x_i + G_i\Delta u_i, \tag{7.53}$$

$$y_i = C_{yi}\Delta x_i + D_{yi}\xi_i, \tag{7.54}$$

where Δx_i is the state vector, Δu_i is the control input, y_i is the measured output, z_i is the controlled output, ξ_i is known as the uncertainty input, ζ_i is known as the uncertainty output, and r_i describes the effect of other subsystems $S_1, \ldots, S_{i-1}, S_{i+1}, \ldots, S_N$ on subsystem S_i. The block diagram of the uncertain system is shown in Fig. 7.42.

The matrices in the system model (7.51)–(7.54), for the test system shown in Fig. 7.39, are given below:

$$E_i = \text{diag}\left(\frac{1}{2H_{m_i}}, \frac{1}{2H_{G_i}}, 0, \frac{1}{2H_{G_i}}, \frac{X_i - X_i'}{T_{o_i}'}, \frac{X_i - X_i'}{T_{o_i}'}, \frac{1}{C_i}, T_{m_i}, 0\right), \tag{7.55}$$

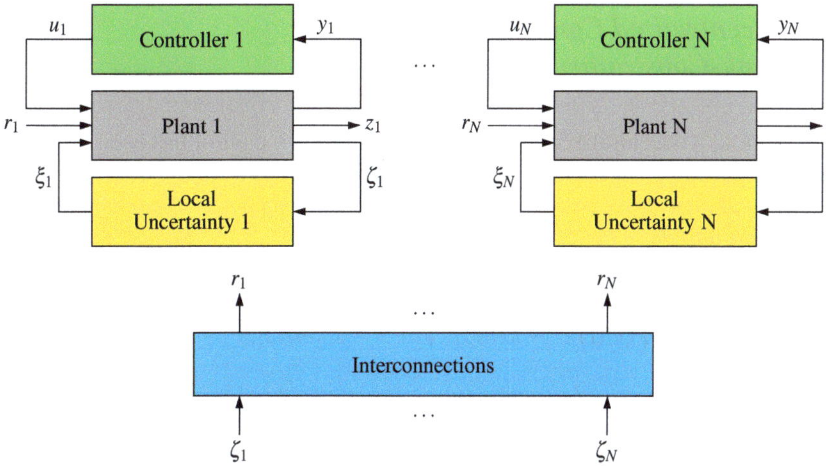

Fig. 7.42 Block diagram of the uncertain system

$$A_i = \left.\frac{\partial f_i}{\partial x_i}\right|_{\substack{x_i=x_{i0}\\ u_i=u_{i0}}}, \; B_i = \left.\frac{\partial f_i}{\partial u_i}\right|_{\substack{x_i=x_{i0}\\ u_i=u_{i0}}}, \; H_i = \sqrt{\Gamma_i}\begin{bmatrix} 0\,0\,0\,1\,0\,0\,0\,0\,0 \\ 0\,0\,0\,0\,1\,0\,0\,0\,0 \\ 0\,0\,0\,0\,0\,1\,0\,0\,0 \\ 0\,0\,0\,0\,0\,0\,1\,0\,0 \end{bmatrix}, \tag{7.56}$$

$$L_i = \mathrm{diag}\left(E_i, 0, \frac{1}{2H_i}, \frac{X_{d_i} - X'_{d_i}}{T'_{do_i}}\right), \; G_i = \begin{bmatrix} 1\,1 \\ 1\,1 \\ 1\,1 \\ 1\,1 \end{bmatrix}. \tag{7.57}$$

and $\xi_i = \tilde{\phi}_i \tilde{H}_i \triangle x_i + \tilde{\psi}_i \tilde{G}_i \triangle u_i, N_i \triangle x_j = L_i r_i,$ where $r_i = \sum_{\substack{j=1\\ j\neq i}}^{n} \tilde{\Xi}_i \zeta_j$.

The value of Γ_i is chosen such that

$$\|\phi_i\|^2 \leq 1, \quad \text{and} \quad \|\Xi_i\|^2 \leq 1. \tag{7.58}$$

The reformulation satisfies the norm bound constraints [70]

$$\|\xi_i\|^2 \leq \|\zeta_i\|^2, \quad \text{and} \quad \|r_i\|^2 \leq \sum_{j\neq i} \|\zeta_j\|^2. \tag{7.59}$$

For this controller design, we consider

$$C_i = [0, 0, 0, 1, 0, 0, 0, 0, 0]^T, \; D_i = 10^{-4}\,[1, 1]^T, \tag{7.60}$$

$$C_{yi} = [0, 0, 0, 0, 0, 0, 0, 1, 0; 0, 0, 0, 0, 0, 0, 0, 0, 1]^T, \tag{7.61}$$

$$D_{yi} = 10^{-4}\,[0, 0, 0, 1, 1, 1, 1, 0, 0; 0, 0, 0, 1, 1, 1, 1, 0, 0]^T. \tag{7.62}$$

7.8.3 Decentralized Controller Design Using Rank-Constrained LMIs

In this section, a decentralized minimax output-feedback control has been designed using LMIs [70, 71]. The robust control design methodology developed in [70] makes use of integral quadratic constraints (IQC) to characterise the magnitude of uncertain perturbations and interconnection between subsystems,

$$E\int_0^{t_l}\left(\|\zeta_i(t)\|^2 - \|\xi_i(t)\|^2\right)dt > -x_{i0}^T M_i x_{i0}, \tag{7.63}$$

$$E\int_0^{t_l}\left(\sum_{n=1,n\neq i}^{N}\|\zeta_n(t)\|^2 - \|r_i(t)\|^2\right)dt > -x_{i0}^T \bar{M}_i x_{i0}, \tag{7.64}$$

where $i = 1, 2, \ldots, N$, N is the number of subsystems, $M_i = M_i^T > 0, \bar{M}_i = \bar{M}_i^T > 0$, $\{t_l\}_{l=1}^{\infty}$, $t_l \rightarrow +\infty$, is a sequence of time instants and E is the expectation operator. It is immediate that the constraints (7.59) can be rewritten in the form of (7.63) and (7.64) with arbitrarily chosen small $x_{i0}^T M_i x_{i0}$ and $x_{i0}^T \bar{M}_i x_{i0}$.

This minimax linear quadratic technique minimises the following performance cost over all permissible integral quadratic constraints (IQCs):

$$J_{wc}(u) \triangleq \int_0^{\infty}\sum_{i=1}^{N}\|z_i(t)\|^2 dt. \tag{7.65}$$

In this design, we consider the norm bounded constraints, as in (7.59), instead of the more general IQCs. This means that the designed controllers are suboptimal for norm-bounded constraints. The control algorithm is to find the infimum of the following function over the set $\mathcal{T}$:

$$J(\tau, \theta) = \sum_{i=1}^{N} x_{i0}^T\left[X_i + \tau_i M_i + \theta_i \bar{M}_i\right]x_{i0}, \tag{7.66}$$

where $\mathcal{T} = \{\{\tau_i\ \theta_i\} \in R^{2N}, \tau_i > 0, \theta_i > 0\}$, $M_i > 0$ and $\bar{M}_i > 0$ are two sets of symmetrical matrices, and matrices X_i and Y_i are the solutions to the following pair of coupled generalized algebraic Riccati equations and algebraic Riccati inequalities [70]:

$$A_i^T X_i + X_i A_i + \bar{C^T}_i \bar{C}_i - X_i\left[B_i R_i^{-1} B_i^T - \bar{B}_{2_i}\bar{B}_{2_i}^T\right]X_i = 0, \tag{7.67}$$

$$A_i^T Y_i + Y_i A_i + Y_i \bar{B}_{2_i}\bar{B}_{2_i}^T Y_i - \left[C_{y_i}^T W_i^{-1} C_{y_i} - \bar{C}_i^T \bar{C}_i\right] < 0, \tag{7.68}$$

where $R_i = \bar{D}_i^T \bar{D}_i$, $W_i = \bar{D}_{y_i} \bar{D}_{y_i}^T$, $\bar{\theta}_i = \sum_{n=1,n\neq i}^{N} \theta_n$,

$$\bar{C}_i = \begin{bmatrix} C_i \\ (\tau_i + \bar{\theta}_i)^{1/2} H_i \end{bmatrix}, \quad \bar{D}_i = \begin{bmatrix} D_i \\ (\tau_i + \bar{\theta}_i)^{1/2} G_i \end{bmatrix},$$
$$\bar{B}_{2_i} = \begin{bmatrix} \tau_i^{-1/2} E_i \; \theta_i^{-1/2} L_i \end{bmatrix}, \quad \bar{D}_{y_i} = \begin{bmatrix} \tau_i^{-1/2} D_{y_i} \; 0 \end{bmatrix}. \tag{7.69}$$

The controller u_i^* with the τ^*, θ^* is given by [70]:

$$\dot{x}_{c_i} = \{A_i - \left[B_i R^{-1} B_i^T - \bar{B}_{2_i} \bar{B}_{2_i}^T\right] X_i\} x_{c_i}$$
$$+ [Y_i - X_i]^{-1} C_{y_i}^T W_i^{-1} \left[y_i - C_{y_i} x_{c_i}\right], \tag{7.70}$$
$$u_i^* = -R_i^{-1} B_i^T X_i x_{c_i}. \tag{7.71}$$

The solutions are required to satisfy the following conditions: $\tau_i > 0, \theta_i > 0, X_i \geq 0$, $Y_i \geq 0$ and $Y_i > X_i$.

The controller u^* guarantees the following minimax property

$$J_{wc}(u^*) \leq J(\tau^*, \theta^*) = \inf_{\mathcal{T}} J(\tau, \theta) \tag{7.72}$$

The suboptimal control design used in this case involves solving the optimisation problem given on the right-hand side of (7.72). Generally, it is difficult to provide a systematic way to perform such optimisation. In this case, the idea is to replace the problem $\inf_\tau J(\tau, \theta)$ with an equivalent optimisation problem involving rank-constrained LMIs [72]. From (7.68), we get

$$A_i^T X_i + X_i A_i + \bar{C^T}_i \bar{C}_i - X_i \left[B_i R_i^{-1} B_i^T - \bar{B}_{2_i} \bar{B}_{2_i}^T\right] X_i < 0, \tag{7.73}$$

and, by multiplying the left and right sides of (7.73) with $\tilde{X}_i = X_i^{-1}$, we get

$$\tilde{X}_i A_i^T + A_i \tilde{X}_i + \tilde{X}_i \bar{C^T}_i \bar{C}_i \tilde{X}_i - \left[B_i R_i^{-1} B_i^T - \bar{B}_{2_i} \bar{B}_{2_i}^T\right] < 0. \tag{7.74}$$

Introducing matrices, F_i of appropriate dimensions, without changing the feasibility of (7.74), we add a quadratic term of F_i to the left-hand side of (7.74) as:

$$\tilde{X}_i A_i^T + A_i \tilde{X}_i + \tilde{X}_i \bar{C^T}_i \bar{C}_i \tilde{X}_i - \left[B_i R_i^{-1} B_i^T - \bar{B}_{2_i} \bar{B}_{2_i}^T\right]$$
$$+ \left[F_i^T + B_i R^{-1}\right] R_i \left[F_i^T + B_i R_i^{-1}\right]^T < 0, \tag{7.75}$$

which is equivalent to

$$\tilde{X}_i A_i^T + A_i \tilde{X}_i + \tilde{X}_i \bar{C^T}_i \bar{C}_i \tilde{X}_i + \bar{B}_{2_i} \bar{B}_{2_i}^T$$

$$+ F_i^T R_i F_i + B_i F_i + F_i^T B_i^T < 0. \tag{7.76}$$

Using (7.69), the terms of (7.76) can be represented as:

$$\begin{aligned}
\bar{B}_{2_i}\bar{B}_{2_i}^T &= \tau_i^{-1}E_iE_i^T + \theta_i^{-1}L_iL_i^T, \\
\tilde{X}_i\bar{C}^T{}_i\bar{C}_i\tilde{X}_i &= \tilde{X}_i\left[C_i^T C_i + (\tau_i + \bar{\theta}_i)H_i^T H_i\right]\tilde{X}_i, \\
F_i^T R_i F_i &= F_i^T\left[D_i^T D_i + (\tau_i + \bar{\theta}_i)G_i^T G_i\right]F_i.
\end{aligned} \tag{7.77}$$

Let $\tilde{\tau}_i = \tau_i^{-1}$, $\tilde{\theta}_i = \theta_i^{-1}$. By combining (7.76), (7.77) and applying the Schur complement, we obtain the following LMIs with the variables $\tilde{X}_i, F_i, \tilde{\theta}_i, \tilde{\tau}_i$:

$$\begin{bmatrix} N_i & X_iC_i^T & F_i^T D_i^T & Q_i \\ \star & -I & 0 & 0 \\ \star & \star & -I & 0 \\ \star & \star & \star & -\Theta_i \end{bmatrix} < 0, \tag{7.78}$$

where

$$\begin{aligned}
N_i &= \tilde{X}_iA_i^T + A_i\tilde{X}_i + \tilde{\tau}_iE_iE_i^T + \tilde{\theta}_iL_iL_i^T + B_iF_i + F_i^T B_i^T \\
Q_i &= \left[F_i^T G_i^T + \tilde{X}_iH_i^T, \ldots, F_i^T G_i^T + \tilde{X}_iH_i^T\right] \quad \text{(N entries)} \\
\Theta_i &= \mathrm{diag}\left[\tilde{\tau}_iI, \tilde{\theta}_1I, \ldots, \tilde{\theta}_{i-1}I, \tilde{\theta}_{i+1}I, \tilde{\theta}_NI\right].
\end{aligned} \tag{7.79}$$

Similarly, by substituting (7.69) into (7.67) and applying the Schur complement, we obtain the LMIs with the variables $\tilde{Y}_i, F_i, \theta_i, \tau_i$:

$$\begin{bmatrix} M_i & Y_iE_i & Y_iL_i \\ \star & -\tau_iI & 0 \\ \star & \star & -\theta_iI \end{bmatrix} < 0, \tag{7.80}$$

where

$$M_i = A_i^T Y_i + Y_iA_i - \tau_iC_{yi}^T\left[D_{yi}D_{yi}^T\right]^{-1}C_{yi} + C_i^T C_i + (\tau_i + \bar{\theta}_i)H_i^T H_i.$$

The coupling condition $Y_i > X_i > 0$ is equivalent to

$$\begin{bmatrix} \tilde{X}_i & I \\ I & Y_i \end{bmatrix} > 0. \tag{7.81}$$

Now, we consider the performance of the upper bound on the right-hand side of (7.72). It should be noted that minimising $J(\tau, \theta)$ is equivalent to minimising $(\lambda_1 + \lambda_2 + \cdots + \lambda_n)$ subject to:

$$\lambda_i > x_{i0}^T \left[X_i + \tau_i M_i + \theta_i \bar{M}_i\right] x_{i0}. \tag{7.82}$$

Using the Schur complement again, (7.82) is equivalent to the following LMIs:

$$\begin{bmatrix} \lambda_i & \Pi & x_{i0}^T M_i^{1/2} & x_{i0}^T \tilde{M}_i^{1/2} \\ \star & \tilde{X}_i & 0 & 0 \\ \star & \star & \tilde{\tau}_i I & 0 \\ \star & \star & \star & \tilde{\theta}_i I \end{bmatrix} < 0, \tag{7.83}$$

here $\Pi = \left[\pi_1^{1/2} x_{i0}^T, \ldots, \pi_k^{1/2} x_{i0}^T\right]$. Also, the conditions $\tilde{\tau}_i > 0$, $\tau_i > 0$, $\tilde{\tau}_i \tau_i = 1$, $\tilde{\theta}_i > 0$, $\theta_i > 0$, $\tilde{\theta}_i \theta_i = 1$ are equivalent to the rank-constrained LMIs:

$$\begin{bmatrix} \tilde{\tau}_i & I \\ I & \tau_i \end{bmatrix} \geq 0, \quad \text{rank} \begin{bmatrix} \tilde{\tau}_i & I \\ I & \tau_i \end{bmatrix} \leq 1, \tag{7.84}$$

$$\begin{bmatrix} \tilde{\theta}_i & I \\ I & \theta_i \end{bmatrix} \geq 0, \quad \text{rank} \begin{bmatrix} \tilde{\theta}_i & I \\ I & \theta_i \end{bmatrix} \leq 1. \tag{7.85}$$

We now consider the following linear cost optimisation problem in the variables λ_i, $\tilde{X}_i$, Y_i, F_i, $\tilde{\theta}_i$, $\tilde{\tau}_i$ and τ_i, θ_i [72]:

$$J_{LMI}^* \triangleq \inf(\lambda_1 + \cdots + \lambda_n), \tag{7.86}$$

subject to (7.78), (7.80), (7.81), (7.83), (7.84) and (7.85).

Note that this problem is a problem of minimising a linear cost subject to rank constrained LMIs. The numerical problem is solved by using the rank constrained LMI solver LMIRank [73].

7.8.4 Controller Design Algorithm

The controller, in this section, is designed for severe faults so it can, in principle, also ensure stability against other disturbances. From simulations we estimated the operating region Ω formed by corner points $\left[\bar{s}_i, \bar{E}'_{dr_i}, \bar{E}'_{qr_i}, \bar{\omega}_{m_i}, \bar{\omega}_{G_i}, \bar{\gamma}_i, \bar{v}_{dc_i}, \bar{v}_{tm_i}, \bar{v}_{sc_i}\right]^T$ and $\left[\underline{s}_i, \underline{E}'_{dr_i}, \underline{E}'_{qr_i}, \underline{\omega}_{m_i}, \underline{\omega}_{G_i}, \underline{\gamma}_i, \underline{v}_{dc_i}, \underline{v}_{tm_i}, \underline{v}_{sc_i}\right]^T$ centred at equilibrium point for severe faults with the values $\bar{s}_i - \underline{s}_i = 2 \times 0.225$ pu, $\bar{E}'_{dr_i} - \underline{E}'_{dr_i} = 2 \times 0.242$ pu, $\bar{E}'_{qr_i} - \underline{E}'_{qr_i} = 2 \times 0.225$ pu, $\bar{\omega}_{m_i} - \underline{\omega}_{m_i} = 2 \times 0.395$ pu, $\bar{\omega}_{G_i} - \underline{\omega}_{G_i} = 2 \times 0.337$ pu, $\bar{\gamma}_i - \underline{\gamma}_i = 2 \times 25°$, $\bar{v}_{dc_i} - \underline{v}_{dc_i} = 2 \times 0.334$ pu, $\bar{v}_{tm_i} - \underline{v}_{tm_i} = 2 \times 0.235$ pu, $\bar{v}_{sc_i} - \underline{v}_{sc_i} = 2 \times 0.248$ pu, $\bar{m}_i - \underline{m}_i = 2 \times 0.328$ and $\bar{\alpha}_i - \underline{\alpha}_i = 2 \times 28°$, $i = 1, 2$. The design process is described in the following steps:

(i) From simulations, select coherent groups of generating units and represent them by equivalent models.
(ii) Perform modal analysis and determine the critical modes. Analyse the participation vectors for the critical modes and identify the states related to them.
(iii) From the simulations of the faulted system (undergoing a large perturbation during the LVRT transient), obtain the range in variations of all state variables and form a volume, Ω, with corner points given by $(x_{0_p} - x_{f_p})$ and $(x_{0_p} + x_{f_p})$, $p = 1, \ldots, 7$, where $2x_{f_p}$ is the largest variation in the pth state variable about its equilibrium value, x_{0_p}.
(iv) Obtain $\Gamma_i^* = \max_{x_i^{*p} \in \Omega_i} \{\Gamma_i : ||\phi_i||^2 < 1 \,|,\ |\psi_i||^2 < 1\}$, as given in (7.58); the process to obtain Γ_i^* involves obtaining the maximum value of $||\phi_i||$ and $||\Xi_i||$ over the operating range of interest.
(v) Check if there exists a feasible controller with $\Gamma_i = \Gamma_i^*$, i.e., scalars τ_i and θ_i exist such that there is a feasible solution to LMIs, as described in Sect. 8.5.
(vi) Compare the control region with the operating region required to provide the LVRT capability.
(vii) If we obtain a feasible controller in the above step, increase the range of the operating region if step (vi) is not satisfied or, if we have arrived at the largest possible range then perform an optimal search over the scalar parameters τ_i and θ_i, to get the infimum in (7.86). If there is no feasible solution with the chosen $\Gamma_i = \Gamma_i^*$, reduce the range and go to step (iv).

For the given system, we are able to obtain feasible controllers with values of $\Gamma_1 = 0.968$ and $\Gamma_2 = 0.976$. The controller is stabilising for all variation of states in the polytope region Ω formed by corner points $\left[\bar{s}_i, \bar{E}'_{dr_i}, \bar{E}'_{qr_i}, \bar{\omega}_{m_i}, \bar{\omega}_{G_i}, \bar{\gamma}_i, \bar{v}_{dc_i}, \bar{v}_{tm_i}, \bar{v}_{sc_i}\right]^T$ and $\left[\underline{s}_i, \underline{E}'_{dr_i}, \underline{E}'_{qr_i}, \underline{\omega}_{m_i}, \underline{\omega}_{G_i}, \underline{\gamma}_i, \underline{v}_{dc_i}, \underline{v}_{tm_i}, \underline{v}_{sc_i}\right]^T$ with the following values:

$\bar{s}_i = s_{i0} + 0.243$ pu , $\underline{s}_i = s_{i0} - 0.243$ pu, $\bar{E}'_{dr_i} = E'_{dr_{i0}} + 0.347$ pu, $\bar{E}'_{dr_i} = E'_{dr_{i0}} - 0.347$ pu, $\bar{E}'_{qr_i} = E'_{qr_{i0}} + 0.315$ pu, $\underline{E}'_{qr_i} = E'_{qr_{i0}} - 0.315$ pu, $\bar{\omega}_{m_i} = \omega_{m_{i0}} + 0.428$ pu, $\underline{\omega}_{m_i} = \omega_{m_{i0}} - 0.428$ pu, $\bar{\omega}_{G_i} = \omega_{G_{i0}} + 0.437$ pu, $\underline{\omega}_{G_i} = \omega_{G_{i0}} - 0.437$ pu, $\bar{\gamma}_i = \gamma_{i0} + 36°$, $\bar{\gamma}_i = \gamma_{i0} - 36°$, $\bar{v}_{dc_i} = v_{dc_{i0}} + 0.365$ pu, $\underline{v}_{dc_i} = v_{dc_{i0}} - 0.365$ pu, $\bar{v}_{tm_i} = v_{tm_{i0}} + 0.269$ pu, $\underline{v}_{tm_i} = v_{tm_{i0}} - 0.269$ pu, $\bar{v}_{sc_i} = v_{sc_{i0}} + 0.275$pu, $\bar{m}_i = m_{i0} - 0.467$, $\bar{\alpha}_i = \alpha_{i0} + 23°$ and $\underline{\alpha}_i = \alpha_{i0} - 23°$, $i = 1, 2$. This range of the variation of the state variables is larger than the range for several large disturbances as noted earlier in this subsection. The above bound for $\phi(\zeta)$ is obtained at a point interior to the region, i.e., $s_i^* = 0.185$ pu, $E'^*_{dr_i} = 0.85$ pu, $E'^*_{qr_i} = 0.825$ pu, $\omega^*_{m_i} = 1.45$ pu, $\omega^*_{G_i} = 1.42$ pu, $\gamma_i^* = 25.5°$, $v^*_{dc_i} = 0.86$ pu, $v^*_{sc_i} = 0.845$ pu, $m_i^* = 0.425$ pu and $\alpha_i^* = 22.5°$, $i = 1, 2$. Although the designed controller is not globally stabilising but we know that it is stabilising over a large operating region which covers most faulted system operation.

From the Bode plot of the designed ninth-order controller for subsystem 1, shown in Fig. 7.43, it can be seen that the controller has a feasible low frequency gain and suitable cut-off frequency. Bode plots for other controllers have similar characteristics.

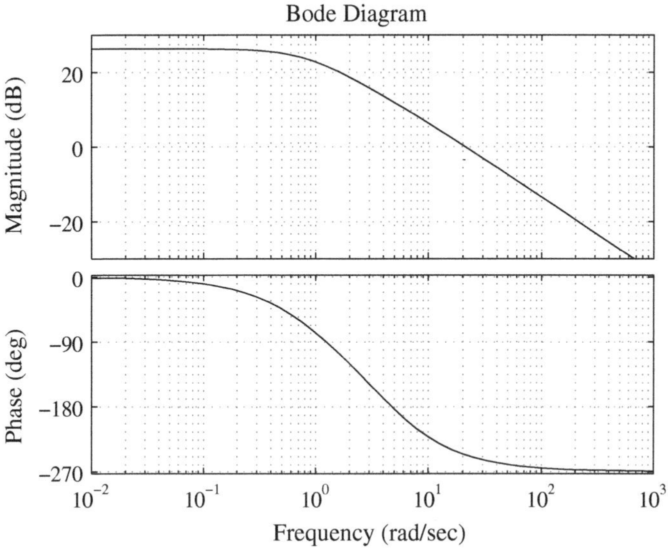

Fig. 7.43 STATCOM/ESS controller for subsystem 1—Bode plot of transfer function of modulation index versus terminal voltage

7.8.5 Controller Performance Evaluation

The FRT capability of a wind generator is expressed in this book as voltage and transient stability margins. The voltage stability margin is defined as the difference between the operating voltage and the critical voltage (V_{cr}). The transient stability margin is given as the difference between the speed after a specified fault duration and the critical speed (CS) of the generator. The critical speed is given by the intersection between the torque-speed curve for the specified system and the mechanical torque [54]. The critical voltage can be obtained from the P–V curves [55]. The stability analysis of a power system may consider the determination of its critical clearing time (CCT), for a given fault, in order to find the maximum value of the CCT for which the system is still stable. In this research, the CCT is first estimated by using the following equations and then exact value is determined from simulations in which it is obtained by increasing the fault time interval until the system loses its stability [74].

$$t_c = \frac{1}{T_m} 2H_m(s_c - s_0), \tag{7.87}$$

where s_c is the CS of a generator.

In this book, the controllers is designed using Matlab and then simulations has been carried out with the nonlinear dynamic models. The performance of the designed controller is tested for the following cases:

Table 7.8 Performance comparison: (*a*) proposed controller; and (*b*) PI-based STATCOM/ESS

STATCOM/ESS (MVA)	(a) Proposed Controller			(b) PI Control		
150 MVA/95F	CS (pu)	CCT (s)	V_{cr} (pu)	CS (pu)(s)	CCT (s)	V_{cr} (pu)
150 MVA/95F	1.35	0.165	0.605	1.315	0.140	0.625

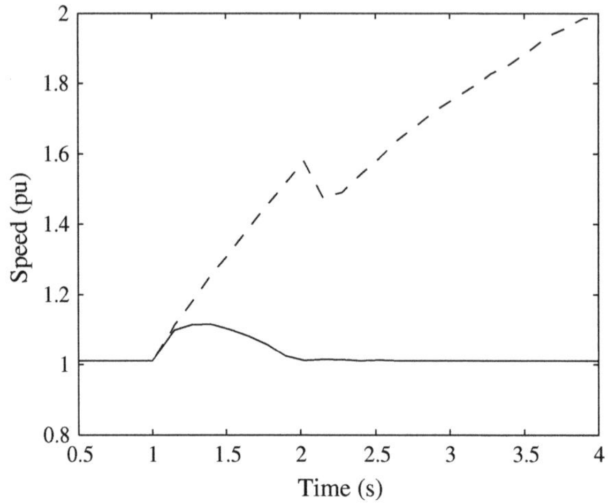

Fig. 7.44 Speed response for three-phase fault at bus 11 (*Solid line* designed and *dashed line* PI-based STATCOM/ESS controller)

7.8.6 Enhancement of Voltage and Transient Stability Margins

A simulation study is performed to emulate the system in Fig. 7.39 with the purpose of evaluating the transient and voltage stability limits achievable with the proposed and PI-based STATCOM/ESS controller. The CCT and CS, as shown in Table 8.1, for a three-phase fault with 150 MVA STATCOMs and 95 F, 19 kJ supercapacitors are 0.165 s and 1.35 pu with the designed controller, compared with 0.140 s and 1.315 pu with a properly tuned PI controller [69]. In this case, the gain of the tuned (trial and error method) PI controller is obtained as $K_P = 0.28$ and $K_I = 20.45$. It can be concluded that an appropriate combination of active and reactive power control by STATCOM/ESS is an effective way of improving the stability and enhancing the fault ride-through capability of the relevant induction-generator-based wind turbines.

In order to evaluate the performance of the designed controller, in the face of system nonlinearity and operating conditions, detailed simulations are performed for a symmetrical three-phase fault at bus 11 which is subsequently cleared after 150 ms. Figures 7.44 and 7.45 show the speeds response and terminal voltages, respectively, of the wind farm WF_1 with the conventional PI controller and the proposed STATCOM/ESS controller.

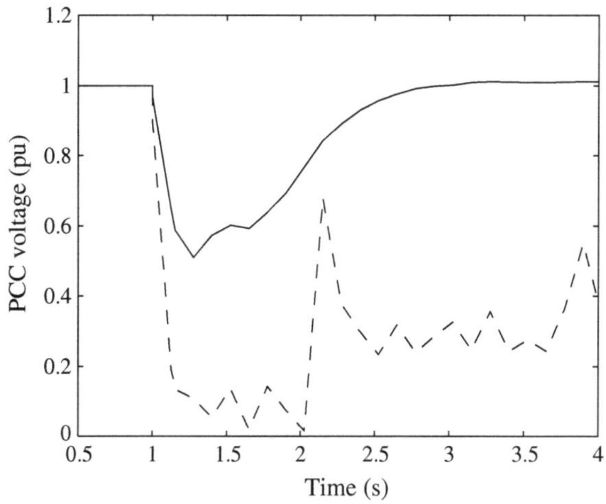

Fig. 7.45 Terminal Voltage for three-phase fault at bus 11 (*Solid line* designed and *dashed line* PI-based STATCOM/ESS controller)

During the fault, the wind generator accelerates, since it is no longer able to generate enough electromagnetic torque to balance the mechanical torque coming from the wind which is obviously unaffected by the grid fault. When the fault is cleared, the generator speed with the proposed control is about 1.15 pu whereas that with the PI control is 1.6 pu. The reclosing time, speed and voltage using the PI controller is greater that the corresponding CCT, CS and critical voltage. With the PI controller it can be seen that the terminal voltage cannot be restored and the induction generator continues to accelerate until the system loses stability.

The robust response of the proposed robust STATCOM/ESS control during a significant voltage sag allows the voltage at the PCC to be kept limited, thus resulting in system stability. The applied control efforts for the designed WF_1 STATCOM controller, for the three-phase fault, are shown in Figs. 7.46 and 7.47. It also avoids a large increase in turbine speed, thereby reducing the likelihood of the turbine exceeding the critical speed and thus increasing the fault ride-through capability.

7.8.7 Real and Reactive Power Output During Low Voltages

In this case, a severe condition for voltage recovery analysis is considered in which the line between the buses 7 and 8 in Fig. 7.39 is opened. Figure 7.48 illustrates the real power output due to the outage of the line 7–8 for 150 ms from which it can be seen that during faulted period the wind farm WF_2, using the proposed control, supplies 82 % of its rated power, but with the PI control, it is 72 %.

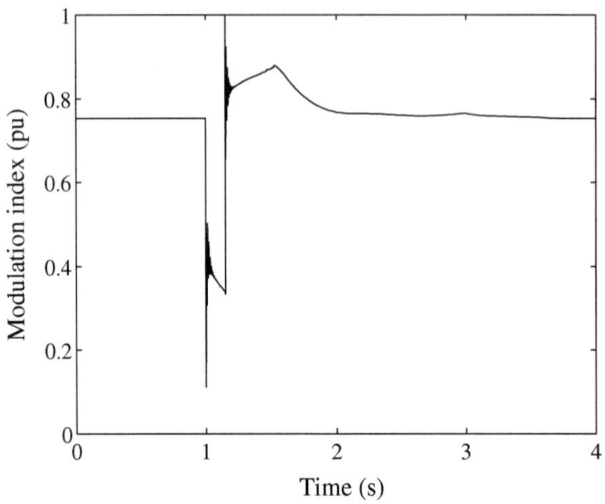

Fig. 7.46 Control effort (m) for three-phase fault at bus 11 with designed controller

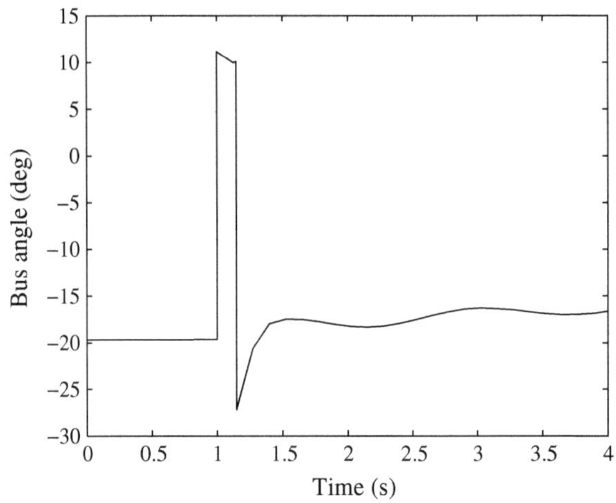

Fig. 7.47 Control effort (α) for three-phase fault at bus 11 with designed controller

Figure 7.49 shows the reactive power absorbed by the wind farm WF_2 and the reactive power output of the designed STATCOM/ESS controller at bus 32. It can be seen that the reactive power output of the controller exactly matches the requirement of the wind farm WF_2. From Fig. 7.48, it can be seen that the PI controller causes oscillatory behaviours at the post fault. The designed controller provides improved performance in terms of oscillations, settling time and damping.

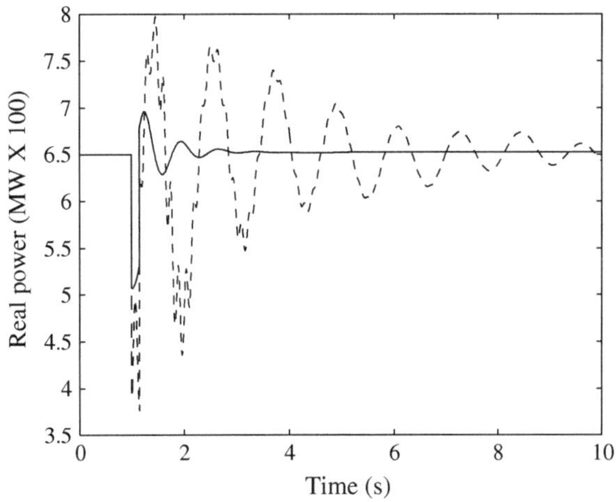

Fig. 7.48 Real power output for temporary outage of line 7–8 (*Solid line* designed and *dashed line* PI-based STATCOM/ESS controller)

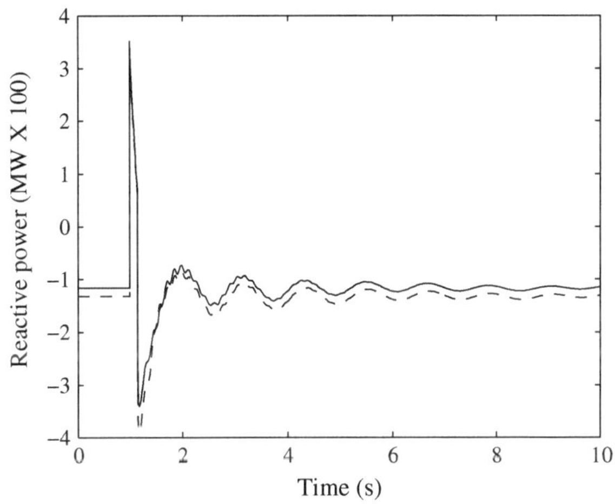

Fig. 7.49 Reactive power output for temporary outage of line 7–8 (*Solid line* reactive power absorption by WF_2 and *dashed line* reactive power supply by designed STATCOM/ESS controller)

7.8.8 Comparisons with Standard LVRT Requirements

Interconnection standards vary from country to country and among individual provinces or states, depending on local grid characteristics and utility specific requirements. Regarding the LVRT, Federal Energy Regulatory Commission requires wind farms to remain in service during any three-phase fault resulting in transmission voltage as low as zero volts, as measured at the high-voltage PCC to the grid, and that

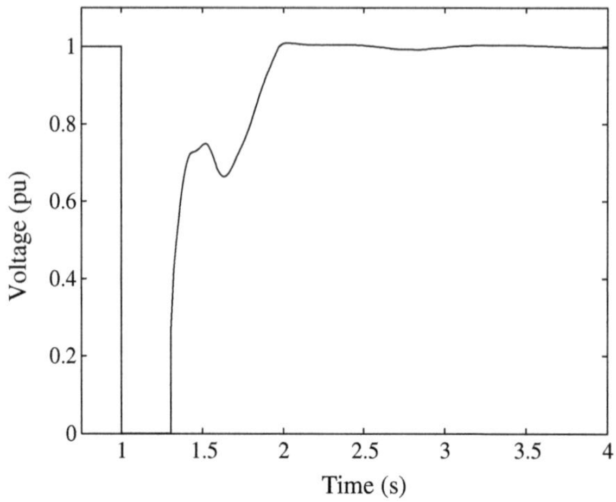

Fig. 7.50 Terminal voltage with zero-voltage for 300 ms

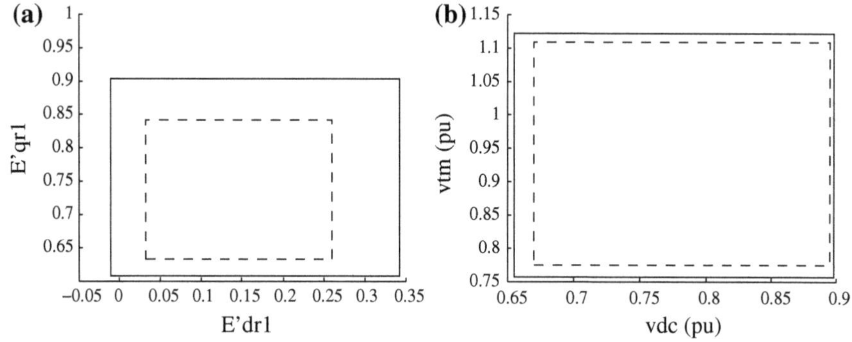

Fig. 7.51 Operating and control region (Ω) (*Solid line* control region (Ω) and *dashed line* operating region during LVRT transient). **a** D-axis versus q-axis transient emf. **b** Capacitor voltage versus terminal voltage

is normally cleared without separating the wind farm from the transmission system. Typical FRT requirements demand that a wind farm remains connected to the grid for voltage levels as low as 5 % of the nominal voltage (for up to 300 ms) [53].

Figure 7.50 shows the terminal voltage of the wind farm WF_2 with the the proposed STATCOM/ESS controller with zero voltage for the duration of 300 ms from which it is clear that the proposed control can meet the standard requirement of the LVRT capability. The reason of providing stability during the LVRT transient is more visible from Fig. 7.51a, b from which it is clear that control region provided by the designed controller is greater that the area required for the LVRT. The same is true for other state variables.

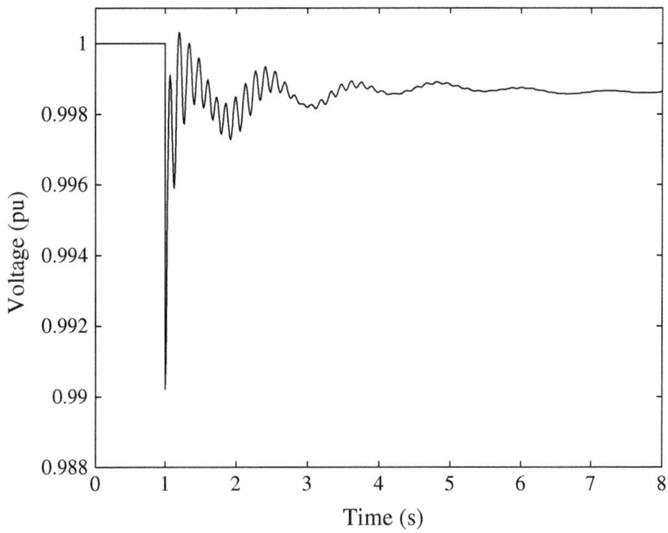

Fig. 7.52 PCC voltage (WF_1) with 10 % increase in load

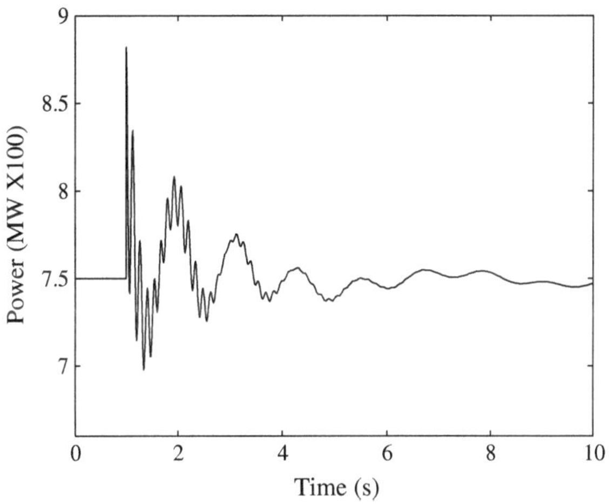

Fig. 7.53 Real power output (WF_1) with 10 % increase in load

7.8.9 Performance Under Different Operating Conditions

Although the controller is designed for rated operating conditions, the designed controller performs well in different loading conditions. This is due to the wide operating regions of the controller which is obtained by using the control algorithm proposed in this research. Figures 7.52 and 7.53 show the PCC voltage and real power output due to the 10 % increase in load from which it is clear that the controller stabilises the system at different equilibrium point.

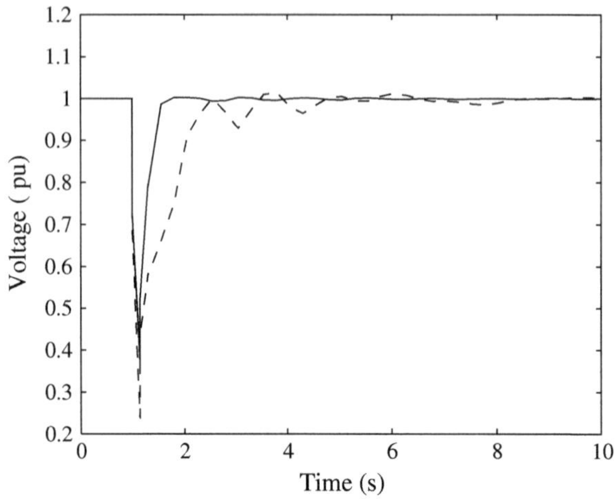

Fig. 7.54 Voltage response for three-phase fault at line 16–17 (*Solid line* designed STATCOM/ESS controllers and *dashed line* STATCOM without ESS controllers)

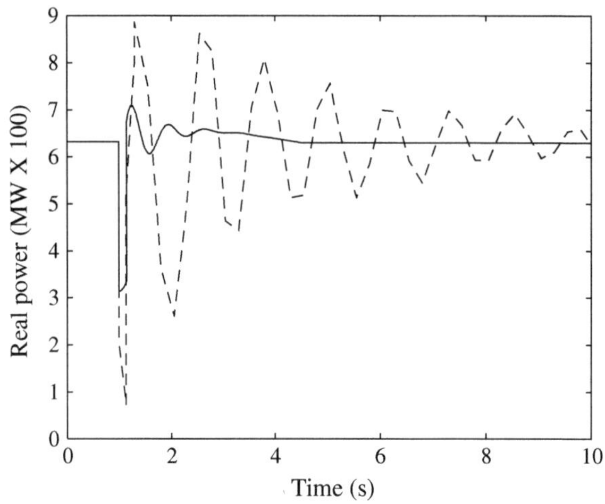

Fig. 7.55 Real power output for three-phase fault at line 16–17 (*Solid line* designed STATCOM/ESS controllers and *dashed line* STATCOM without ESS controllers)

7.8.10 Impact of Adding Supercapacitors

To test the effects of adding supercapacitor, a simulation is performed with (i)STATCOM/ESS, and (ii) STATCOM only. A three phase fault is applied at middle of the line 16–17 at 1 s and the fault is cleared after 140 ms by opening the line 16–17. This line is again restored after 150 ms.

Figures 7.54 and 7.55 show the voltage and real power output of WF_2 from which, it is clear that although the addition of supercapacitor does not produce significant difference in terminal voltage response, however, it damps the oscillation in output power quickly. The active power is controlled using energy storage type supercapacitor and this is effective to enhance the transient stability of the rest of the system.

7.9 Chapter Summary

This chapter presents the effects of increases in wind generation and STATCOM ratings on the FRT capability of a wind farm. It is observed that the critical speed and voltage, as well as the FRT capability, decrease with the integration of FSIGs. On the other hand, a system with a higher-rating STATCOM can have a longer FRT capability. The simulation results illustrate that a fast pitching control strategy activated by power system disturbances can lead to significant improvements in the dynamic stability of the wind turbine and the voltage stability of the connecting network. By reducing the wind turbine mechanical input torque by pitching to reduce the available wind power throughout the duration of the fault the wind turbine is protected against over-speed and remains stable and connected to the power system.

Four different control schemes have been developed to control the STATCOM. Simulation studies have shown that the STATCOM significantly reduces the voltage fluctuations caused by wind speed variations and enhances the fault ride-through capability of the WTGS. The STATCOM/ESS exhibits increased flexibility over the traditional STATCOM with improved damping capabilities due to the additional degree of control freedom provided by the active power capabilities.

References

1. Blaabjerg F, Chen Z, Kjaer SB (2005) Power electronics as efficient interface in dispersed power generation systems. IEEE Trans Power Syst 19(5):1184–1194
2. IEA Wind Energy Annual Report 2005 (2006) International Energy Agency (IEA). Technical report
3. Holttinen H, Pedersen J (2003) The effect of large-scale wind power on a thermal system operation. In: Proceedings of 4th international workshop on large-scale integration of wind power and transmission networks for offshore wind farms, Stockholm, Sweden, pp E1–E7
4. Chen Z, Blaabjerg F (2009) Wind farm a power source in future power systems. Renew Sustain Energy Rev 13(6–7):1288–1300
5. Johnson GL (2004) Wind energy systems. Prentice-Hall, Manhattan
6. Rathi MR, Mohan N (2005) A novel robust low voltage and fault ride through for wind turbine application operating in weak grids. In: 31st annual conference of IEEE Industrial Electronics Society, pp 2481–2486
7. Roy NK, Pota HR, Hossain MJ (2013) Reactive power management of distribution networks with wind generation for improving voltage stability. Renew Energy 58:85–94
8. Hossain MJ, Pota HR, Ramos RA (2011) Robust STATCOM control for the stabilisation of fixed speed wind turbines during low voltages. Renew Energy 36(11):2897–2905
9. Chen Z, Hu Y, Blaabjerg F (2007) Stability improvement of induction generator-based wind turbine systems. IET Renew Power Gener 1(1):81–93

10. Roy NK, Pota HR, Hossain MJ (2013) A cost-effective approach for fast voltage recovery of power distribution networks with wind generators. Int J Sustain Energy. doi:10.1080/14786451.2013.767803
11. Molinas M, Suul JA, Undeland T (2006) Wind farms with increased transient stability margin provided by a STATCOM. In: IEEE 5th international conference on power electronics and motion, control, pp 1–7
12. Molinas M, Suul JA, Undeland T (2007) Improved grid interface of induction generators for renewable energy by use of STATCOM. In: International conference on clean electrical, power, Capri, pp 215–222
13. Muyeen SM, Mannan MA, Ali MH, Takahashi R, Murata T, Tamura J (2005) Stabilization of grid connected wind generator by STATCOM. In: International conference on power electronics and drives systems, vol 2. Kuala Lumpur, Malaysia, pp 1584–1589
14. Qi L, Langston J, Steurer M (2008) Applying a STATCOM for stability improvement to an existing wind farm with fixed-speed induction generators. In: IEEE power and energy society general meeting—conversion and delivery of electrical energy in the 21st Century, Pittsburgh, PA, pp 1–6
15. Molinas M, Suul JA, Undeland T (2008) Low voltage ride through of wind farms with cage generators: STATCOM versus SVC. IEEE Trans. Power Electron 23(3):1104–1117
16. Mahmud MA, Pota HR, Hossain MJ (2013) Nonlinear DSTATCOM controller design for distribution network with distributed generation to enhance voltage stability. IEEE J Photovoltaics. doi:10.1109/JPHOTOV.2013.2281721
17. Aziz T, Hossain MJ, Saha TK, Mithulananthan N (2013) Var planning with tuning of statcom in a DG integrated industrial system. IEEE Trans Power Delivery 28(2):875–885
18. Griffo A, Lauria D (1990) Series connected PWM GTO current/source convertor with symmetrical phase angle control. IEE Proc B, Electric Power Appl 137(1):205–212
19. Blaabjerg F, Teodorescu R, Liserre M, Timbus AV (2006) Overview of control and grid synchronization for distributed power generation systems. IEEE Trans Ind Electron 53(5):1398–1409
20. Hossain MJ, Pota HR, Ugrinovski V, Ramos RA (2009) Robust STATCOM control for the enhancement of fault ride-through capability of fixed-speed wind generators. In: IEEE multi-conference on systems and control, Saint Petersburg, Russia, pp 1505–1510
21. Shen C, Yang Z, Crow ML, Atcitty S (2000) Control of STATCOM with energy storage device. In: IEEE power Engineering Society winter meeting, vol 4. Singapore, pp 2722–2728
22. Kraan I, Bongers PMM (1993) Control of a wind turbine using several linear robust controllers. In: Proceedings of the 32nd IEEE conference on decision and control, vol 2. San Antonlo, Texas, pp 1928–1929
23. Jauch C, Islam SM, Sorensen P, Jensen BB (2007) Design of a wind turbine pitch angle controller for power system stabilisation. Renew Energy 32(14):2334–2349
24. Hossain MJ, Pota HR, Ugrinovskii V, Ramos RA (2009) Excitation control for large disturbances in power systems with dynamic loads. In: IEEE Power and Energy Society general meeting, Calgary, Canada, pp 1–8
25. Hossain MJ, Pota HR, Ramos RA (August 2012) Improved LVRT capability of fixed-speed wind turbines with decentralized STATCOM/ESS control. IET Gener Transm Distrib 6(8):719–730
26. Muyeen SM, Takahashi R, Murata T, Tamura J, Ali MH (2009) Application of STATCOM/BESS for wind power smoothening and hydrogen generation. Electric Power Syst Res 79(2):365–373
27. Kuiava R, Ramos RA, Bretas NG (2009) Control design of a STATCOM with energy storage system for stability and power quality improvements. In: IEEE international conference on industrial technology, Victoria, Australia, pp 897–902
28. Abdin ES, Xu W (2000) Control design and dynamic performance analysis of a wind turbine-induction generator unit. IEEE Trans Energy Convers 15(1):91–96
29. Akhmatov V, Knudsen H, Nielsen AH, Pedersen JK, Poulsen NK (2003) Modeling and transient stability of large wind farms. Int J Electric Power Energy Syst 25(2):123–144

30. Bianchi F, Mantz R, Christiansen C (2005) Gain scheduling control of variable speed wind energy conversion systems using quasi-LPV models. Control Eng Pract 13(2):247–255
31. Peng L (2004) Study on the voltage stability of induction motor load. IEEE PES Power Syst Conf Exposition 1:10–13
32. Chedid R, Mrad F, Basma M (1999) Intelligent control of a class of wind energy conversion systems. IEEE Trans Energy Convers 14(4):1597–1604
33. Li P, Keung PK, Ooi BT (2009) Development and simulation of dynamic control strategies for wind farms. IET Renew Power Gener 3(2):180–189
34. Connor B, Leithead WE, Grimble M (1994) LQG control of a constant speed horizontal axis wind turbine. In: Proceedings of the third IEEE conference on control applications, vol 1. Glasgow, pp 251–252
35. Shaked U, Soroka E (1985) On the stability robustness of the continuous time LQG optimal control. IEEE Trans Autom Control 30(9):1039–1043
36. Sakamoto R, Senjyu T, Kaneko T, Urasaki N, Takagi T, Sugimoto S, Sekine H (2006) Output power leveling of wind turbine generator by pitch angle control using H_∞ control. In: IEEE PES power systems conference and exposition, Atlanta, GA, pp 2044–2049
37. Senjyu T, Kikunaga Y, Yona A, Sekine H, Saber AY, Funabashi T (2008) Coordinate control of wind turbine and battery in wind power generator system. In: IEEE Power and Energy Society general meeting, vol 2. Pittsburgh, PA, pp 1–7
38. Qiao W, Harley RG (2007) Power quality and dynamic performance improvement of wind farms using a STATCOM. In: IEEE power electronics specialists conference, Orlando, FL, pp 1832–1838
39. Guo Y, Hill DJ, Wang Y (2000) Nonlinear decentralized control of large-scale power systems. Automatica 36(9):1275–1289
40. Trinh H, Aldeen M (1991) Decentralized load-frequency control of interconnected power systems. In: International conference on advances in power system control, operation and management, Hong Kong, pp 815–820
41. Yeung L, Gouping L, Nobakhti A, Ho D, Wang H (2007) Large scale system decentralized controller design based on locally imposed decomposition constraints. In: American control conference, New York, USA, pp 1473–1478
42. Hossain MJ, Pota HR, Ugrinovskii V, Ramos RA (2010) Decentralized control to augment LVRT capability of wind generatorswith statcom/ess. In: IEEE Power and Energy Society general meeting, Minnesota, USA, pp 1–8
43. Kamwa I, Grondin R, Hebert Y (2001) Wide-area measurement based stabilizing control of large power systems-a decentralized/hierarchical approach. IEEE Trans Power Syst 16(1):136–153
44. Ilic M, Zaborszky J (2000) Dynamics and control of large electric power systems. Wiley, New York
45. Siljak DD, Zecevic AI, Neskovic G (2004) Robust decentralized exciter control with linear feedback. IEEE Trans Power Syst 19(2):1096–1103
46. Siljak DD, Stipanovic DM, Zecevic AI (2002) Robust decentralized turbine/governor control using linear matrix inequalities. IEEE Trans Power Syst 17(2):715–722
47. Suul JA, Molinas M, Undeland T (2010) STATCOM-based indirect torque control of induction machines during voltage recovery after grid faults. IEEE Trans Power Electron 25(5):1240–1250
48. El-Moursi MS, Bak-Jensen B, Abdel-Rahman MH (2010) Novel STATCOM controller for mitigating SSR and damping power system oscillations in a series compensated wind park. IEEE Trans Power Electron 25(2):429–441
49. Rahimi M, Parniani M (2010) Efficient control scheme of wind turbines with doubly fed induction generators for low-voltage ride-through capability enhancement. IET Renew Power Gener 4(3):242–252
50. Hossain MJ, Pota HR, Ugrinovskii V, Ramos RA (2010) Simultaneous STATCOM and pitch angle control for improved LVRT capability of fixed-speed wind turbines. IEEE Trans Sustain Energy 1(3):142–152

51. Guo Y, Hill DJ, Wang Y (2001) Global transient stability and voltage regulation for power systems. IEEE Trans Power Syst 16(4):678–688
52. Hossain MJ, Pota HR, Ugrinovskii V, Ramos RA (2009) A robust STATCOM control to augment lvrt capability of fixed-speed induction generators. In: 48th IEEE conference on decision and control, Shanghai, China, pp 1–8
53. Bary D (2004) Increasing renewable accessibility in ireland. 9th World Energy Congr, 1–10
54. Akhmatov V, Knudsen H, Bruntt M, Nielsen A, Pedersen JK, Poulsen NK (2000) A dynamic stability limit of grid-connected induction generator. In: International conference on power and energy systems, pp 235–244
55. Cutsem TV, Vournas C (1998) Voltage stability of electric power systems. Kluwer Academic, Norwell
56. Ugrinovskii VA, Petersen IR (2001) Minimax LQG control of stochastic partially observed uncertain systems. SIAM J Control Optim 40(4):1189–1226
57. Petersen IR, Ugrinovskii VA, Savkin AV (2000) Robust control design using H_∞ methods. Springer, London
58. Giroux P, Sybille G, Le-Huy H (2001) Modeling and simulation of a distribution STATCOM using simulink's power system blockset. In: 27th annual conference of the IEEE Industrial Electronics Society, pp 990–996
59. Garulli A, Tesi A, Vicino A (1999) Robustness in identification and control. Springer, New York
60. Fernandez L, Garcia C, Saenz JR, Jurado F (2009) Equivalent models of wind farms by using aggregated wind turbines and equivalent winds. Energy Convers Manage 50(3):691–704
61. Savkin AV, Petersen IR (1995) Nonlinear versus linear control in the absolute stabilisability of uncertain linear systems with structured uncertainty. IEEE Trans Autom Control 40(1):122–127
62. Moheimani SR, Savkin AV, Petersen IR (1995) A connection between H_∞ control and the absolute stabilizability of discrete-time uncertain linear systems. Automatica 31(8):1193–1195
63. Hiskens IA. Power system test cases. http://psdyn.ece.wisc.edu/IEEE_benchmarks/
64. Germond AJ, Podmore R (1978) Dynamic aggregation of generating unit models. IEEE Trans Power Apparatus Syst PAS-97(4):1060–1069
65. Pal BC, Coonick AH, Macdonald DC (2000) Robust damping controller design in power systems with superconducting magnetic energy storage devices. IEEE Trans Power Syst 15(1):320–325
66. Werner H, Korba P, Yang TC (2003) Robust tuning of power system stabilizers using LMI-techniques. IEEE Trans Control Syst Technol 11(1):147–152
67. Ackermann T (2005) Wind power in power systems. Wiley, England
68. Hossain MJ, Pota HR, Ugrinovski V (2008) Short and long-term dynamic voltage instability. In: 17th IFAC world congress, Seoul, Korea, pp 9392–9397
69. Srithorn P, Sumner M, Yao L, Parashar R (2008) A STATCOM with supercapacitors for enhanced power system stability. In: 4th IET conference on power electronics, machines and drives. London, pp 96–100
70. Li L, Ugrinovskii VA, Orsi R (2007) Decentralized robust control of uncertain markov jump parameter systems via output feedback. Automatica 43(11):1932–1944
71. Li L, Petersen IR (2007) A rank constrained LMI algorithm for the robust H_∞ control of an uncertain system via a stable output feedback controller. In: 46th IEEE conference on decision and control New Orleans, LA, pp 5423–5428
72. Li L, Ugrinovskii VA (2007) On necessary and sufficient conditions for H_∞ output feedback control of markov jump linear systems. IEEE Trans Autom Control 52(7):1287–1292
73. Orsi R (2005) LMIRank : software for rank constrained LMI problems. http://rsise.anu.edu.au/robert/lmirank/
74. Hemeida AM (2009) Improvement of voltage stability and critical clearing time for multi-machine power systems using static var compensator. ICGST-ACSE 9(2):41–47

Chapter 8
LVRT Capability of DFIGs in Interconnected Power Systems

Abstract This chapter presents a new control scheme for the enhancement of the low-voltage ride-through (LVRT) capability of doubly-fed induction generators (DFIGs). LVRT capability is provided by extending the range of operation of the controlled system to include typical post-fault conditions. Simultaneous controllers are designed for both converters using a linear quadratic (LQ) decentralized control strategy. The nonlinear behavior of power systems is represented in this paper by an uncertain term derived from the Cauchy remainder of the Taylor series expansion. A genetic algorithm is used to calculate the bound on the uncertainty and the robust controller resulting from this design provides acceptable performances to enhance voltage and transient stability margins and thereby to limit the oscillations, peak value of rotor current and DC-link voltage fluctuations which are needed to operate a wind farm during severe faults. The performance of the designed controller is demonstrated by large disturbance simulations on a test system.

8.1 Introduction

In the past, most wind farms were disconnected whenever there was a fault at the point of common coupling (PCC) or in the vicinity of the wind farm [1]. However, with the improvement in technology and the introduction of DFIG units, wind farms are no longer required to be disconnected as they have an inherent capability to control the active and reactive power. This capability can be utilized in designing a robust control to meet the current grid code requirements. In the grid code requirements, LVRT is one of the important and pertinent issues with wind farms based on DFIG units [2]. The stator of a DFIG is connected through a transformer and the rotor through a back to back power converter to the grid.

Due to partial-scale power converters, wind turbines based on DFIGs are very sensitive to grid disturbances, especially voltage dips during grid faults [3–5]. Faults in a power system, even far away from the location of a wind turbine, can cause a

J. Hossain and H. R. Pota, *Robust Control for Grid Voltage Stability: High Penetration of Renewable Energy*, Power Systems,
DOI: 10.1007/978-981-287-116-9_8, © Springer Science+Business Media Singapore 2014

voltage dip at the connection point of the generator. Such an abrupt drop in voltage will cause over-current in the rotor windings and over-voltage in the DC buses of the power converters. Typically, a rotor's current limit is 2 pu and the DC-link voltage limit is 1.2 times its nominal value (1 pu) [6, 7]. In addition, a voltage dip will cause over-speeding of the wind turbine which will threaten its safe operation. The existing LVRT strategies can be divided into two main types: active control strategies and passive schemes which incorporate additional hardware protection devices.

Several LVRT control schemes that involve hardware modifications [8], such as the crowbar approach [9, 10], virtual resistance strategy [11], dynamic braking resistance [12] and DFIG converter control, have been proposed in the literature. Some of the active control strategies for LVRT are: an indirect matrix converter control scheme for the voltage source converter (VSC) to enhance the dynamic stability of DFIGs in [13]; a coordinated control approach for fulfilling the requirements of a LVRT capability in DFIG-based wind turbines in [6, 14]; a robust control technique in [15]; current sensing control [16]; advanced control approach [17], internal model control [18] and predictive control approach for improving system stability in [19–21]. However, these approaches do not consider the inherent nonlinearity of power systems and interactions among wind farms. As the main difficulty of DFIG converter control during a large disturbance is the system's nonlinear behavior, it is essential that both this and its interconnection effects are considered in the design of the controllers for multi-machine power systems. Also, it is necessary to quantify the operating region around the equilibrium point for which the system maintains closed-loop stability.

In order to improve the LVRT capacity of DFIGs, nonlinear control strategies have also been used [22, 23]. Rather than using a linearised model, as is common in the design of a converter control design, nonlinear models, with nonlinear feedback linearisation techniques, are used; this has the advantage that they are not dependent on the operating conditions as are linearised models. Although nonlinear controllers can significantly improve a power system's transient stability, they usually have more complicated structures and are harder to implement in practice. Linear controllers are preferred as they are easy to design and implement.

Linear controllers have a limited range of operation which normally does not include post-fault voltage conditions, and the difficulty of providing a LVRT capability is due to the nonlinearities in the power system model [24, 25]. During faults, and in post-fault operation, a system's state can be significantly far from the desired equilibrium point. In most situations, a post-fault uncontrolled system has an unstable post-fault trajectory. A solution to the LVRT problem is to design globally stabilizing controllers [26]. However, they often require a full-state feedback and are not sufficiently robust to handle modeling uncertainties. The next option is to use a linear controller which is robust to changes in the linear model with varying operating conditions—a necessary outcome of the underlying nonlinear model.

As mentioned above, robust controllers exist for power systems but few have been able to systematically provide robustness against large deviations required for LVRT. In this chapter, we present a method which can be used to design a linear controller which is sufficiently robust to accommodate post-fault low-voltage conditions.

The design of a robust decentralized controller has been presented, which enhances dynamic voltage and transient stability where instability is caused by severe system faults. The extended robustness in the proposed method is provided by the exact linearisation of the nonlinear model using the mean value and Cauchy remainder.

The organization of this chapter is as follows: Sect. 8.2 provides the mathematical modeling of the power system devices under consideration; the test system and control objectives are presented in Sect. 8.3; Sect. 8.4 describes the linearisation technique and the process for obtaining the bounds on the nonlinear terms; Sect. 8.5 discusses the decentralized LQ output-feedback controller design procedure using a rank-constrained LMI technique; Sect. 8.6 presents the control design algorithm; and, in Sect. 8.7, the performance of the controller is demonstrated through a series of simulation results. Concluding remarks are given in Sect. 8.8.

8.2 Power System Model

Dynamic models of the devices used in this chapter are presented in this section. A single-axis third order synchronous generator model and an IEEE-ST1A type excitation system are used in this chapter [27]. The nonlinear model of the wind turbines, made up of the various block shown in Fig. 8.1, is based on a static model of the aerodynamics, a two mass model of the drive train, a third order model of the generator, the grid side converter (GSC) with DC-link capacitor, the pitch controller and the rotor side converter (RSC).

The aerodynamic torque applied to the rotor for the ith turbine by the effective wind speed passing through the rotor is given as [28]:

$$T_{ae_i} = \frac{\rho_i}{2\omega_{m_i}} A_{wt_i} c_{p_i}(\lambda_i, \theta_i) V_{w_i}^3, \tag{8.1}$$

where $i = 1, \ldots, n$, and n is the number of wind turbines and $\lambda_i = \frac{\omega_{m_i} R_i}{V_{w_i}}$.

The drive train attached to the wind turbine converts the aerodynamic torque T_{ae_i} on the rotor into the torque on the low speed shaft, which is scaled down through the gearbox to the torque on the high-speed shaft. A two-mass drive train model of a wind turbine generator system (WTGS) is used in this chapter as the drive train modeling can satisfactorily reproduce the dynamic characteristics of the WTGS. The dynamic of the ith shaft is represented as [28]:

$$\dot{\omega}_{m_i} = \frac{1}{2H_{m_i}} \left[T_{ae_i} - K_{s_i}\gamma_i - D_{m_i}\omega_{m_i} \right], \tag{8.2}$$

$$\dot{\omega}_{G_i} = \frac{1}{2H_{G_i}} \left[K_{s_i}\gamma_i - T_{e_i} - D_{G_i}\omega_{G_i} \right], \tag{8.3}$$

$$\dot{\gamma}_i = 2\pi f (\omega_{m_i} - \frac{1}{N_{g_i}}\omega_{G_i}). \tag{8.4}$$

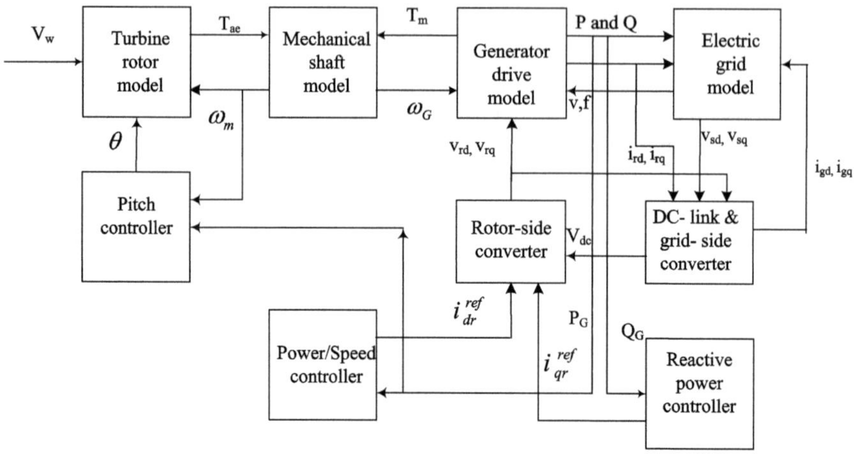

Fig. 8.1 Block diagram of DFIG wind turbine system

The transient model of a DFIG is described by the following equations [29]:

$$\dot{s}_i = \frac{1}{2H_{G_i}}\left[T_{e_i} - T_{m_i}\right], \tag{8.5}$$

$$\dot{E}'_{qr_i} = -\frac{1}{T'_{o_i}}\left[E'_{qr_i} - (X_i - X'_i)i_{ds_i}\right] - s_i\omega_s E'_{dr_i} - \omega_s v'_{dr_i}, \tag{8.6}$$

$$\dot{E}'_{dr_i} = -\frac{1}{T'_{o_i}}\left[E'_{dr_i} + (X_i - X'_i)i_{qs_i}\right] + s_i\omega_s E'_{qr_i} + \omega_s v'_{qr_i}, \tag{8.7}$$

$$v_{ds_i} = R_{s_i}i_{ds_i} - X'_i i_{qs_i} + E'_{dr_i}, \tag{8.8}$$

$$v_{qs_i} = R_{s_i}i_{ds_i} + X'_i i_{qs_i} + E'_{qr_i}, \tag{8.9}$$

$$v_{t_i} = \sqrt{v_{ds_i}^2 + v_{qs_i}^2}, \tag{8.10}$$

where

$$X'_i = X_{s_i} + X_{m_i}X_{r_i}/(X_{m_i} + X_{r_i}), \tag{8.11}$$

$$X_i = X_{s_i} + X_{m_i}, \tag{8.12}$$

$$T'_{o_i} = (L_{r_i} + L_{m_i})/R_{r_i}, \tag{8.13}$$

$$T_{e_i} = E_{dr_i}i_{ds_i} + E_{qr_i}i_{qs_i}, \tag{8.14}$$

$$v'_{dr_i} = \frac{v_{dr_i}X_{m_i}}{(X_{m_i} + X_{r_i})}, \tag{8.15}$$

$$v'_{qr_i} = \frac{v_{qr_i}X_{m_i}}{(X_{m_i} + X_{r_i})}. \tag{8.16}$$

The DC-link dynamics is given by:

$$C_i v_{dc_i} \dot{v}_{dc_i} = -\frac{v_{dc_i}^2}{R_{loss_i}} - P_{r_i}(t) - P_{g_i}(t) \tag{8.17}$$

where resistor R_{loss_i} represents the total conducting and switching losses of the converter. Also, $P_{r_i}(t)$ is the instantaneous input rotor power, and $P_{g_i}(t)$ is the instantaneous output power of the GSC which are given by:

$$P_{r_i} = v_{rd_i} i_{rd_i} + v_{rq_i} i_{rq_i}, \tag{8.18}$$

$$P_{g_i} = v_{gd_i} i_{gd_i} + v_{gq_i} i_{gq_i}. \tag{8.19}$$

Using this model of DFIG a control strategy is proposed in the next section for a large interconnected test system.

8.3 Test System and Control Task

A one-line diagram of the New England system [30] is shown in Fig. 8.2. To test the performance of the proposed LVRT controllers, this New England is modified by connecting four wind farms–each rated at 50 MW, at buses 7, 11, 23 and 27. Each wind turbine supplies 2 MW power at rated condition and each wind farm has twenty five turbines. The modified system network consists of ten thermal power plants and four wind farms. The conventional generation, wind generation, and the total load in this system are 5993.41 MW, 200 MW and 6150.5 MW, respectively. The parameters of the wind turbine are given in [7]. Aggregated wind generator [31] and synchronous generator models [32] are used in this research for designing controllers. One decentralised controller is designed for each wind farm.

Interconnection standards vary from country to country and among individual provinces or states, depending on local grid characteristics and utility specific requirements. In this book the standard, jointly recommend by the North American Electric Reliability Council (NAERC) and the American Wind Energy Association (AWEA) [33] as shown in Fig. 8.3, is used. This standard demands that if the voltage remains at a level greater than 15 % of the nominal voltage for a period that does not exceed 0.625 seconds, the plant must stay online. Ireland also follows a similar standard [34]. Although this standard is followed in this research work, the designed controller fulfils the some other grid codes and zero voltage ride-through requirements as well.

The function of the RSC control is to limit the rotor fault current and to increase the damping of the stator flux, and consequently, to enhance the ride-through capability. In normal conditions, the rotor current limiter (RCL) which is also known as crowbar protection, shown in Fig. 8.4, is inactive. Once a voltage dip is detected, the crowbar is activated, the rotor-side converter operates to control active and reactive power.

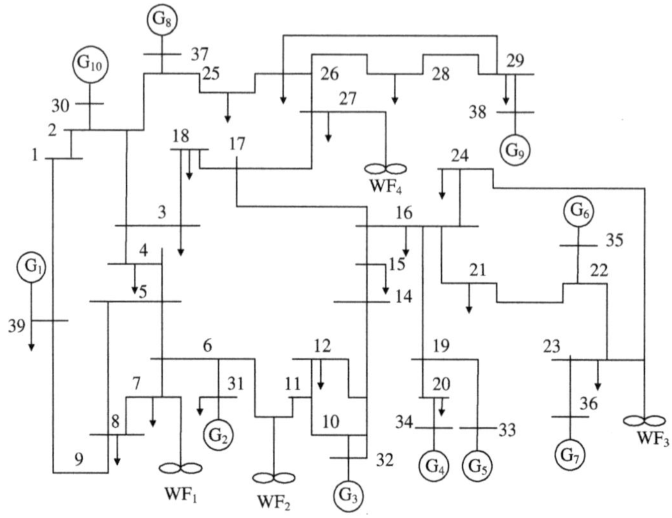

Fig. 8.2 Single line diagram of the test system

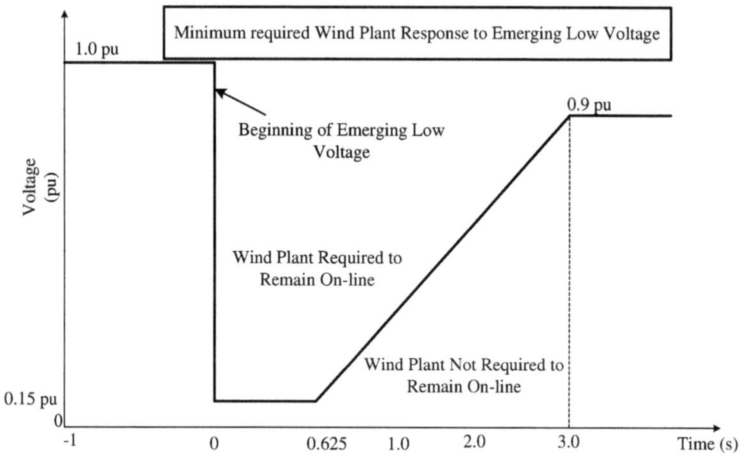

Fig. 8.3 Voltage limit criteria according to grid-code

Once the fault is cleared, the crowbar remains in service for another 150 ms to limit the DFIG transients generated due to voltage increase at the time of clearing the fault [6]. However, in this research the performance of the designed controller are verified without firing the crowbar as the use of a crowbar is not permitted for all grids.

The GSC and RSC control strategies are shown in Fig. 8.5. The main objective of the GSC control is to keep the DC voltage constant. The RSC requires a constant DC voltage to operate which is supplied by the GSC. GSC supplies the real power

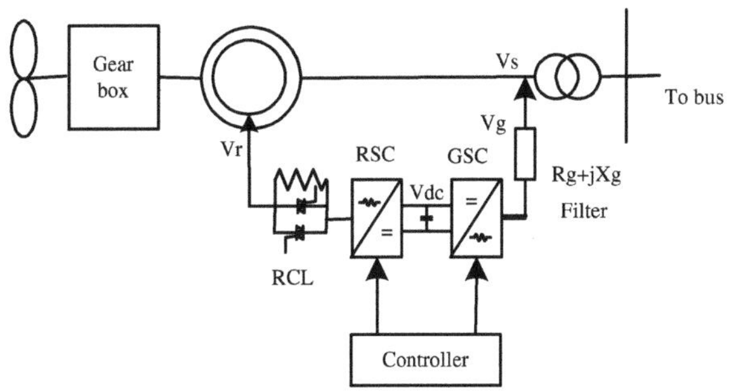

Fig. 8.4 Schematic diagram of a DFIG

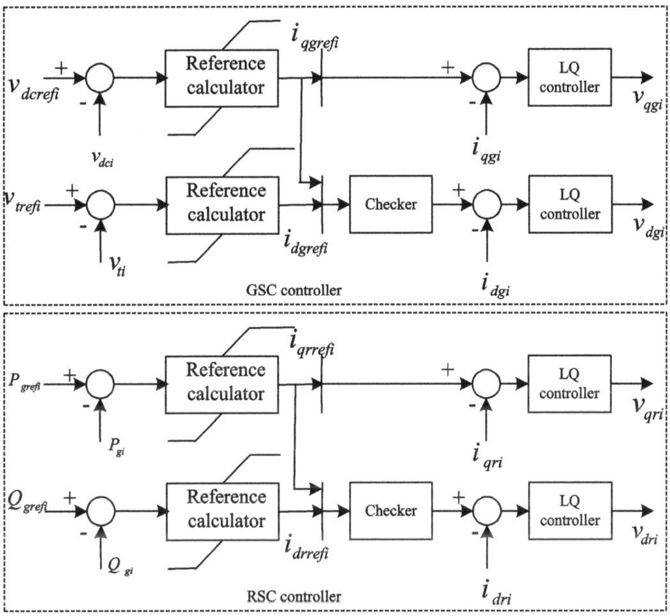

Fig. 8.5 Structure of the converter controller

demand of the RSC through the DC link. As GSC operates to maintain a relatively constant voltage RSC will receive almost constant power supply. RSC also controls real and reactive power flow between the grid and the GSC. The q-axis current is used to control the DC voltage and d-axis current regulates the reactive power or terminal voltage. The currents i_{dg_i} and i_{qg_i} are obtained from Δv_{dc_i} and Δv_{t_i} by PI controllers as shown in Fig. 8.5. Always the priority is to control the active power, so the Checker block checks the available capacity for i_{dg_i} with respect to i_{qg_i} while the rating of GSC is 25 % of the DFIG. The RSC controller controls the active and reactive power with i_{qr_i} and i_{dr_i}. As during the voltage dip it is difficult to keep the DC-link voltage constant, the LVRT control can be provided by limiting the

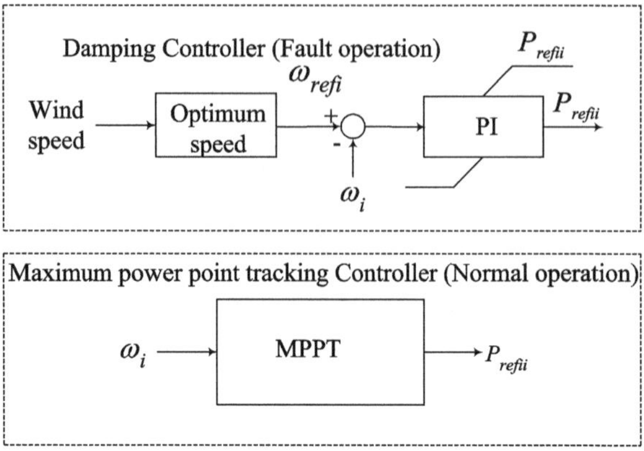

Fig. 8.6 Determination of active power set point

DC-voltage fluctuation. Since the DC link dynamics are nonlinear, the conventional linear control cannot properly limit the DC voltage under severe voltage dips. In the proposed ride-through approach, the DFIG contributes to grid stability enhancement by controlling both active and reactive power taking into account the nonlinearity of DFIGs.

The reference signals for the DC and terminal voltage, vd_{cref_i} and v_{tref_i} are set to a constant value (1 pu), independent of the wind turbine operation mode. In fact, those values are firmly dependent on the size of the converter, the stator-rotor voltage ratio and the modulation factor of the power converter. The reference set point Q_{gref_i} for the RSC can be set to either zero or a value depending on whether or not the DFIG is required to contribute reactive power. In this chapter, the performance of the controller is tested for both zero and nonzero values of reactive power contribution. During normal operation the active power set point P_{gref_i} is defined by maximum power point tracking (MPPT) control as shown in Fig. 8.6. However, during faulted condition, it is determined by the damping controller which damps out the torsional oscillations excited in the drive train.

Proper placement and selection of feedback signals and controllers allows control objectives to be met much more easily. The most effective locations and best feedback signals for the controllers are found by the method of comparing the residues, which are the products of modal controllability and observability gramians [35]. The modal controllability indicates the degree of influence of the given input to the mode in question. The modal observability is a measure of the modal information contained in a feedback signal. They are independent and hence can be computed separately. The selection of the controller locations are made by comparison of the residues at all locations. The best input and output signals are those which give the largest residue for a given mode.

The test system is divided into six subsystems based on the coherent groups of generating units: (1) wind farms 1, (2) wind farm 2, (3) wind farm 3, (4) wind farm 4, (5) G_2–G_7, and (6) G_1, G_8–G_{10}. A converter controller is a part of subsystems 1–4, and PSSs are parts of subsystems 5 and 6. The PSSs are designed using the standard process given in [36]. For subsystems 1–4, the state vector is $\Delta x_i = \left[\Delta\omega_{m_i}, \Delta\omega_{G_i}, \Delta\gamma_i, \Delta s_i, \Delta E'_{dr_i}, \Delta E'_{qr_i}, \Delta V_{dc_i}\right]^T$, i=1–4. For (PSSs), $u_i = \Delta V_{s_i}$ and $y_i = \Delta\omega_i$, where V_{s_i} is the PSS output signal and ω_i is the rotor speed of the synchronous generator.

The structure of the controller is shown in Fig. 8.5. The input and output signals for controllers are depicted in Fig. 8.5. With the above control strategy and dynamic model of DFIG, the problem formulation for control design is explained in the next section.

8.4 Problem Formulation

A linearised model of the power system is usually obtained by expanding the equations, about an equilibrium point, in a Taylor series and retaining only the linear terms. In this chapter, to quantify the neglected higher order terms, we propose the use of a linearisation scheme which retains the contributions of the higher order terms in the form of the Cauchy remainder. The wind generator is a nonlinear system, which is modeled as a linear part plus a nonlinear part, the nonlinear term being the Cauchy remainder term in the Taylor series expansion and of the equations used to model the wind farm. Bounds derived for this Cauchy remainder term are used to define an uncertain linear model for which a robust control design is performed. In this chapter, in the design of the linear controller, the Cauchy remainder is incorporated as an uncertain term thus quantifying the deviations from the equilibrium point.

Each wind generator is considered as a subsystem of the interconnected system shown in Fig. 8.7 and can be represented as follows:

$$S_i : \Delta\dot{x}_i = A_i\Delta x_i + B_i\Delta u_i + E_i\xi_i + L_i r_i, \tag{8.20}$$
$$z_i = C_i\Delta x_i + D_i\Delta u_i, \tag{8.21}$$
$$\zeta_i = H_i\Delta x_i + G_i\Delta u_i, \tag{8.22}$$
$$y_i = C_{yi}\Delta x_i + D_{yi}\xi_i, \tag{8.23}$$

where Δx_i is the state vector, Δu_l the control input, y_l the measured output, z_l the controlled output, ξ_i the uncertainty input, ζ_i the uncertainty output, and r_i describes the effect of other subsystems $S_1, \ldots, S_{i-1}, S_{i+1}, \ldots, S_N$ on subsystem S_i.

The procedure for obtaining the matrices in (8.20)–(8.23) and the bounding uncertainty is described below for the wind turbine subsystem.

Let (x_{i0}, u_{i0}) be an arbitrary point in the region of interest. Using the mean-value theorem, the wind farm subsystem (subsystems 1–4) can be rewritten as follows [37]:

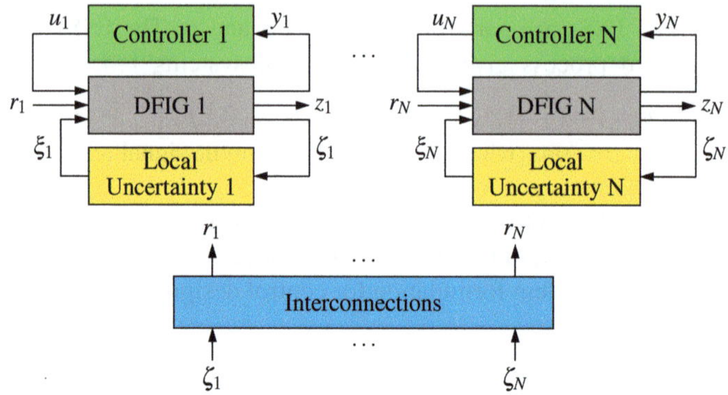

Fig. 8.7 Block diagram of the uncertain system

$$\begin{aligned}\dot{x}_i &= f_i(x_{i_0}, u_{i_0}) + P_i(x_i - x_{i0}) + Q_i(u_i - u_{i_0}) \\ &\quad + \sum_{\substack{j=1\\ j\neq i}}^{n} N_i(x_j - x_{j_0})\end{aligned} \tag{8.24}$$

where the Cauchy remainder terms are:

$$P_i = \begin{bmatrix} \left.\frac{\partial f_{i1}}{\partial x_i}\right|_{\substack{x_i = x_i^{*1}\\ u_i = u_i^{*1}}} \\ \vdots \\ \left.\frac{\partial f_{i7}}{\partial x_i}\right|_{\substack{x_i = x_i^{*7}\\ u_i = u_i^{*7}}} \end{bmatrix}, \; Q_i = \begin{bmatrix} \left.\frac{\partial f_{i1}}{\partial u_i}\right|_{\substack{x_i = x_i^{*1}\\ u_i = u_i^{*1}}} \\ \vdots \\ \left.\frac{\partial f_{i7}}{\partial u_i}\right|_{\substack{x_i = x_i^{*7}\\ u_i = u_i^{*7}}} \end{bmatrix},$$

$$N_i = \begin{bmatrix} \left.\frac{\partial f_{i1}}{\partial x_j}\right|_{\substack{x_i = x_i^{*1}\\ u_i = u_i^{*1}}} \\ \vdots \\ \left.\frac{\partial f_{i7}}{\partial x_j}\right|_{\substack{x_i = x_i^{*7}\\ x_i = u_i^{*7}}} \end{bmatrix}.$$

Here x_i is a 1×7 column vector, $f_i = [f_{i1}, \ldots, f_{i7}]^T$ is also a 1×7 column vector made up of the right-hand-side of Eqs. (8.2)–(8.7) and (8.17); (x_i^{*p}, u_i^{*p}), $p = 1, \ldots, 7$, denote points lying on the line segment connecting points (x_i, u_i) and (x_{i_0}, u_{i_0}). Equation (8.24) is an exact reformulation of the system in which the first term is the value of f_i at the equilibrium point (normally zero) and the rest of the terms are Cauchy remainders. The Cauchy remainder terms can also be linear but they have to be evaluated on a point lying on the segment joining the equilibrium point and the current operating point, and *not* on the equilibrium point.

It should be noted that the exact values of x_i^{*p} and u_i^{*p} are not available, and due to this, instead of the exact expressions for P_i, Q_i and N_i in system (8.24), their bounds, obtained numerically, are used in the control design.

Letting (x_{i0}, u_{i0}) be an equilibrium point and defining $\Delta x_i \triangleq x_i - x_{i0}$ and $\Delta u_i \triangleq u_i - u_{i0}$, it is possible to rewrite (8.24) as follows

$$\begin{aligned}
\Delta \dot{x}_i &= \dot{x}_i - \dot{x}_{i0}, \\
&= P_i(x_i - x_{i0}) + Q_i(u_i - u_{i0}) + \sum_{\substack{j=1\\ j\neq i}}^{n} N_i(x_j - x_{j0}), \\
&= A_i \Delta x_i + B_i \Delta u_i + (P_i - A_i)\,\Delta x_i \\
&\quad + (Q_i - B_i)\,\Delta u_i + \sum_{\substack{j=1\\ j\neq i}}^{n} N_i \Delta x_j,
\end{aligned} \tag{8.25}$$

where $A_i = \left.\frac{\partial f_i}{\partial x_i}\right|_{\substack{x_i=x_{i0}\\ u_i=u_{i0}}}$ and $B_i = \left.\frac{\partial f_i}{\partial u_i}\right|_{\substack{x_i=x_{i0}\\ u_i=u_{i0}}}$ are the Jacobian matrices evaluated about the point $\{x_{i0}, u_{i0}\}$.

The system (8.25) is of the form which allows for an application of the LQ control design technique [38]. To apply this technique, we rewrite system (8.25) in terms of the block diagram shown in Fig. 8.7.

We introduce a signal ξ_i such that the following relationship is satisfied:

$$(P_i - A_i)\,\Delta x_i + (Q_i - B_i)\,\Delta u_i = E_i \xi_i, \tag{8.26}$$

where

$$E_i = \operatorname{diag}\left(\frac{1}{2H_{m_i}}, \frac{1}{2H_{G_i}}, 1, \frac{1}{2H_{G_i}}, \frac{X_i - X_i'}{T_{0_i}'}, \frac{X_i - X_i'}{T_{0_i}'}, \frac{1}{C_i R_{loss_i}}\right),$$

We can write (8.26) as:

$$\xi_i = E_i^{-1}\,(P_i - A_i)\,\Delta x_i + (Q_i - B_i)\,\Delta u_i \tag{8.27}$$

$$= E_i^{-1}\left[P_i - A_i \;\; Q_i - B_i\right]\begin{bmatrix}\Delta x_i\\ \Delta u_i\end{bmatrix} \tag{8.28}$$

$$= E_i^{-1}\left[P_i - A_i \;\; Q_i - B_i\right]\sqrt{\Gamma_i}\frac{1}{\sqrt{\Gamma_i}}\begin{bmatrix}\Delta x_i\\ \Delta u_i\end{bmatrix} \tag{8.29}$$

$$= E_i^{-1}\sqrt{\Gamma_i}\left[P_i - A_i \;\; Q_i - B_i\right]\zeta_i \tag{8.30}$$

where ζ_i defined as

$$\zeta_i = \frac{1}{\sqrt{\Gamma_i}}\begin{bmatrix}\Delta x_i\\ \Delta u_i\end{bmatrix}.$$

We introduce a signal r_i such that

$$r_i = L_i^{-1} \sum_{\substack{j=1 \\ j \neq i}}^{n} [N_j \ 0] \sqrt{\Gamma_j} \frac{1}{\sqrt{\Gamma_j}} \begin{bmatrix} \Delta x_j \\ \Delta u_j \end{bmatrix} \tag{8.31}$$

$$= L_i^{-1} \sum_{\substack{j=1 \\ j \neq i}}^{n} \sqrt{\Gamma_j} [N_j \ 0] \zeta_j \tag{8.32}$$

where L_i is a 7×7 identity matrix.

The scaling factor $\sqrt{\Gamma_i}$ is chosen such that

$$\sqrt{\Gamma_i} || E_i^{-1} [P_i - A_i \ Q_i - B_i] || \leq 1, \tag{8.33}$$

and

$$\sqrt{\Gamma_j} || L_i^{-1} [N_j \ 0] || \leq 1, \quad \forall i, j = 1, \ldots n \ (i \neq j), \tag{8.34}$$

over the entire region of interest. The selection of Γ_j is done numerically as explained in Sect. 8.6.

From Eqs. (8.33) and (8.34), we recover the norm bound constraints [39],

$$\|\xi_i\|^2 \leq \|\zeta_i\|^2, \quad \text{and} \quad \|r_i\|^2 \leq \sum_{j \neq i} \|\zeta_j\|^2. \tag{8.35}$$

The bounds given by (8.35) can be used with the LQ output-feedback control design method to obtain a controller for the underlying nonlinear system. The expressions of $P_i - A_i$ and $Q_i - B_i$ can be determined following the procedure given in [27].

The process is similar for the other subsystems. Robustness properties of the LQ output-feedback controller ensure that this controller stabilizes the nonlinear system (8.20)–(8.23) for all instances of linearisation errors. From the reformulation presented in Appendix-I, we can quantify the deviations from the equilibrium points and compare with the estimated region for the LVRT obtained from simulations. The new formulation presented in this section is used with the LQ output-feedback control theory to design decentralized controllers for the DFIG using the linear matrix inequality (LMI) technique.

8.5 Decentralized Control Design Using Rank Constrained LMIs

A decentralized LQ output-feedback control has been designed for this research work using an LMI technique [39]. This LQ technique minimizes the following performance cost over all permissible integral quadratic constraints (IQCs):

$$J_{wc}(u) \triangleq \int_0^{\infty} \sum_{i=1}^{N} \|z_i(t)\|^2 dt. \tag{8.36}$$

In this design measured and controlled outputs are the same, i.e., $y_i = z_i$. The control algorithm is to find the infimum of the following function over the set $\mathscr{T}$:

$$J(\tau, \Theta) = \sum_{i=1}^{N} x_{i0}^T \left[\bar{X}_i + \tau_i M_i + \Theta_i \bar{M}_i\right] x_{i0}, \tag{8.37}$$

where $\mathscr{T} = \{\{\tau_i\ \Theta_i\} \in R^{2N}, \tau_i > 0, \Theta_i > 0\}$, $M_i > 0$ and $\bar{M}_i > 0$ are two sets of symmetric matrices, and matrices $\bar{X}_i$ and $\bar{Y}_i$ are the solutions to the following pair of coupled generalized algebraic Riccati equations and algebraic Riccati inequalities [39]:

$$A_i^T \bar{X}_i + \bar{X}_i A_i + \bar{C^T}_i \bar{C}_i - \bar{X}_i \left[B_i R_i^{-1} B_i^T - \bar{B}_{2_i} \bar{B}_{2_i}^T\right] \bar{X}_i = 0, \tag{8.38}$$

$$A_i^T \bar{Y}_i + \bar{Y}_i A_i + \bar{Y}_i \bar{B}_{2_i} \bar{B}_{2_i}^T \bar{Y}_i - \left[C_{y_i}^T W_i^{-1} C_{y_i} - \bar{C}_i^T \bar{C}_i\right] < 0, \tag{8.39}$$

where $R_i = \bar{D}_i^T \bar{D}_i$, $W_i = \bar{D}_{y_i} \bar{D}_{y_i}^T$, $\bar{\Theta}_i = \sum_{n=1, n\neq i}^{N} \Theta_n$,

$$\bar{C}_i = \begin{bmatrix} C_i \\ (\tau_i + \bar{\Theta}_i)^{1/2} H_i \end{bmatrix}, \quad \bar{D}_i = \begin{bmatrix} D_i \\ (\tau_i + \bar{\Theta}_i)^{1/2} G_i \end{bmatrix},$$

$$\bar{B}_{2_i} = \left[\tau_i^{-1/2} E_i\ \Theta_i^{-1/2} L_i\right], \quad \bar{D}_{y_i} = \left[\tau_i^{-1/2} D_{y_i}\ 0\right]. \tag{8.40}$$

The controller u_i^* with the τ^*, Θ^* is given by [39]:

$$\dot{x}_{c_i} = \{A_i - \left[B_i R^{-1} B_i^T - \bar{B}_{2_i} \bar{B}_{2_i}^T\right] \bar{X}_i\} x_{c_i} + \left[\bar{Y}_i - \bar{X}_i\right]^{-1} C_{y_i}^T W_i^{-1} \left[y_i - C_{y_i} x_{c_i}\right], \tag{8.41}$$

$$u_i^* = -R_i^{-1} B_i^T \bar{X}_i x_{c_i}. \tag{8.42}$$

The solutions are required to satisfy the following conditions: $\tau_i > 0$, $\Theta_i > 0$, $\bar{X}_i \geq 0$, $\bar{Y}_i \geq 0$ and $\bar{Y}_i > \bar{X}_i$. The controller u^* guarantees the following LQ property

$$J_{wc}(u^*) \leq J(\tau^*, \Theta^*) = \inf_{\mathscr{T}} J(\tau, \Theta) \tag{8.43}$$

The solution of the optimization problem using the LMI technique is discussed in [40]. A control algorithm for the LVRT enhancement using the above formulation is presented in the next section.

8.6 Control Design Algorithm

The controller is designed for severe faults so it can, in principle, also ensure stability against other disturbances. Prior to the controller design, we carry out several large disturbance simulations to get an idea of the region of interest. The maximum value of $|| [P_i - A_i|| \, ||Q_i - B_i] ||$ is obtained over this region and not globally. If the maximum value is evaluated over the entire uncertainty region, the calculation burden will be very high and it will lead to a conservative controller. The design process is described in the following steps:

- From the simulations of the faulted system (undergoing a large perturbation during LVRT transients), obtain the range of variation of all the state variables and form a volume, Ω, with corner points given by $(x_{f_{p_i}} - x_{0_{p_i}})$ and $(x_{f_{p_i}} + x_{0_{p_i}})$, $p = 1, \ldots, 7$, where $2x_{f_{p_i}}$ is the largest variation in the pth state variable about its equilibrium value, $x_{0_{p_i}}$.
- Obtain $\Gamma_i^* = \max_{x_i^{*p} \in \Omega}$ such that (8.33) and (8.34) hold, using a genetic algorithm [41, 42] .
- Check if there exists a feasible controller with $\Gamma_i = \Gamma_i^*$, i.e., scalars τ_i and Θ_i exist such that there is a feasible solution to LMIs, as described in [40]. If there is no feasible solution with the chosen Γ_i^*, reduce the volume Ω, by choosing different corner points, and repeat the process.
- Perform an optimal search over the scalar parameters τ_i and Θ_i, to get the infimum in (8.43).

For the given system, we are able to obtain a solution with values of $\Gamma_1 = 0.865$, $\Gamma_2 = 0.925$, $\Gamma_2 = 0.956$, and $\Gamma_4 = 0.975$. The controller is stabilizing for all variations of states in the polytope region Ω_i formed by the corner points as: $\left[\bar{\omega}_{m_i}, \bar{\omega}_{G_i}, \bar{\gamma}_i, \bar{s}_i, \bar{E}'_{dr_i}, \bar{E}'_{qr_i}, \bar{v}_{dc_i}\right]^T$ and $\left[\underline{\omega}_{m_i}, \underline{\omega}_{G_i}, \underline{\gamma}_i, \underline{s}_i, \underline{E}'_{dr_i}, \underline{E}'_{qr_i}, \underline{v}_{dc_i}\right]^T$ with the following values: $\bar{\omega}_{m_i} = \omega_{mi0} + 0.465$ pu, $\underline{\omega}_{m_i} = \omega_{mi0} - 0.465$ pu, $\bar{\omega}_{G_i} = \omega_{Gi0} + 0.427$ pu, $\underline{\omega}_{G_i} = \omega_{Gi0} - 0.427$ pu, $\bar{\gamma}_i = \gamma_{i0} + 35°$, $\bar{\gamma}_i = \gamma_{i0} - 35°$, $\bar{s}_i = s_{i0} + 0.253$ pu , $\underline{s}_i = s_{i0} - 0.253$ pu, $\bar{E}'_{dr_i} = E'_{dri0} + 0.325$ pu, $\bar{E}'_{dr_i} = E'_{dri0} - 0.325$ pu, $\bar{E}'_{qr_i} = E'_{qri0} + 0.335$ pu, $\underline{E}'_{qr_i} = E'_{qri0} - 0.335$ pu, $\bar{v}_{dc_i} = v_{dci0} + 0.225$, $\bar{v}_{dc_i} = v_{dci0} - 0.225$pu, $\bar{v}_{dg_i} = v_{dgi0} + 0.314$, $\underline{v}_{dg_i} = v_{dgi0} - 0.314$ and $\bar{v}_{qg_i} = v_{qgi0} + 0.427$, $\underline{v}_{qg_i} = v_{qgi0} - 0.427$, $\bar{v}_{dr_i} = v_{dri0} + 0.327$, $\underline{v}_{dr_i} = v_{dri0} - 0.327$. Although the designed controller is not globally stabilizing we know that it is stabilizing over a large operating region which covers most faulted system operation. From the two cross-sections of the polytope Ω shown in Fig. 8.17, it can be seen that the region of controller operation is larger than the region of faulted system trajectories.

Figures 8.8 and 8.9 show the Bode plot of the designed and the PI (V_{dc_1} versus V_{gd_1}) controller of WF$_1$. The Bode plot of the designed 7th-order controller in Fig. 8.8 shows that it has an adequate low-frequency gain (25 dB) and the band-

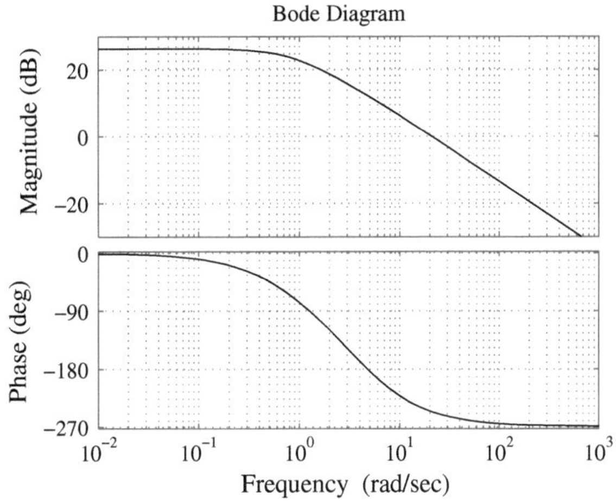

Fig. 8.8 Designed controller for subsystem 1-Bode plot of transfer function of V_{dc} versus V_{gd}

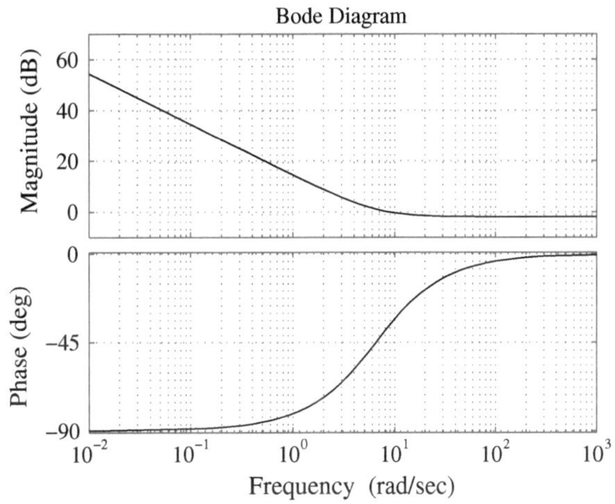

Fig. 8.9 PI controller for subsystem 1-Bode plot of transfer function of V_{dc} versus V_{gd}

width is between 10 and 20 rad/s. Bode plots for the other controllers show similar characteristics. The bandwidth of the designed controller is larger than that of the PI controller. It is desirable to have a high gain at low frequency and a rapidly decreasing gain after the gain crossover frequency; these are provided by the robust controller. The above control algorithm is applied to the New England test system. Large disturbance simulations are used in the next section to demonstrate the performance of the designed controller.

8.7 Controller Performance Evaluation

The LVRT capability of a wind generator is expressed as voltage and transient stability margins. The voltage stability margin is defined as the difference between the operating voltage and the critical voltage. The transient stability margin is the difference between the speed after a specified fault duration and the critical speed (CS) of the generator. The critical speed is given by the intersection between the torque-speed curve for the specified system and the mechanical torque [43]. The critical voltage can be obtained from the P–V curves [44]. The stability analysis of a power system may consider the determination of its critical clearing time (CCT), for a given fault, in order to find the maximum value of the CCT for which the system is stable. In this chapter, the CCT is first estimated by using (8.44) and then the exact value is determined from simulations by increasing the fault time interval until the system loses its stability [43].

$$t_c = \frac{1}{T_m} 2H_m(s_c - s_0), \tag{8.44}$$

where t_c is the critical clearing time, s_c the critical slip and s_0 the slip at equilibration point of a generator.

8.7.1 Enhancement of Voltage and Transient Stability Margins

A simulation study is performed for the 10 machine 39 bus system in Fig. 8.2 with the purpose of evaluating the transient and voltage stability limits achievable with the proposed and conventional PI-based vector control scheme. The critical clearing time (CCT) and critical speed (CS), as shown in Table 8.1, for a three-phase fault are 0.425 s and 1.436 pu, respectively, with the designed controller as compared with 0.340 s and 1.345 pu with a properly tuned PI controller [45, 46].

The PI controller is of the first order $[K_P + K_I/s]$ and it has the same input and output signals as those of the designed controller given in Fig. 8.5. In practice, PI parameters are often tuned by trial-and-error approaches, sometimes starting with initial values given by, for example, the Ziegler-Nichols method [45]. We adopt this strategy for getting the optimal settling time and peak overshoot. In this case, the gain of the tuned (trial and error method) PI controller in WF_1 is obtained as for GSC $K_{P_{GSC}} = 0.8$ and $K_{I_{GSC}} = 5.25$ and for RSC $K_{P_{RSC}} = 0.6$ and $K_{I_{RSC}} = 8.5$. It can be concluded that an appropriate combination of active and reactive power control by both converters is an effective way of improving the stability and enhancing the fault ride-through capability of the relevant induction-generator-based wind turbines.

Table 8.1 Performance comparison: (*a*) proposed controller; and (*c*) PI-based vector control

(a) Proposed controller			(c) PI control		
CS (pu)	CCT (s)	$V_{critical}(pu)$	CS (pu)	CCT (s)	$V_{critical}(pu)$
1.436	0.425	0.825	1.345	0.340	0.875

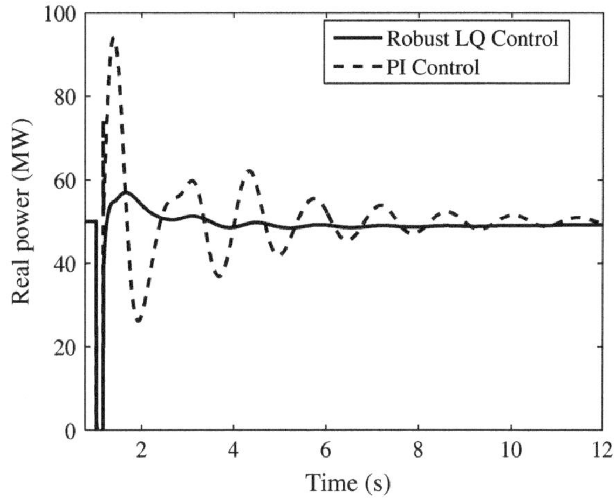

Fig. 8.10 Real Power of WF_1 for three-phase fault at bus 6

8.7.2 Comparison of Robust LQ and PI Controllers

In order to evaluate the performance of the designed controller, in the face of large change in operating conditions, detailed simulations are performed for a symmetrical three-phase fault at bus 6 which is subsequently cleared after 150 ms. Figures 8.10 and 8.11 show the real power and terminal voltage response of the wind farm WF_1 with the proposed LVRT control and the conventional PI-based vector controller.

From Fig. 8.10, it can be seen that the PI controller results in an oscillatory post-fault behavior. The designed controller provides improved performance in terms of damped oscillations, lower settling time and better damping. The proposed ride-through methods also reduce the oscillations and settling time of the DFIG transient response, and consequently, enhance the DFIG voltage dip behavior. It also reduces the rotor instantaneous power oscillations fed to the DC-link capacitor and in turn limits the DC-link voltage fluctuations.

Next a three phase fault is applied on bus 12 at 1 s and cleared after 300 ms. Figures 8.12 and 8.13 show the responses of DC link voltage and rotor current of the wind farm WF_2. From Fig. 8.12, by using the proposed DC-link control technique, the fluctuation range of the DC-link voltage is reduced and limited to 1.16 pu.

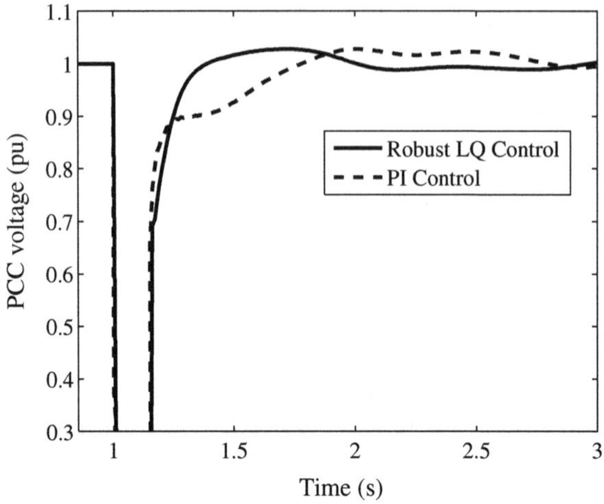

Fig. 8.11 PCC voltage of WF_1 for three-phase fault at bus 6

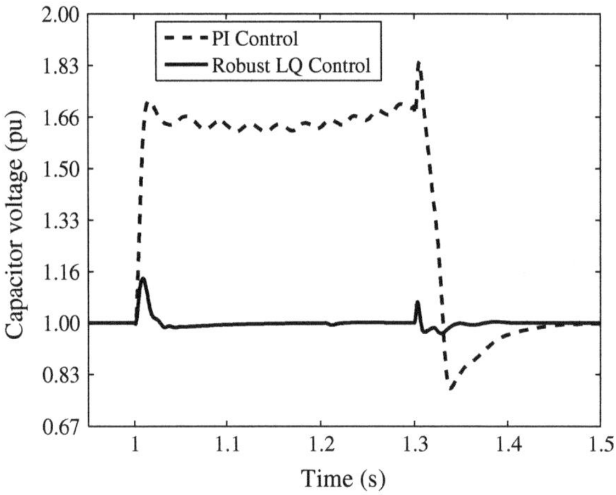

Fig. 8.12 DC-link voltage of WF_2 for three-phase fault at bus 12

The DC-link voltage with the conventional control of Fig. 8.12 reaches the maximum value of 1.83 pu during the fault, and thus the DC-link capacitor would be under a large voltage stress and may be damaged. Figure 8.13 shows the rotor current from which it is clear that the peak rotor fault current at the instants of fault occurring and clearing is below the rotor-side converter tolerable limit, i.e., 2 pu with both controllers. However, it is clear that the performance of the proposed approach is more effective than that of the conventional control.

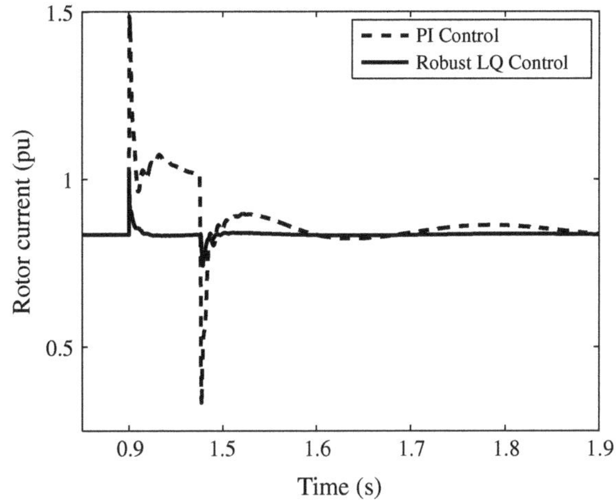

Fig. 8.13 Rotor current of WF_2 for three-phase fault at bus 12

8.7.3 Severe Low Impedance Fault Close to Wind Farm

In this case a low impedance (0.5 Ω) is applied close to the PCC (bus 26) of wind farm, WF_4. When a disturbance or fault occurs, the voltages at the terminals of the wind turbine drop significantly, causing the electromagnetic torque and electric power output of the generator to be greatly reduced. However, given that the mechanical input torque is almost constant when typical non-permanent faults occur in a wind farm, this leads to an acceleration of the machine rotor. Figure 8.14 shows the speed deviation for the same fault from which it is clear that, although the speed is temporarily increased, it returns to its pre-fault value after the fault is cleared. Figure 8.15 shows the three-phase stator current from which it is clear that current returns back to the previous value within 0.3 s. From the above simulation results it can be concluded that the designed controller provides robust performance and ensures stability under varying operating conditions.

8.7.4 Comparisons with LVRT Standards

As mentioned earlier LVRT requirements in the North American Electric Reliability Council (NAERC) demand that a wind farm remains connected to the grid for voltage levels as low as 15 % of the nominal voltage for 625 ms [33]. However, the performance of the designed controller is demonstrated for zero voltage and it ensures stability for both zero voltage ride-through and conventional 85 % LVRT.

Figure 8.16 shows the terminal voltage of the wind farm WF_1 using the proposed control with zero voltage for the duration of 625 ms from which it is clear that

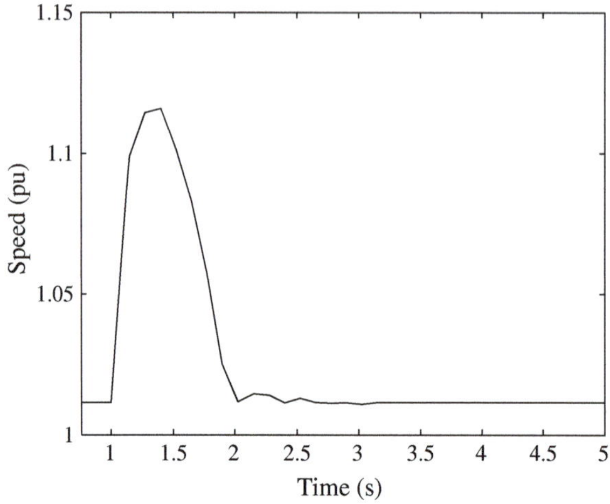

Fig. 8.14 Speed deviation of WF_4 for three-phase fault at bus 26

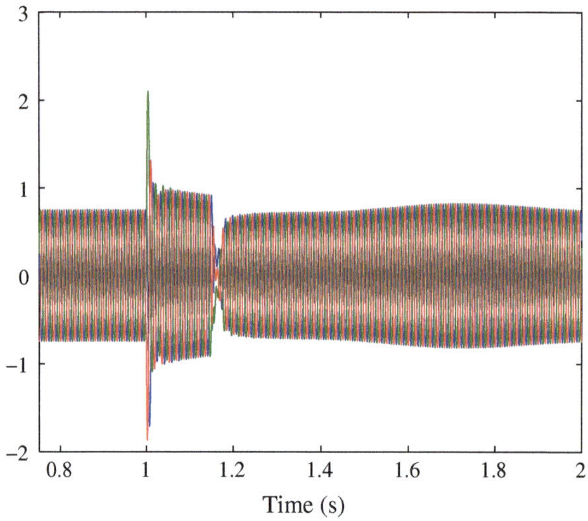

Fig. 8.15 Stator current of WF_4 for three-phase fault at bus 26

the proposed control can meet the standard requirement of the LVRT capability. The reason for providing stability during the LVRT transient is clear from Fig. 8.17 which shows that the control region covered by the designed controller is larger than the area required for the LVRT. Similar comparison holds true for other state variables. In Fig. 8.17, the dotted area represents the maximum variation in the state variable, E'_{dr_1} (d-axis transient emf) and E'_{qr_1} (q-axis transient emf) of wind farm 1 at bus

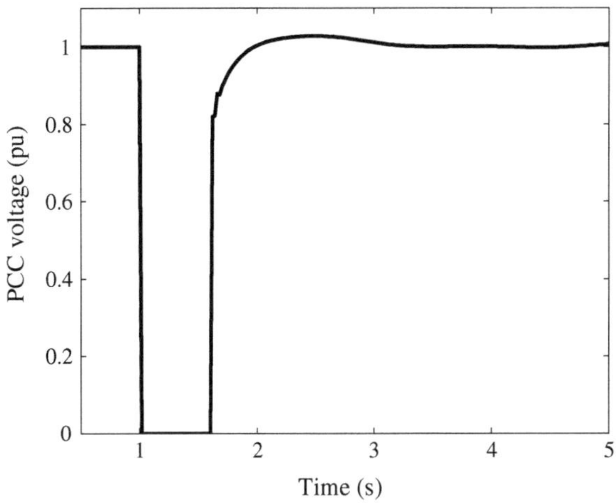

Fig. 8.16 Terminal voltage with zero-voltage for 625 ms

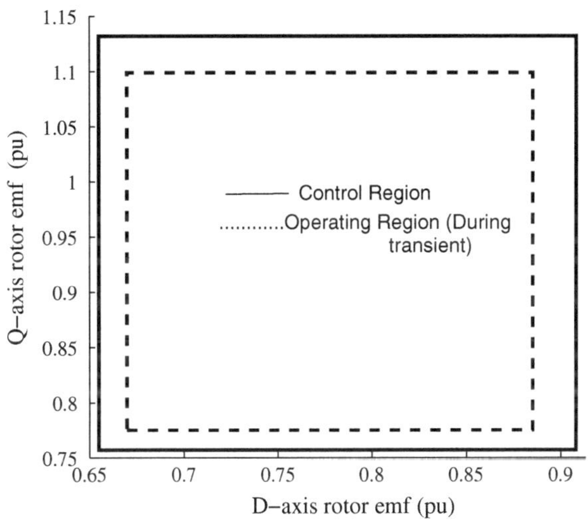

Fig. 8.17 Operating and control region (*Solid line* control region and dashed line operating region during LVRT transient

7 for a severe three-phase fault at the point of common coupling (PCC) of WF_1 as shown in Fig. 8.2. This variation is obtained from the nonlinear simulation. The solid line represents the maximum change in the above mentioned state variable for which the designed controller ensures stability, this region is obtained from the presented control design algorithm in Sect. 8.6.

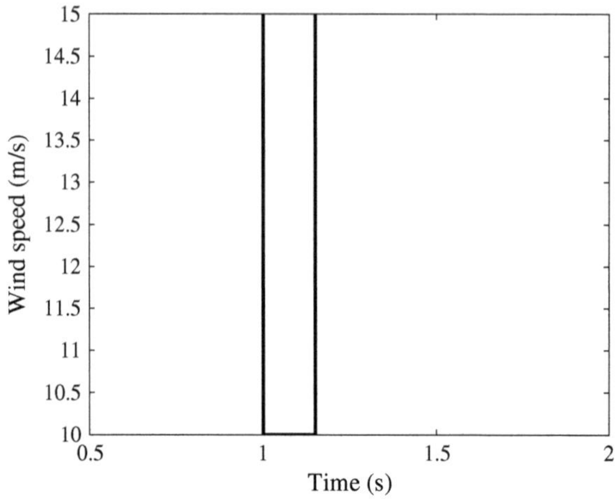

Fig. 8.18 Change in wind speed

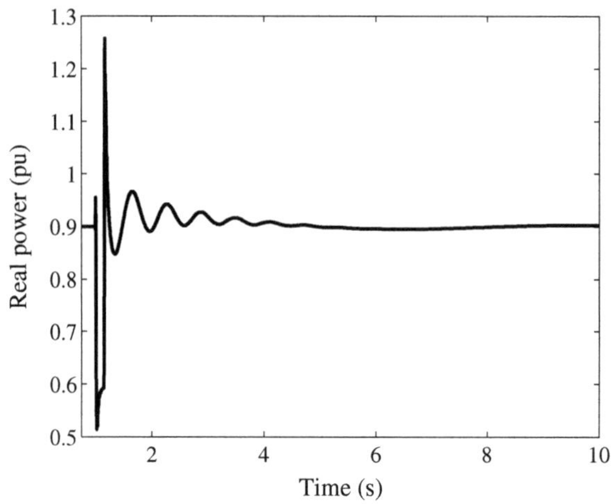

Fig. 8.19 Real power output of WF_4 due to change in wind speed

8.7.5 Turbine Response to a Change in Wind Speed

When the SCIG is subjected to a sudden drop in speed, active power drops and reactive powers increase substantially. By changing active power and providing an adequate reactive support to the system, the proposed control scheme ensures a stable response to these sudden changes in operating conditions related to speed variations. To illustrate this stabilising capability of the proposed controller, a simulation in

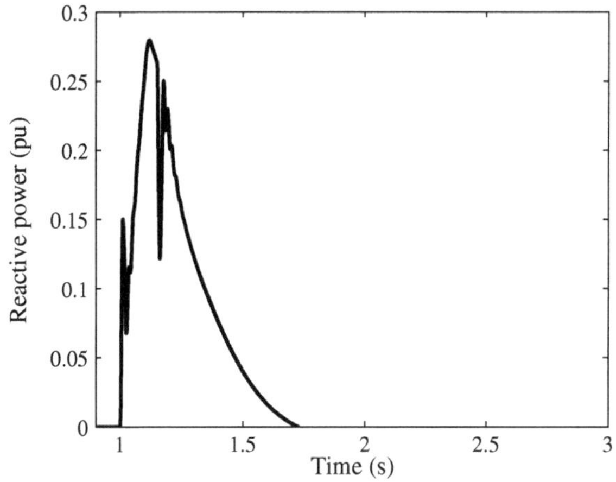

Fig. 8.20 Reactive power output of WF_4 due to change in wind speed

which the initial wind speed is 15 m/s and then stepped down to 10 m/s. Figure 8.18 shows the applied variation of the wind speed, and Fig. 8.19, refReactivePowerdfig show the response to this variation. In this simulation zero reactive power set point is used. The designed DFIG controllers increase the reactive power supply to keep the voltage constant which is visible from Fig. 8.20, while the speed controller prevents the generator speed becoming too high.

8.7.6 Robustness Under Different Operating Conditions

To verify the robust performance of the controller, its effectiveness is evaluated under load uncertainty. Although the controller is designed for rated operating conditions, the designed controller performs well in different loading conditions. Figures 8.21 and 8.22 show the real and reactive power output of WF_1 due to the 10 % increase in load from which it is clear that the controller stabilizes the system at different equilibrium point.

8.7.7 Dynamic Interactions Among Several DFIG Controllers

In this research work, both local and interconnection uncertainties are taken into account in the design of the robust controller for each subsystem. Therefore, they do not produce any adverse effects to reduce the stability margin. To investigate the

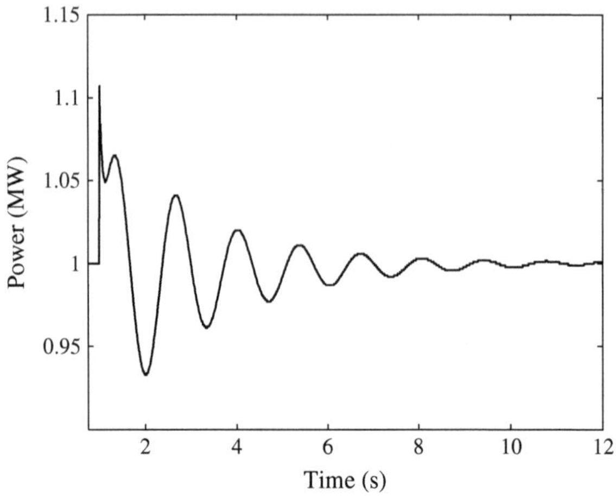

Fig. 8.21 Real power output of WF_1 with 10 % increase in load

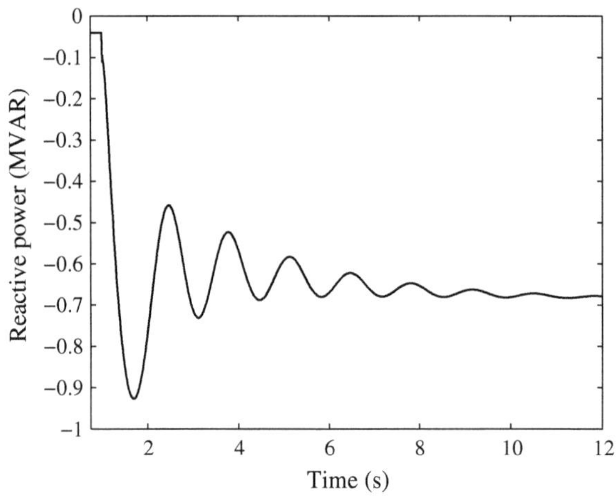

Fig. 8.22 Reactive power output of WF_1 with 10 % increase in load

interactions among several DFIG controllers, simulations are carried out considering: (1) only wind farms (aggregated model); and (2) a detailed model of each DFIG embedded with the designed controller. Figures 8.23 and 8.24 show the terminal voltages and real power outputs of WF_1 and a DFIG with a severe three-phase fault at bus 7, respectively. It is clear that interactions among the DG units with the designed controllers do not deteriorate the dynamic performance of the system.

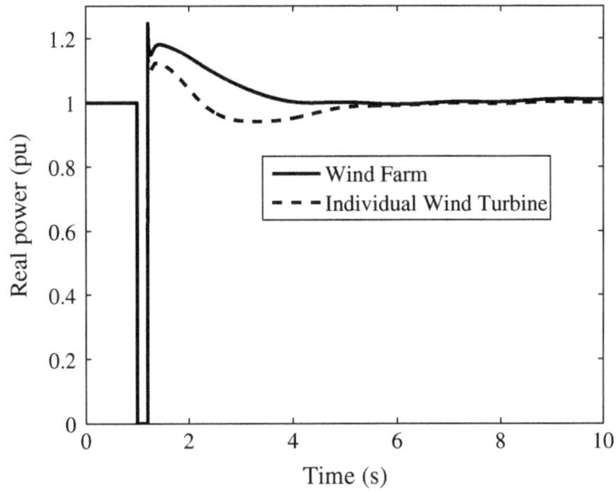

Fig. 8.23 Real power output for three-phase fault at bus 6

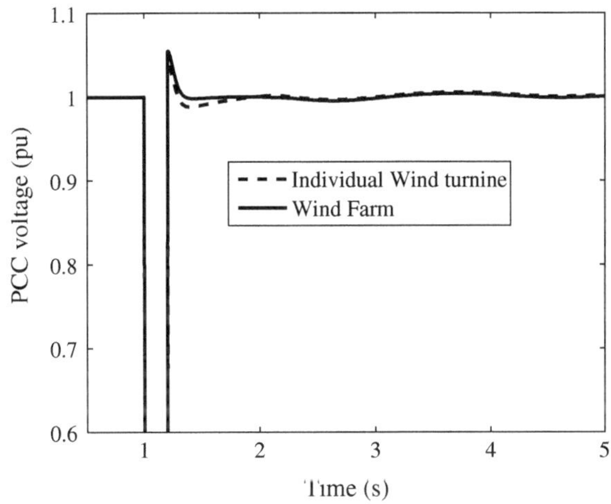

Fig. 8.24 PCC voltage for three-phase fault at bus 6

8.7.8 Asymmetrical Faults

The performance of the designed controller is also validated against unbalanced faults, i.e., single line to ground fault. The duration of the fault is five cycles and the original system is restored after the clearance of the fault. Figures 8.25 and 8.26 show the real and reactive power output of WF_3 due to a single line to ground fault on bus 23. From the above results it is clear that the proposed robust controller ensures stability in the presence of both symmetrical and unsymmetrical fault.

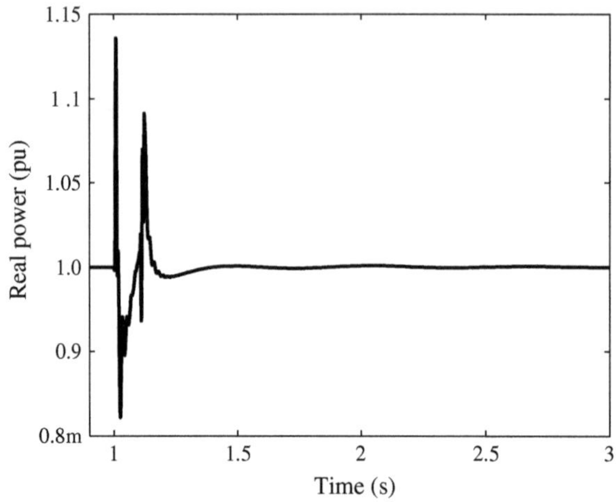

Fig. 8.25 Real power output of WF_3 with single line to ground fault

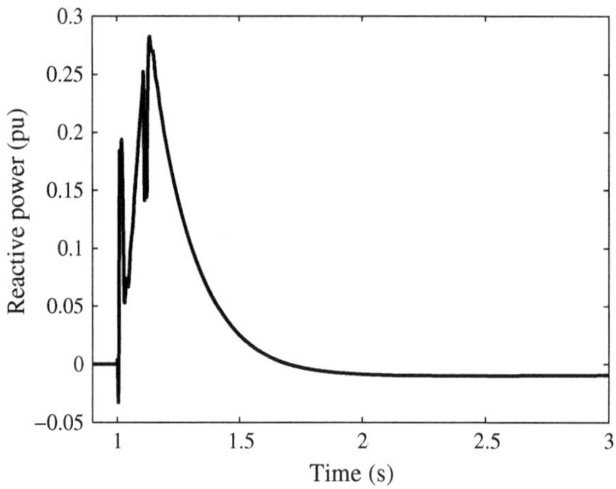

Fig. 8.26 Reactive power output of WF_3 with single line to ground fault

8.8 Conclusions

In this chapter a new robust decentralized output-feedback control has been proposed to enhance the LVRT capability of DFIGs. A systematic procedure to design the controller has been discussed. The designed controller guarantees stability if the system post-fault operating point is in the region for which the controller is designed. The control for a grid-side converter has been designed to stabilize both internal and external dynamics and limit the DC voltage fluctuations. Also, the control for a rotor-side converter improves the damping of transients during voltage dip. Simulation results show that despite the nonlinear coupling between different types of generators and significant operating condition variations, the proposed controller can greatly enhance the damping and voltage stability as well as LVRT capability of wind turbines. The future aim of this study is to implement the proposed algorithm in real-time applications.

References

1. Alberdi M, Amundarain M, Garrido AJ, Garrido I, Maseda FJ (2011) Fault-ride-through capability of oscillating-water-column-based wave-power-generation plants equipped with doubly fed induction generator and airflow control. IEEE Trans Ind Electron 58(5):1501–1517
2. de Luna A, Araujo Lima FK, Santos D, Rodriguez P, Watanabe EH, Arnaltes S (2011) Simplified modeling of a DFIG for transient studies in wind power applications. IEEE Trans Ind Electron 58(1):9–20
3. Verma V, Hossain MJ, Saha TK, Chakraborty C (2012) Performance of MRAS-based speed estimators for grid connected doubly fed induction machines during voltage dips. In: IEEE power and energy society general meeting, CA, pp 1–8
4. Abdel-Baqi O, Nasiri A (2011) Series voltage compensation for DFIG wind turbine low-voltage ride-through solution. IEEE Trans Energy Convers 26(1):272–280
5. Lopez J, Gubia E, Olea E, Ruiz J, Marroyo L (2009) Ride through of wind turbines with doubly fed induction generator under symmetrical voltage dips. IEEE Trans Ind Electron 56(10):4246–4254
6. Rahimi M, Parniani M (2010) Efficient control scheme of wind turbines with doubly fed induction generators for low-voltage ride-through capability enhancement. IET Renew Power Gener 4(3):242–252
7. Perdanad A (2008) Dynamic models of wind turbines. Ph.D. thesis, Chalmers University of Technology, Sweden
8. Joshi N, Mohan N (2009) A novel scheme to connect wind turbines to the power grid. IEEE Trans Energy Convers 24(2):504–510
9. Tohidi S, Oraee H, Zolghadri MR, Shao S, Tavner P (2013) Analysis and enhancement of low-voltage ride-through capability of brushless doubly fed induction generator. IEEE Trans Ind Electron 60(3):1146–1155
10. Kasem AH, El-Saadany EF, El-Tamaly HH, Wahab MAA (2008) An improved fault ride-through strategy for doubly fed induction generator-based wind turbines. IET Renew Power Gener 2(4):201–214
11. Hu S, Lin X, Kang Y, Zou X (2011) An improved low voltage ride through control strategy of doubly-fed induction generator during grid faults. IEEE Trans Power Electron 26(12):3653–3665
12. Ibrahim AO, Nguyen TH, Lee DC, Kim SC (2011) A fault ride-through technique of DFIG wind turbine systems using dynamic voltage restorers. IEEE Trans Energy Convers 26(3):871–882
13. Pena R, Cardenas R, Reyes E, Clare J, Wheeler P (2011) Control of a doubly fed induction generator via an indirect matrix converter with changing DC voltage. IEEE Trans Ind Electron 58(10):4664–4674
14. Hu J, He Y, Xu L, Williams BW (2009) Improved control of DFIG systems during network unbalance using PIR current regulators. IEEE Trans Ind Electron 56(2):439–451
15. da Costa JP, Pinheiro H, Degner T, Arnold G (2011) Robust controller for DFIGs of grid-connected wind turbines. IEEE Trans Ind Electron 58(9):4023–4038
16. Karimi S, Gaillard A, Poure P, Saadate S (2009) Current sensor fault-tolerant control for WECS with DFIG. IEEE Trans Ind Electron 56(11):4660–4670

17. Zhang S, Tseng KJ, Choi SS, Nguyen TD, Yao DL (2012) Advanced control of series voltage compensation to enhance wind turbine ride-through. IEEE Trans Power Electron 27(2):763–772
18. Campos-Gaona D, Moreno-Goytia EL, Anaya-Lara O (2013) Fault ride-through improvement of dfig-wt by integrating a two-degrees-of-freedom internal model control. IEEE Trans Ind Electron 60(3):1143–1145
19. Abad C, Rodriguez MA, Poza J (2008) Three-level NPC converter-based predictive direct power control of the doubly fed induction machine at low constant switching frequency. IEEE Trans Ind Electron 56(10):4417–4429
20. Xu L, Zhi D, Williams BW (2009) Predictive current control of doubly fed induction generators. IEEE Trans Ind Electron 56(10):4143–4153
21. Rodriguez J, Pontt J, Silva A, Correa P, Lezana P, Cortes P, Ammann U (2007) Predictive current control of a voltage source inverter. IEEE Trans Ind Electron 54(1):495–503
22. Mullane A, Lightbody G, Yacamini R (2005) Wind-turbine fault ride-through enhancement. IEEE Trans Power Syst 20(4):1929–1937
23. Wu F, Zhang XP, Ju P, Sterling MJH (2008) Decentralized nonlinear control of wind turbine with doubly fed induction generator. IEEE Trans Power Syst 23(2):613–621
24. Hossain MJ, Saha TK, Mithulananthan N, Pota HR (2013) Control strategies for augmenting LVRT capability of dfigs in interconnected power systems. IEEE Trans Ind Electron 60(6):2510–2522
25. Hossain MJ, Pota HR, Ugrinovskii V, Ramos RA (2010) Simultaneous STATCOM and pitch angle control for improved LVRT capability of fixed-speed wind turbines. IEEE Trans Sustain Energy 1(3):142–152
26. Guo Y, Hill DJ, Wang Y (2001) Global transient stability and voltage regulation for power systems. IEEE Trans Power Syst 16(4):678–688
27. Hossain MJ, Pota HR, Ugrinovskii V, Ramos RA (2010) Voltage mode stabilisation in power systems with dynamic loads. Int J Electr Power Energy Syst 32(8):911–920
28. Ackermann T (2005) Wind power in power systems. Wiley, London
29. Lei Y, Mullane A, Lightbody G, Yacamini R (2006) Modeling of the wind turbine with a doubly-fed induction generator for grid integration studies. IEEE Trans Energy Convers 21(1):257–264
30. Hiskens IA: Power system test cases. http://psdyn.ece.wisc.edu/IEEE_benchmarks/
31. Fernandez L, Garcia C, Saenz JR, Jurado F (2009) Equivalent models of wind farms by using aggregated wind turbines and equivalent winds. Energy Convers Manage 50(3):691–704
32. Germond AJ, Podmore R (1978) Dynamic aggregation of generating unit models. IEEE Trans Power Apparatus Syst PAS-97(4):1060–1069
33. Interconnection for wind energy. Tech. rep., Order No. 661, 70 Fed. Reg. 34993 (FERC Stats. & Regs. 31,186 (2005) ("Final Rule" or "Order No. 661")
34. Wu Q, Xu Z, Ostergaard J (2010) Grid integration issues for large scale wind power plants (wpps). In: IEEE power and energy society general meeting, Minneapolis, Minnesota, pp 1–6
35. Pal BC, Coonick AH, Macdonald DC (2000) Robust damping controller design in power systems with superconducting magnetic energy storage devices. IEEE Trans Power Syst 15(1):320–325
36. Werner H, Korba P, Yang TC (2003) Robust tuning of power system stabilizers using LMI-techniques. IEEE Trans Control Syst Technol 11(1):147–152
37. Khalil HK (1992) Nonlinear systems. Prentice-Hall, New York
38. Petersen IR, Ugrinovskii VA, Savkin AV (2000) Robust control design using H_∞ methods. Springer, London
39. Li L, Ugrinovskii VA, Orsi R (2007) Decentralized robust control of uncertain markov jump parameter systems via output feedback. Automatica 43(11):1932–1944
40. Hossain MJ, Pota HR, Ugrinovskii VA, Ramos RA (2010) Decentralized control to augment LVRT capability of wind generators with STATCOM/ESS. In: IEEE power and energy society general meeting, Minneapolis, Minnesota, pp 1–8

41. Hossain MJ, Pota HR, Mahmud MA, Ramos RA (2010) Impacts of large-scale wind generators penetration on the voltage stability of power systems. In: IEEE power and energy society general meeting, Michigan, pp 1–8
42. Fogel DB, Back T, Michalewicz Z (2000) Evolutionary computation: basic algorithms and operators. Institute of Physics Publishing
43. Molinas M, Suul JA, Undeland T (2008) Low voltage ride through of wind farms with cage generators: STATCOM versus SVC. IEEE Trans Power Electron 23(3):1104–1117
44. Cutsem TV, Vournas C (1998) Voltage stability of electric power systems. Kluwer Academic, Norwell
45. Mallesham G, Mishra S, Jha AN (2009) Maiden application of Ziegler-Nichols method to AGC of distributed generation system. In: IEEE/PES power systems conference and exposition, Seattle, pp 1–7
46. Rahimi M, Parniani M (2010) Efficient control scheme of wind turbines with doubly fed induction generators for low voltage ride-through capability enhancement. IET Renew Power Gener 4(3):242–252

Chapter 9
Interactions of PV Units in Distribution Networks

Abstract In this chapter it is shown that there exist oscillations in distribution networks with physically close photovoltaic (PV) units due to control interaction among them. The main reason for oscillations being the changing "gain" of the PV units due to the nonlinear dynamics and the decentralized tuning of the controllers for the nominal operating condition. The decentralized robust control design, presented in this chapter, ensures a fast and well-damped response for varying operating conditions with physically close PV units. The control design process uses an estimate of the change in model due to variations in solar radiance, changing electric load, and physically close PV units. A minimally conservative method is used to capture the estimate as norm bounded unmodeled dynamics. Simulations on a test distribution system demonstrate the performance of the proposed robust controller for different system contingencies.

9.1 Introduction

The integration level of PV units in low and medium voltage networks has increased significantly in recent years [1]. As the penetration level of PV units in distribution system increases it is essential to develop a good understanding of its potential impact on the dynamic operation of power systems [2]. Although sporadic integration of relatively small size PV units at lower penetration levels is not a concern, higher penetration of small size PV plants might affect the voltage profile and dynamic performance of the overall system. The existing PV units in low voltage networks do not contribute to voltage control as they inject real power only and operate at unity power factor. It is expected that in near future the PV units with sophisticated inverters will allow the injection of reactive power at the point of common coupling (PCC) along with other distributed energy resource (DER) units. In such cases it must be ensured that these PV units are integrated without any undesirable effect.

J. Hossain and H. R. Pota, *Robust Control for Grid Voltage Stability: High Penetration of Renewable Energy*, Power Systems,
DOI: 10.1007/978-981-287-116-9_9, © Springer Science+Business Media Singapore 2014

As the use of power electronics based renewable energy injection is increasing at the distribution level, there is a strong possibility that future voltage and reactive power control will be at the primary control level with smaller time constants rather than discrete and at tertiary level control [3]. Different techniques are used for inverter control as the inverters are produced by different manufacturers. There are many control loops for voltage control in DERs. Multiple inverters connected to the distribution system with uncoordinated controllers and minimal impedance between different DERs may interact negatively and degrade the stability of the system.

Dynamic interactions have been studied in transmission systems, however interactions in distribution systems could be completely different as there are many differences between transmission and distribution systems. Higher resistance to reactance ratio, low and medium voltage levels, small number of devices to control the voltage and different types of loads connected to distribution systems are some of these differences. The presence of PV units near a load and the strong coupling of the dynamic load with power electronic controls complicate the interaction problem. It is important to understand this issue as it can influence the dynamic operation of the entire system.

A recent work reports on a preliminary study on dynamic interactions of voltage regulating units and PV voltage regulators [4]. In [4], only the interactions of voltage regulators among different types of devices, e.g., on-load tap changers (OLTCs), shunt capacitors and PV voltage regulators are analyzed. In [5], the interactions of DER controllers in low voltage distribution networks are analyzed and a simple tuning algorithm to minimize the negative interactions is proposed. However, the interaction among the PV inverter controllers is not analyzed in [5]. The controller proposed in [5] works well for a fixed set of operating points, but the performance for changing operating conditions cannot be ensured.

A centralized control scheme based on optimal power flow is designed by collecting the system-wide information and sending the set-points to each unit in [6, 7]. For a distribution network which has numerous PV units, it is quite expensive to implement centralized control. Decentralized control strategies based on droop control, power flow control and maximum power point tracking (MPPT) have been proposed [6–12] which use local information and are easy to implement. A decentralized control scheme is proposed in [13] for optimal voltage regulation in a complex distribution system. A decentralized control scheme for DER units based on disturbance rejection and tracking is presented in [14, 15]. However, most of the controllers are designed either using classical linear control theory or proportional-integral (PI) control techniques which are unable to guarantee the robustness under changing operating conditions.

It is important to minimise the negative interactions of PV controllers for the secure and reliable operation of the power system. This chapter presents an analysis of the possible negative interaction among PV controllers and presents a design of a robust controller to mitigate unwanted interactions. In the robust control design proposed in this article, parts of nonlinear dynamics and control interaction are modeled as disturbances and this ensures non-interacting robustness control.The rest of the chapter is organized as follows: Sect. 9.2 provides the mathematical

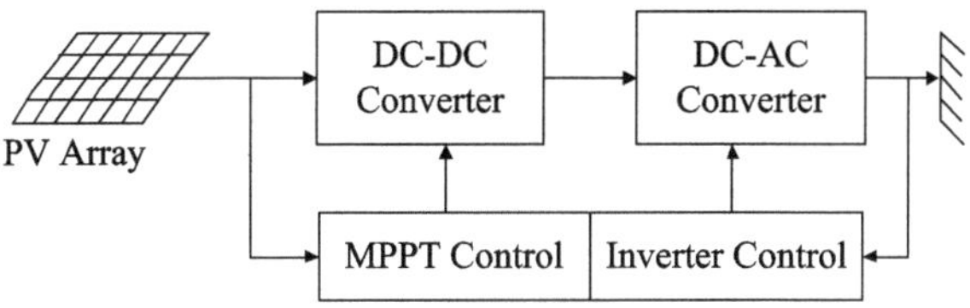

Fig. 9.1 Block Diagram of a PV system

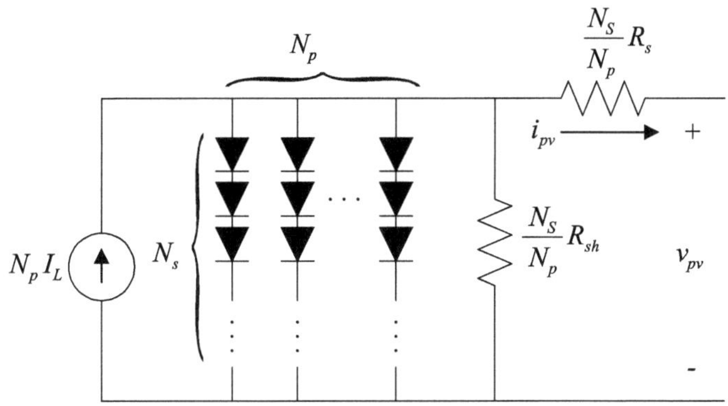

Fig. 9.2 Equivalent circuit of a PV array

modeling of PV units. Interactions among PV units are investigated in Sect. 9.3 via case studies. The decentralized control strategy and uncertainty modeling are presented in Sect. 9.4. The PV control is presented in Sect. 9.5. Section 9.6 presents the control design algorithm and the performance of the controller is demonstrated through a series of simulations. Concluding remarks are provided in Sect. 9.7.

9.2 PV System Model

As shown in Fig. 9.1, PV plants have mainly two parts (a) solar energy conversion and (b) electrical interface with the electrical network (a power electronic converter). A PV array is connected to the grid through a DC-DC converter and a DC-AC inverter. A DC-DC converter enables the transfer of maximum power from the solar module to the inverter. The PV array as shown in Fig. 9.2 is described by its current-voltage characteristics function [16, 17]:

$$i_{\text{pv}i} = N_{\text{p}i}I_{\text{L}i} - N_{\text{p}i}I_{\text{s}i}[\exp\left[\alpha_{p_i}\left(\frac{v_{\text{pv}i}}{N_{\text{s}i}} + \frac{R_{\text{s}i}i_{\text{pv}i}}{N_{\text{p}i}}\right)\right] - 1] - \frac{N_{\text{p}i}}{R_{\text{sh}i}}\left(\frac{v_{\text{pv}i}}{N_{\text{s}i}} + \frac{R_{\text{s}i}i_{\text{pv}i}}{N_{\text{p}i}}\right), \quad (9.1)$$

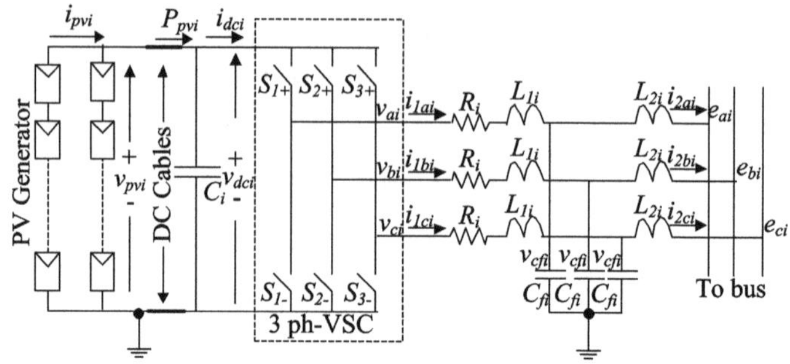

Fig. 9.3 PV system connected to the grid

where I_{L_i} is the light-generated current, I_{s_i} is the reverse saturation current, chosen as 9×10^{-11} A, N_{s_i} is the number of cells in series and N_{p_i} is the number of modules in parallel, R_{s_i} and R_{sh_i} are the series and shunt resistances of the array respectively, i_{pv_i} is the current flowing through the array, and v_{pv_i} is the output voltage of the array. The constant α_{p_i} in Eq. (9.1) is given by

$$\alpha_{p_i} = \frac{q_i}{A_i k_i T_{r_i}} \tag{9.2}$$

where $k_i = 1.3807 \times 10^{-23}$ J K^{-1} is the Boltzmann constant, $q_i = 1.6022 \times 10^{-19}$ C is the charge of an electron, A_i is the p-n junction ideality factor with a value between 1 and 5, and T_{r_i} is the cell reference temperature. The schematic of a grid-connected PV system consisting of switching elements is shown in Fig. 9.3 [18, 19]. A nonlinear model of the three-phase grid connected PV system, shown in Fig. 9.3, can be written as [18, 19]:

$$\begin{aligned}
\dot{i}_{1\mathrm{a}_i} &= -\frac{R_i}{L_{1_i}} i_{1\mathrm{a}_i} - \frac{1}{L_{1_i}} e_{\mathrm{a}_i} + \frac{v_{\mathrm{pv}_i}}{3L_{1_i}} (2K_{\mathrm{a}_i} - K_{\mathrm{b}_i} - K_{\mathrm{c}_i}) \\
\dot{i}_{1\mathrm{b}_i} &= -\frac{R_i}{L_{1_i}} i_{1\mathrm{b}_i} - \frac{1}{L_{1_i}} e_{\mathrm{b}_i} + \frac{v_{\mathrm{pv}_i}}{3L_{1_i}} (-K_{\mathrm{a}_i} + 2K_{\mathrm{b}_i} - K_{\mathrm{c}_i}) \\
\dot{i}_{1\mathrm{c}_i} &= -\frac{R_i}{L_{1_i}} i_{1\mathrm{c}_i} - \frac{1}{L_{1_i}} e_{\mathrm{c}_i} + \frac{v_{\mathrm{pv}_i}}{3L_{1_i}} (-K_{\mathrm{a}_i} - K_{\mathrm{b}_i} + 2K_{\mathrm{c}_i}) \\
\dot{v}_{\mathrm{cfa}_i} &= \frac{1}{C_{\mathrm{f}_i}} \left(i_{1\mathrm{a}_i} - i_{2\mathrm{a}_i}\right), \quad \dot{v}_{\mathrm{cfb}_i} = \frac{1}{C_{\mathrm{f}_i}} \left(i_{1\mathrm{b}_i} - i_{2\mathrm{b}_i}\right) \\
\dot{v}_{\mathrm{cfc}_i} &= \frac{1}{C_{\mathrm{f}_i}} \left(i_{1\mathrm{c}_i} - i_{2\mathrm{c}_i}\right), \quad \dot{i}_{2\mathrm{a}_i} = \frac{1}{L_{2_i}} \left(v_{\mathrm{cfa}_i} - e_{\mathrm{a}_i}\right) \\
\dot{i}_{2\mathrm{b}_i} &= \frac{1}{L_{2_i}} \left(v_{\mathrm{cfb}_i} - e_{\mathrm{b}_i}\right), \quad \dot{i}_{2\mathrm{c}_i} = \frac{1}{L_{2_i}} \left(v_{\mathrm{cfc}_i} - e_{\mathrm{c}_i}\right)
\end{aligned} \tag{9.3}$$

where K_{a_i}, K_{b_i}, and K_{c_i} are the binary input switching signals. By applying KCL at the node where the DC-link is connected, we get

$$\dot{v}_{pv_i} = \frac{1}{C_i}\left(i_{pv_i} - i_{dc_i}\right). \tag{9.4}$$

The input current of the inverter i_{dc_i} can be written as [19]

$$i_{dc_i} = i_{a_i}K_{a_i} + i_{b_i}K_{b_i} + i_{c_i}K_{c_i}. \tag{9.5}$$

Now Eq. (9.4) can be rewritten as:

$$\dot{v}_{pv_i} = \frac{1}{C_i}i_{pv_i} - \frac{1}{C_i}\left(i_{a_i}K_{a_i} + i_{b_i}K_{b_i} + i_{c_i}K_{c_i}\right). \tag{9.6}$$

Equations (9.3) and (9.6) can be transformed into dq frame using the angular frequency ω_i of the grid as:

$$\begin{aligned}
L_{1_i}\dot{i}_{1d_i} &= -R_i i_{1d_i} + \omega_i L_{1_i} i_{1q_i} - v_{cfd_i} + K_{d_i} v_{pv_i} \\
L_{1_i}\dot{i}_{1q_i} &= -R_i i_{1q_i} - \omega_i L_{1_i} i_{1d_i} - v_{cfq_i} + K_{q_i} v_{pv_i} \\
L_{2_i}\dot{i}_{2d_i} &= +\omega_i L_{2_i} i_{2q_i} + v_{cfd_i} - E_{d_i} \\
L_{2_i}\dot{i}_{2q_i} &= -\omega L_2 i_{2d} + v_{cfq} - E_q \\
C_{f_i}\dot{v}_{cfd_i} &= \omega_i C_{f_i} v_{cfq_i} + C_{f_i}\left(i_{1d_i} - i_{2d_i}\right) \\
C_{f_i}\dot{v}_{cfq_i} &= -\omega_i C_{f_i} v_{cfd_i} + C_{f_i}\left(i_{1q_i} - i_{2q_i}\right) \\
C_i\dot{v}_{pv_i} &= i_{pv_i} - i_{1d_i}K_{d_i} - i_{1q_i}K_{q_i}
\end{aligned} \tag{9.7}$$

The synchronization scheme for $abc \rightarrow dq$ transformation is chosen such that the q-axis of the dq frame is aligned with the grid voltage vector, $E_{q_i} = 0$, and the real and reactive power delivered to the grid can be written as $P_i = \frac{3}{2}E_{d_i}I_{d_i}$ and $Q_i = -\frac{3}{2}E_{d_i}I_{q_i}$.

For the PV system, the main control objective is to (a) regulate the DC-link voltage to control the power extracted from the PV array and (b) regulate the terminal voltage by controlling the voltage source converter (VSC). The DC-link voltage reference is usually obtained from the maximum power point tracking (MPPT) scheme. The DC-link control as presented in [20] is used in this chapter. The VSC control for the PV units, considered in the case study in the next section to study the interaction, is shown in Fig. 9.4 [20].

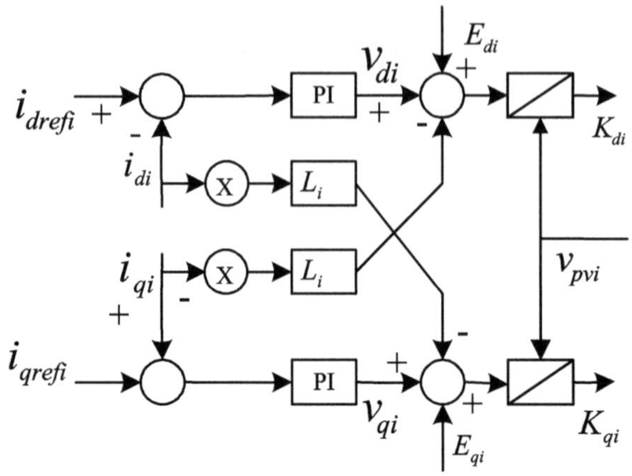

Fig. 9.4 *dq*-frame current control scheme

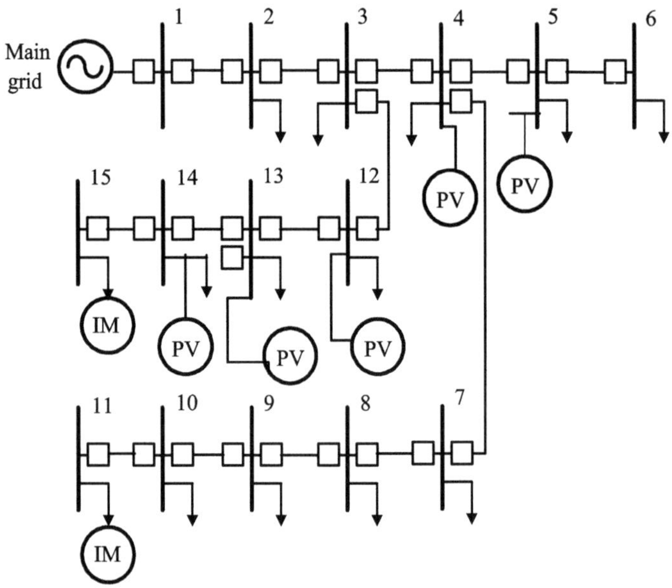

Fig. 9.5 Single line diagram of the test system

9.3 Case Studies

A single-line diagram of the test distribution system, used in this chapter, is shown in Fig. 9.5. The parameters of this system are given in [21]. This test system is a representative distribution feeder of Kumamoto area in Japan [21]. It is a representative of real distribution systems and in the literature it has been used for distribution systems research [21–23]. The system consists of three 6.8 KV feeders which are

Table 9.1 PV system parameters

Parameter	Value
Number of parallel strings in each array	150
Number of modules in each string	80
Module open circuit voltage	43.5 V
Module short circuit current	2.8735 A
Diode ideality factor	1.3
VSC switching frequency	3,000 Hz
DC-link Capacitance	5,000 μF
Filter capacitance	100 μF
Interface resistance	3 mΩ
Interface inductance	100 μH
Filter inductance (L_1 and L_2)	500 μH and 220 μH

connected to the grid through a 69 KV radial line. The total load of the system is 6.301 MW, 0.446 MVAr. Out of the fifteen loads in the distribution system, two are induction motor (IM) loads. The original system is modified by connecting five PV units, each rated at 1.5 MW, at buses 4, 5, 12, 13, and 14. These buses are chosen as they are weak buses which will allow to consider the worst case scenario [23]. Each distribution line is represented by a lumped series RL branch. The maximum real and reactive powers of a PV unit are 1,570 KW and 708 K VAr, respectively. Each PV array consists of 150 parallel strings of 80 modules connected in series. Each module is rated for 43.5V@2.8735 A and the PV array is rated for 3,480V@431 A.

The value of the DC-link capacitor is 5,000 μF and the nominal voltage is 1,582 V. The following levels of solar irradiance are considered to test controllers for robustness to the change in the operating point: 1.0 KW m^{-2}, 900, 800, 700 and 600 W m^{-2}. The PV unit is connected to the PCC using a step up transformer. The parameters for the PV system are shown in Table 9.1 [17]. The rating of the inverter used in this chapter is given as: nominal power 1.5 MW, output voltage (low voltage) 330 VAC and maximum output current per phase is 2,624 A.

Multi-PV Unit Interaction Analysis

There are several measures for quantifying the control interaction in multi-variable systems. The most commonly used measure for interaction study is the relative gain array (RGA) [24, 25]. This measure suffers from some disadvantages, and the major one is that it only considers one frequency at a time and it often provides limited limited into how to design multi-variable controllers. Gramian based interaction measures known as Hankel interaction index array (HIIA) have been introduced to overcome most of the limitations of the RGA technique. The HIIA method overcomes the limitations of the RGA technique when dealing with systems that have interactions varying non-monotonically with frequency [26]. An analytical approach based on normal form analysis technique has been used for interaction measures in power systems [27, 28]. A better understanding of the complex nonlinear interaction phenomenon can be obtained using this index, and it can provide some guidance to design coordinated controllers. Recently, small-signal analysis complemented by nonlinear time-domain simulations has been widely used in power systems for studying control interactions [29–31]. In this chapter the commonly used eigenvalue analysis, quantitative index based on normal form theory and nonlinear simulation are used to study the controller interaction among nearby PV units. Small-signal analysis is chosen as it provides an insight into the problem and quantifies the contribution of the states in a particular mode. The linear analysis is verified using nonlinear simulation under faulted conditions and it is an accepted way to test the performance of nonlinear systems with controllers designed for the linearised models.

In the following three approaches for quantifying dynamic interactions are applied to the distribution system with multiple PV units: (a) small-signal analysis, (b) normal form technique (index-based) and (c) nonlinear simulation.

9.3.1 Small-Signal Analysis

To get an insight into the interaction phenomenon, small signal analysis is carried out. For this analysis, the nominal system is represented by the equations: $\Delta\dot{x} = A\Delta x + B\Delta u$ and $y = C\Delta x + Dy$, where A is the system matrix, B is control or input matrix, C is the output matrix and D is the feed-forward matrix. Eigenvalues of the A matrix are calculated and the dominant mode is the mode with the eigenvalues closest to the imaginary axis with the lowest damping factor. Different scenarios are considered in order to find out the possible interaction among several PV controllers. The scenarios are: (i) one PV unit is connected at bus 13, (ii) two PV units are connected at buses 12 and 13, and (iii) three PV units at buses 12, 13 and 14.

Figure 9.6 shows the position of the dominant eigenvalues with single, two, and three PV units where '.' represents the eigenvalues for a single unit, '*' for two units and '+' for three units. From Fig. 9.6, it can be seen that the addition of the second and the third PV units significantly affect the position of the eigenvalues and the modes are moved towards the right-hand side. From this analysis it can be concluded that the modes are coupled. This process is repeated for four and five units and a further shift of the dominant modes towards the right-hand side is observed.

The effect of changing gain parameters on system interaction is also analyzed. In this study two PV units at buses 12 and 13 are considered. The gain of the PV unit at bus 13 is varied while the gain of the other unit is unchanged. The effect of changing the proportional gain for q-axis current control (shown in Fig. 9.4) of the VSC for the PV unit at bus 13 on system eigenvalues is analyzed. Figure 9.7 shows the dominant eigenvalues with two different gain parameters where '*' and $\star$ represent the eigenvalues for the PV unit at bus 13 with the gain parameters 0.1

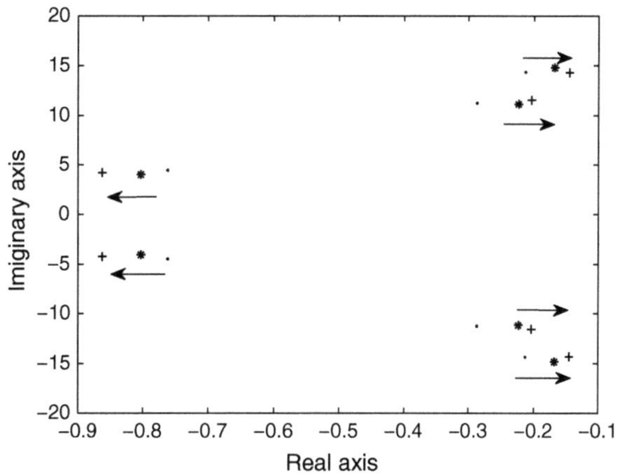

Fig. 9.6 Movement of dominant complex modes of the system with single PV due to additional PV units (marker '*dot*' single, '*asterisk*' two and '*plus sign*' three PV units

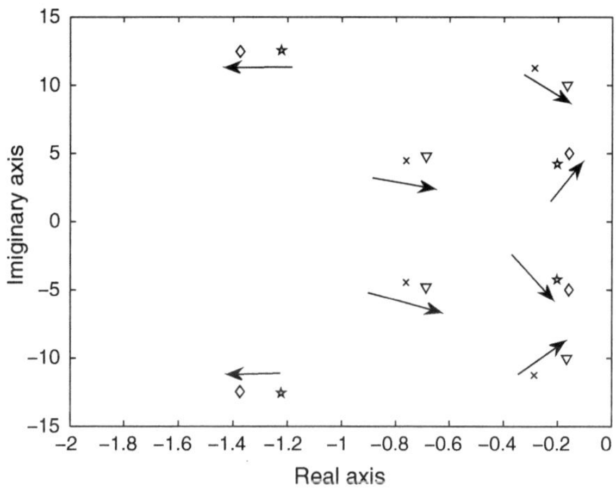

Fig. 9.7 Two PV case (bus 12 and 13): Movement of dominant complex modes of the system with two PV units due to change in gain parameter of one PV unit (bus 13) from 0.1 to 0.5 pu unit (PV bus 13: '*star*' gain 0.1, '*open diamond*' gain 0.5; PV bus 12, '*multiplicationsign*' gain 0.1, '*inverted triangle*' gain 0.5 pu)

pu and 0.5 pu, respectively, whereas '+' and × mark eigenvalues for the PV unit at bus 12 with gain 0.1 pu. It can be seen that increasing the gain of PV unit at bus 13, affects the position of its own eigenvalues and also the position of the eigenvalues of the other unit.

9.3.2 Interaction Index Based on Normal Form Theory

Interaction among generators connected at different nodes in a distribution system is due to power transfer via distribution feeders. The power transfer is a nonlinear function of the states and thus the relative magnitudes of the nonlinear terms can be a good indicator of the interaction among different PV units. An indicator of this relative magnitude is the nonlinear interaction index defined in [32]. The index is obtained by first writing the nonlinear system equations in the normal form as first-order modal equations with a decoupled linear term and quadratic terms including the other modes. The largest coefficient among the quadratic terms is used to obtain the index [32]. In other words the nonlinear interaction index is an indicator of the magnitude of the quadratic terms in the Taylor series expansion. Using normal form representation and initial conditions, the following steps are used to calculate the interaction index [32]:

- Represent the dynamical equations governing the PV units in the general nonlinear form,

$$\dot{X} = f(X), \tag{9.8}$$

where $X = [I_{1d_1}, I_{1q_1}, I_{2d_1}, I_{12q_1}, v_{cfd_1}, v_{cfq_1}, v_{pv_1} \ldots I_{1d_n}, I_{1q_n}, I_{2d_n}, I_{12q_n}, v_{cfd_n}, v_{cfq_n}, v_{pv_n}]^T$, n is the number of PV units and f is an analytic vector field.
- Expand Eq. (9.8) as a Taylor series about the post disturbance stable equilibrium point X_{SEP} and retain up to second order terms:

$$\dot{X} = A_i X + \frac{1}{2} X^T H^i X + \text{higher order terms}, \tag{9.9}$$

where A_i is the ith row of Jacobian A which is equal to $\left[\frac{\delta F}{\delta X}\right]_{SEP}$ and Hessian matrix $H^i = \left[\frac{\delta^2 F_i}{\delta X_j \delta X_k}\right]_{SEP}$
- Represent the second order Taylor series expansion in the Jordan Form variables using $X = UY$ as

$$\dot{y}_{d_j} = \lambda_j + \sum_{k=1}^{n} \sum_{l=1}^{n} C_{kl}^j y_{d_k} y_{d_l}, \tag{9.10}$$

where $C^j = \frac{1}{2} \sum_{p=1}^{n} V_{jp}^T [U^T H^P U] = [C_{kl}^j]$. The normal form transformation is then given by:

$$Y = Z + h2(Z), \tag{9.11}$$

where $h2^j(z) = \sum_{k=1}^{n} \sum_{l=1}^{n} h_{kl}^j z_k z_l$ and $h2_{kl}^J = \frac{C_{kl}^j}{\lambda_k + \lambda_l - \lambda_j}$,

Table 9.2 Nonlinear index

(a) PV units	Index	(b) Solar radiance (W m^{-2})	Index
One	0.578	600	2.56
Two	3.45	700	2.94
Three	7.940	800	3.08
Four	13.650	900	3.25
Five	15.44	1,000	3.45

- Determine the initial condition in the Jordan form variables from the relation $Y_0 = U^{-1}X_0$. Then solve the the following equation to obtain the initial conditions of normal form variables:

$$\min f(Z_0) = Z_0 + h2(Z_0) - Y_0. \tag{9.12}$$

- Finally, the nonlinear interaction index is given by:

$$\text{index}(j) = |\max_{k,l}\left(h2^j_{kl} z_{k0} z_{l0}\right) / z_{j0}| \tag{9.13}$$

where $h2^j_{kl}$ indicates the strength of the nonlinear interaction between modes, i.e., it shows the interaction of mode k and mode l to mode j and $Z_{I_{k0}}$ is the initial condition for the mode k.

The interaction index for two different cases is listed in Table 9.2: (a) increase in the number of PV units under constant solar irradiance (1,000 W m^{-2}) and (b) the change in solar irradiance with two PV units (buses 12 and 13) in the distribution system. The interaction indices given in Table 8.2 are for the dominant control modes and the control mode is calculated from the participation factor analysis. It is also found that the nonlinear participating states of this mode include control states of several PV units. The index is a good indicator of the interaction among PV units as can be seen from the correlation between the the numerical values in Table 9.2 and the simulation plots in Fig. 9.8. The larger is the interaction index the less is the damping in the mode indicating that due to control interaction the modes have shifted towards the right-hand side.

9.3.3 Time-Domain Simulation

Time-domain simulations are carried out to confirm the results from the small-signal and normal formal analyses. Each PV unit is equipped with the controller shown in Fig. 9.4. A three-phase fault is applied at bus 15 at 0.5 s and sequently cleared after 0.1 s. Figure 9.8 shows the power response of the PV unit at bus 13 with single and

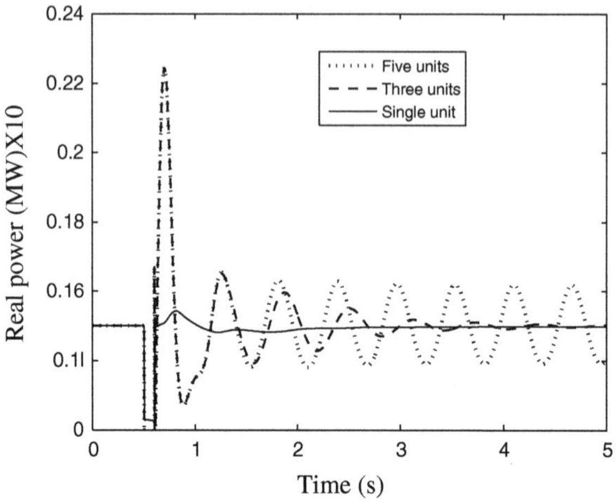

Fig. 9.8 Power response with single and multiple PV units

multiple PV units. Bus 13 is chosen due to its proximity to PV units and dynamic loads. From Fig. 9.8 it is clear that a negative interaction is possible for multiple uncoordinated PV controllers located in close proximity to each other. The results obtained by small-signal and normal form analyses show a good agreement with time-domain simulation results using nonlinear models. As strong interactions among multiple PV units exist, the coordination among controllers must be considered to ensure stable operation. A properly designed robust controller explained later can alleviate this negative interaction and enhance damping performance.

9.4 Problem Formulation for Noninteractive Control

Adequate damping of weak modes is essential to improve the stability and increase the safe penetration limit of PV units in distribution systems. As the PV model given in Eq. (9.7) is nonlinear, the linear controller designed with linear model cannot ensure stability under severe fault and other stressed operating conditions. In this chapter to design the robust controller, the PV system model is reformulated to capture the full nonlinearity of systems in the region of interest. The power angle and control interactions and the design of robust H_∞ controllers is considered in the problem formulation. The PV system described by (9.7) can be represented by the following equations for a given operating condition:

$$\Delta\dot{x}_i = (A_i + \Delta A_i)\,\Delta x_i + (B_i + \Delta B_i)\,\Delta u_i + \sum_{j=1}^{n} G_{ij}\Delta x_j, \qquad (9.14)$$

$$y_i = C_i\Delta x_i + D_i y_i,$$

where Δx_i is the state vector (7×1), Δu_i is the control input (2×1), y_i is the measured output (2×1) of the ith subsystem. In this case,

$\Delta x_i = [\Delta i_{1d_i}, \Delta i_{1q_i}, \Delta i_{2d_i}, \Delta i_{2q_i}, \Delta v_{cfd_i}, \Delta v_{cfq_i} \Delta v_{pv_i}]^T$, $\Delta u_i = [\Delta K_{d_i}, \Delta K_{q_i}]$ and $y_i = [P_i, Q_i]$. The matrices A_i, B_i, C_i and D_i are constant and known, expressing the nominal model. The order of the matrices are: $A_i = 7 \times 7, B_i = 7 \times 2$ and $C_i = 2 \times 7$. The summation on the right-hand side represents interconnections with the neighboring N subsystems. The matrices ΔA_i and ΔB_i represent unmodeled part of the nonlinear terms. These are the higher order terms, known as the Cauchy remainder, in the Taylor series:

$$(L_i - A_i)\,\Delta x_i + (M_i - B_i)\,\Delta u_i = \begin{bmatrix} L_{2i} \\ L_{3i} \end{bmatrix} \times$$

$$F_i(t)\left[E_{2i}\ E_{3i},\right] L_i = \left[\frac{\partial f_{i1}}{\partial x_i}\bigg|_{\substack{x_i = x_i^{*1} \\ u_i = u_i*1}}, \ldots, \frac{\partial f_{iN}}{\partial x_i}\bigg|_{\substack{x_i = x_i^{*N} \\ u_i = u_i*N}} \right]^T,$$

$$M_i = \left[\frac{\partial f_{i1}}{\partial u_i}\bigg|_{\substack{x_i = x_i^{*1} \\ u_i = u_i*1}}, \ldots, \frac{\partial f_{iN}}{\partial u_i}\bigg|_{\substack{x_i = x_i^{*N} \\ u_i = u_i*N}} \right]^T.$$

It is difficult to represent the matrices (L_i and M_i) symbolically but it is straightforward to evaluate them numerically at a given point and this is what we need to do to calculate the bounds on the uncertainty. The matrix A_i is calculated at nominal operating points and L_i at arbitrary points to model the uncertainty. For example, if only the d-axis current $\dot{i}_{1d_i}$ of Eq. (9.7) is considered, then $A_i = \left[-\frac{R_i}{L_i}, \omega_i L_{1_i}, 0, 0, 0, 0, \frac{K_{d_i0}}{L_i}\right]^T$ and $L_i - A_i == \left[0, 0, 0, 0, 0, 0, \frac{K^*_{d_i}}{L_i} - \frac{K_{d_i0}}{L_i}\right]^T$. The complete description of this reformulation technique can be found in [33] and [34]. The interaction terms are restructured as $G_{ij} = L_{1i} F_{ij}(t) E_{1j}$. All the matrices are known except $F_{ij}(t)$ and $F_i(t)$, however they are assumed to satisfy [35]: $\|F_{ij}(t)\| \le f_{ij}$ and $\|F_i(t)\| \le 1$, where the norms represent the maximum singular values of the indicated matrices and f_{ij} is a positive scalar representing the magnitude of the interactions.

The subsystem interaction can be represented as external disturbances. Equation (9.14) can be rewritten as:

$$\begin{aligned} \Delta \dot{x}_i &= A_i \Delta x_i + B_i \Delta u_i + \begin{bmatrix} \sigma_i L_{2i} \\ \gamma_i L_{3i} \end{bmatrix} w_i, \\ y_i &= C_i \Delta x_i + \begin{bmatrix} \sigma_i L_{3i} \\ 0 \end{bmatrix} w_i + D_i y_i, \\ z_i &= \begin{bmatrix} \frac{1}{\sigma_i} E_{2i} \\ E_{1i} \end{bmatrix} x_i + \begin{bmatrix} \frac{1}{\sigma_i} E_{3i} \\ 0 \end{bmatrix} u_i. \end{aligned} \tag{9.15}$$

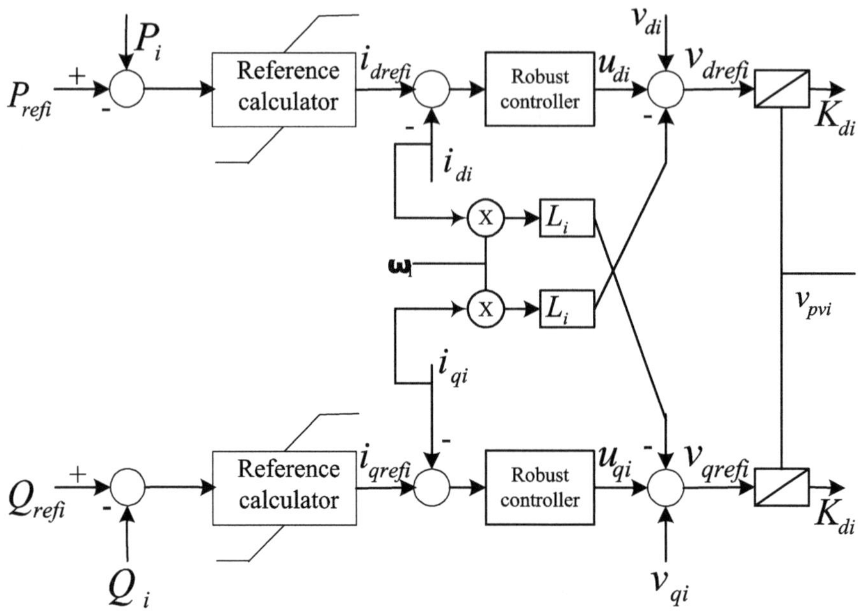

Fig. 9.9 Proposed PV control strategy

where w_i represents disturbance and z_i is the control output. The new formulation presented in this section is used with the H_∞ output-feedback control theory to design decentralized controllers for PV systems.

9.5 PV Control Design

A simulation is carried out to find out the coherent PV units and form subsystems consisting of those PV units in order to reduce the calculation burden. Based on the coherency analysis the system is divided into three subsystems, (a) subsystem 1: PV at buses 13 and 14, (b) subsystem 2: PV at bus 12 and (c) subsystem 3: PV at buses 4 and 5. One decentralized controller is designed for each subsystem which only requires local information. The proposed inverter control strategy is shown in Fig. 9.9. In the voltage control mode of PV units, P_i and Q_i are controlled by the amplitude of the VSC terminal voltage. The error signals $P_{\mathrm{ref}_i} - P_i$ and $Q_{\mathrm{ref}_i} - Q_i$ are fed to the controller which produces d- and q-axis components of the VSC current at their respective reference values (i_{dref_i}, i_{qref_i}), which are again processed to get reference voltage components (v_{dref_i}, v_{qref_i}). These two signals are then divided by v_{pv_i} to generate K_{d_i} and K_{q_i} for the pulse width modulation (PWM). Real power reference for the proposed control is determined by the maximum power point control [36] and reactive power reference from the steady-state solution.

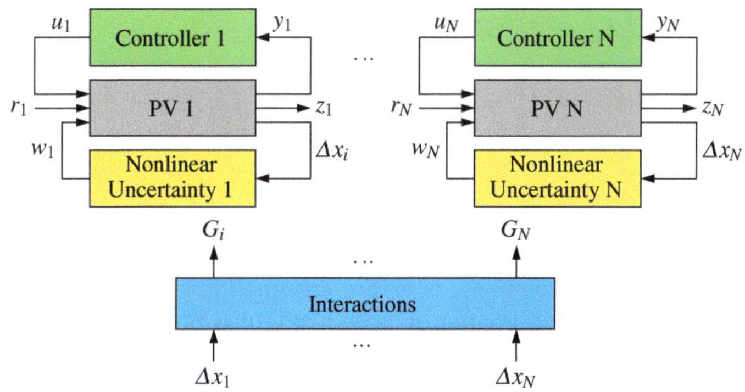

Fig. 9.10 Block diagram of the uncertain system

The complete dynamic system is linearised around the equilibrium point. From linearised small-signal model matrices, A_i, B_i and C_i are obtained. The most challenging problem is to bound the uncertainty from the nonlinearity in PV system models. The nonlinear converters of PV units are modeled as a linear part plus a nonlinear part, the nonlinear term being the Cauchy remainder term in the Taylor series expansion. Bounds derived for this Cauchy remainder term are used to define an uncertain linear model for which a robust control design is performed. The calculation procedure can be found in [33, 34]. An advantage of this approach over the existing linearisation scheme is the treatment of the nonlinear PV model in a rigorous framework for the control design. The interaction among different subsystems are calculated as $\|G_{ij}\|$ using Gramian based interaction measure [26].

The block diagram of the decentralized controller is shown in Fig. 9.10. A decentralized controller can stabilize the systems (9.15) with unitary H_∞ disturbance attenuation [35, 37]. The linear output-feedback controller is given as [37]:

$$\dot{x}_{c_i} = A_{c_i}x_{c_i} + B_{c_i}y_i(t), \ u_i = C_{c_i}x_{c_i} + D_{c_i}y_i(t), \tag{9.16}$$

where

$$\begin{aligned} A_{c_i} &= A_i + B_iC_{c_i} - B_{c_i}C_i + (\gamma_i^2 L_{1i}L_{1i}^T + \\ &\quad \sigma_i^2 L_{2i}L_{2i}^T)X_i - \left(Y_i^{-1} - X_i\right)^{-1}\left(X_iB_iC_{c_i} + Q_i\right), \\ B_{c_i} &= \frac{\beta_i}{2}\left(Y_i^{-1} - X_i\right)^{-1} C_i^T, \\ C_{c_i} &= \frac{\alpha_i}{2}B_i^T X_i \text{ and } D_{c_i} = 0, \end{aligned} \tag{9.17}$$

and X_i and Y_i are solutions of the following algebraic Riccati equations

$$A_i^T X_i + X_i^T - \alpha_i X_i B_i B_i^T X_i + \gamma_i^2 X_i L_{1i} L_{1i}^T +$$

$$
\begin{aligned}
&E_{1i}^T E_{1i} + \sigma_i^T X_i L_{2i} L_{2i}^T X_i + \frac{1}{\sigma_i^2} E_{2i}^T E_{2i} + Q_i = 0, \\
&A_i Y_i + Y_i A_i^T - \beta_i Y_i C_i^T C_i Y_i + \gamma_i^2 L_{1i} L_{1i}^T + \\
&Y_i E_{1i}^T E_{1i} + \frac{1}{\sigma_i^2} Y_i E_{2i}^T E_{2i} Y_i + \sigma_i^T L_{2i} L_{2i}^T + R_i = 0,
\end{aligned}
\tag{9.18}
$$

where α_i, β_i, γ_i and σ_i are positive numbers, Q_i and R_i are positive definite matrices and $Y_i^{-1} > X_i$. The parameters α_i, β_i and σ_i are small positive numbers which are chosen by successive approximation in such a way that both the Riccati equations have positive definite solutions. The parameter γ_i is determined from the relation, $||C_i(SI - A_i)^{-1}B_i|| < \gamma_i$ and it is generally less than 1. In this chapter we have chosen $\alpha_i = 0.0005$, $\beta_i = 0.0001$, $\sigma_i = 0.0025$ and $\gamma_i = 0.694$. An algorithm to design non-interacting controllers for PV systems using the above formulation is presented in the next section.

9.6 Control Design Algorithm and Performance Evaluation

The proposed control algorithm can be applied to a general distribution system including radial and meshed systems. In this chapter the PV control scheme is based on the reformulation of the nonlinear dynamics of the PV units using the mean-value theorem. With this representation, it becomes easier to explicitly account for the effect of nonlinearities in the system dynamics, which enables us to more accurately represent the system and also provides guaranteed robust performance. It also significantly reduces the conservativeness of the designed controllers. The designed controller guarantees stability under wide varieties of operating conditions.

The design steps of the output-feedback decentralized controllers for the nominal system are given below:

- model the system using Eq. (9.7) and solve it for the steady-state conditions;
- from simulations, select coherent groups of generating units and represent them by equivalent models;
- determine the best input/output variables for the controller from the residues which are the product of system controllability and observability Gramians;
- linearise the system with the best input and output signals and determine nominal system matrices (A_i, B_i, C_i);
- Reformulate the nonlinear system using the technique proposed in Sect. 9.4 and determine the other matrices given in Eq. (9.15);
- quantify the uncertain terms $(L_i - A_i)$ and $(M_i - B_i)$ due to variations in solar irradiance, changing electric loads, and physically close PV units and check that $F_i^T F_i \leq I$ where I is the identity matrix;
- determine the interaction among different subsystems;
- solve the Riccati equations given in (9.18); and
- design the controller given by Eq. (9.16) and (9.17).

Table 9.3 Dominant closed loop eigenvalues

$-1.95 \pm j5.56$	$-3.75 \pm j2.76$	$-2.64 \pm j9.39$	$0.44 \pm j3.47$

Table 9.4 Nonlinear index with robust controllers

PV units	Index	Faulted bus	Index
One	0.4618	12	0.4625
Two	0.4639	13	0.4613
Three	0.4664	14	0.4637
Four	0.4689	15	0.4628
Five	0.4692	16	0.4642

The control algorithm given above is applied to a benchmark distribution test system. The controller is designed offline and applied to the system. The computational burden in calculating the uncertainty bound and controller design is not of concern as it is done offline. The order of the controller is exactly equal to the order of the PV model. Table 9.3 shows only the dominating eigenvalues for the closed-loop system. The critical mode for the closed-loop system is $0.44085 \pm j3.4737$ and the damping ratio is 0.1269 which is well above the minimum acceptable damping ratio of five percent. From the eigenvalue it is clear that the closed-loop system is well-damped. The nonlinear index presented in Sect. 9.3 is also used to validate the performance of the designed controller under different operating conditions. Table 9.4 shows the nonlinear index for a three-phase fault for five cycles at different locations of the network with five PV units. From Table 9.4 it is clear that with the designed controller the variation of the index in the critical mode is small and can be neglected.

The next section contains results of large disturbance simulations, carried out to demonstrate the performance of the designed controller. The simulations are done with the full nonlinear model and robustness is verified with the complete nonlinear model under different operating conditions including severe three-phase fault close to the PV bus. The controller is designed simultaneously for five PV units and the performance is tested under different operating conditions: (a) three-phase faults, (b) sudden load changes, (c) changes in reference points, and (d) rapid changes in solar irradiations.

9.6.1 Three-Phase Fault

Firstly, the performance of the designed controller is verified for a severe three-phase fault at bus 15. The fault is applied at 1 s and subsequently cleared after 150 ms by opening the circuit breaker. Figures 9.11 and 9.12 show the terminal voltage response of the PV unit at bus 13 and its reactive power response. It can be observed that the voltage is reduced during the fault and generated reactive power is

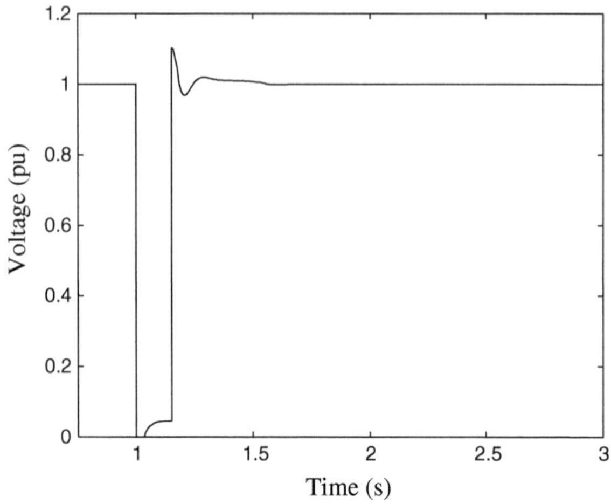

Fig. 9.11 Voltage response (bus 13) for three-phase fault at bus 15

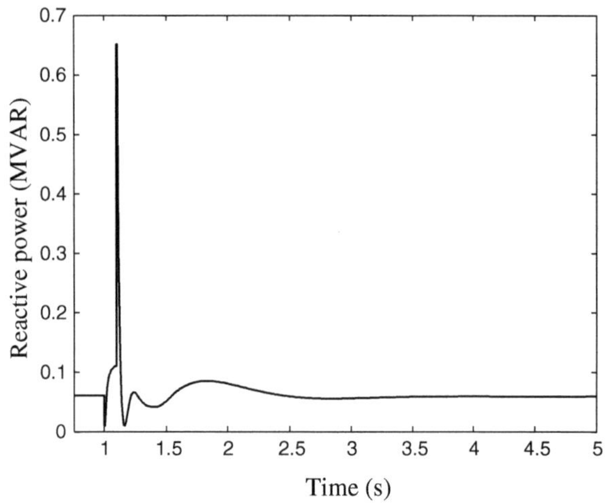

Fig. 9.12 Reactive power of PV unit (bus 13) for three-phase fault at bus 15

increased. The designed controllers reacts promptly when the fault is removed and the system regains synchronous operation. From Figs. 9.11 and 9.12 it is evident that the designed controller provides excellent performance in terms of overshoot, rise time, settling time, oscillation and steady-state error. Figure 9.13 shows the real power response of the PV unit at bus 13. The simulation result presented in Fig. 9.8 for the multi-PV units with the same fault exhibits negative interaction. From Figs. 9.13 and 9.8 it can be concluded that the designed controllers do not have any unwanted negative interaction as they coordinate with each other.

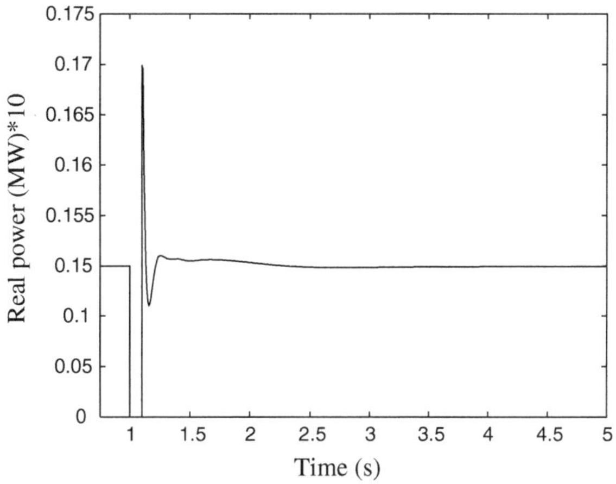

Fig. 9.13 Real power response for PV unit (bus 13) for three-phase fault at bus 15

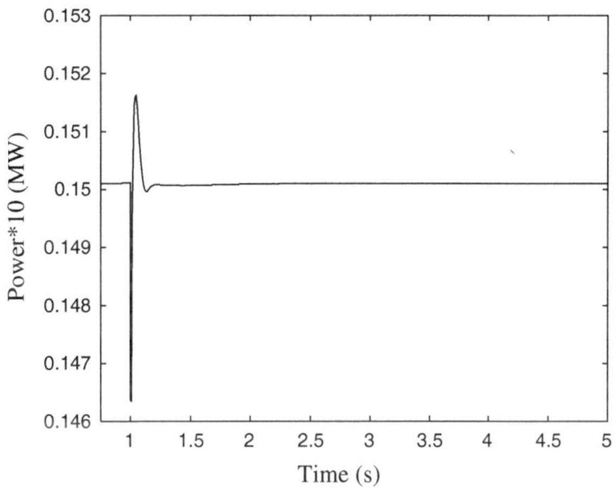

Fig. 9.14 Real power of PV unit (bus 4) for 10 % change in load

9.6.2 Sudden Change in Connected Load

The performance of the controller is also tested for a comparatively less severe disturbance. In this simulation, the two IMs at buses 11 and 15 are represented with a detailed nonlinear model as given in [33]. In this case, a 10 % sudden change in the connected load is applied to verify the dynamic performance of the controller. Figures 9.14 and 9.15 show the real power of the PV unit at bus 4 and its terminal voltage response for ten percent changes in connected loads. From Figs. 9.14 and 9.15, it is obvious that the designed controller ensures stability at different operating conditions with mix of static and dynamic loads and provides strong robustness

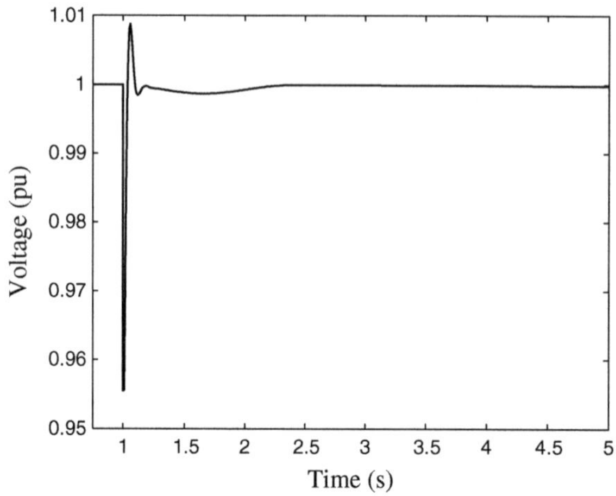

Fig. 9.15 Voltage excursion (bus 4) for 10 % change in load

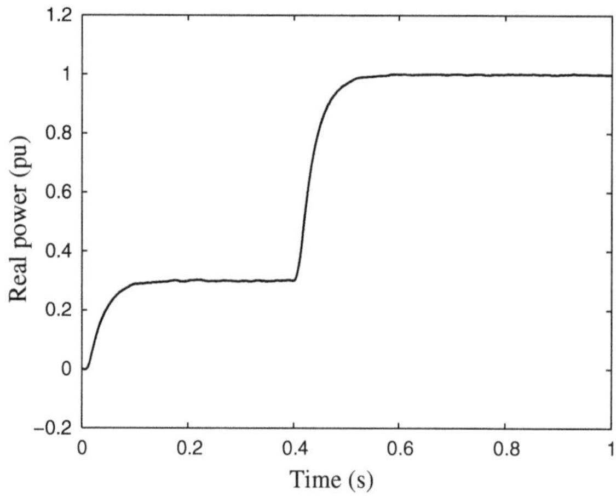

Fig. 9.16 Real power (bus 12) for change in reference power

against sudden load variations. From the above analysis, it is clear that the designed controller ensures stability for both small and large signal disturbances and minimizes the negative interactions.

9.6.3 Change in the Reference Point

It is expected that the designed controller will track the reference value under different operating conditions. In this case, initially the reference power is 0.3 p.u. and at 0.4 s

Table 9.5 Nonlinear index with changing radiation

Solar radiance in W/m^2	1,000	900	800	700	600
Index	0.4692	0.4689	0.4691	0.4681	0.4659

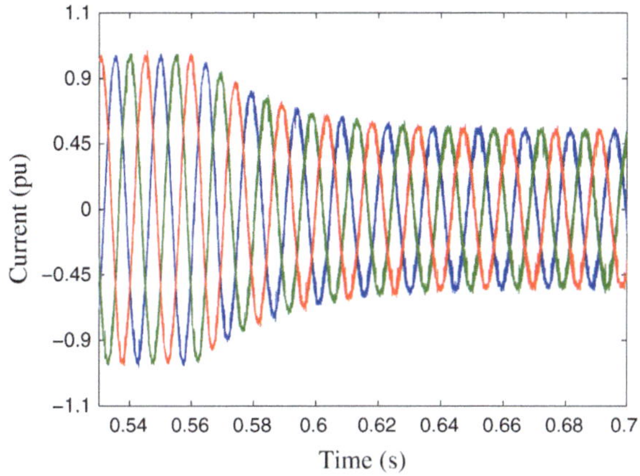

Fig. 9.17 PV system (bus 4) current due to the change in radiation

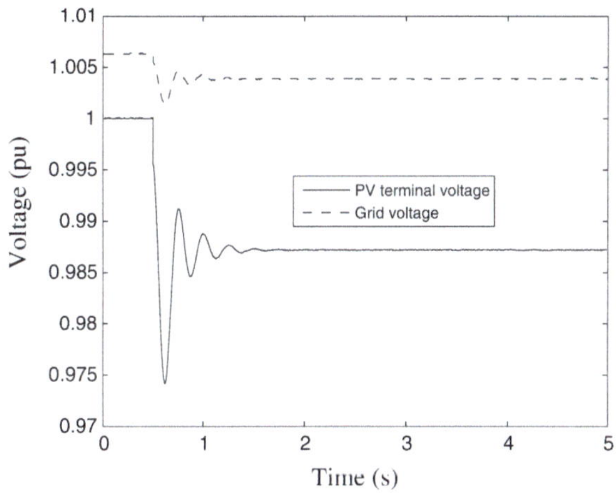

Fig. 9.18 Bus voltage (bus 4) due to the change in radiation

the reference power is increased gradually to 1.0 p.u. by the MPPT controller for the PV unit at bus 12. Figure 9.16 shows the real power of that particular PV unit. From Fig. 9.16 it is clear that the controller follows the reference value exactly and provides excellent tracking performance.

9.6.4 Rapid Change in Sunlight

In a practical PV system, the atmospheric condition changes continuously for which the PV output, the output current, and power of PV unit changes significantly. As sunlight changes, the decentralized robust control makes the PV output converge to new operating points. Table 9.5 shows the index with the changes in solar radiation and by comparing it with the index in second column of Table 9.2, it is evident that the designed controller provides excellent performance in minimizing the interaction. The change in grid current is shown in Fig. 9.17 from which it is seen that the PV system operates at standard atmospheric conditions till 0.5 s. However, the irradiance changes from 1,000 to 700 W m^{-2} at 0.55 s and the weather remains cloudy. At this stage, the amount of power delivered to the grid change and MPPT will select a different MPP. Figure 9.17 shows the performances of the designed dynamic controller with the change in atmospheric conditions. Figure 9.18 shows the voltage response for the PV unit at bus 4 and grid voltage. From Figure 9.18 it is evident that the PV terminal voltage slightly reduces and become stable at another operating point as it is a permanent disturbance. However, the grid voltage does not change significantly. From the results it is clear that the designed controller performs well at varying atmospheric conditions and provides robust performance.

9.7 Conclusions

The analysis in this chapter has shown that distribution systems with PV units in close proximity are likely to have oscillations due to control interactions. It is important to minimize the negative interactions for secure and reliable operation of the power system. To avoid the oscillations, a robust decentralized control design approach can be used for which it is important to identify the coherent PV units as this enables the design process to be less complex and less conservative with lower order controllers. A properly designed robust controller damps small-signal oscillations and maintains a stable operation in post-fault conditions even with severe contingencies. The emerging trend of high penetration of PV units needs novel control design for stable operation. The idea presented in the chapter can be extended to other DER units which are placed in close proximity.

References

1. Liserre M, Sauter T, Hung JY (2010) Future energy systems: integrating renewable energy sources into smart power grids through industrial electronics. IEEE Indus Electron Mag 4(1):18–37
2. Mather B, Neal R (2012) Integrating high penetrations of PV into Southern California: year 2 project update. In: 38th IEEE photovoltaic specialists conference, pp 737–741

3. Casey LF, Schauder C, Cleary J, Ropp M (2010) Advanced inverters facilitate high penetration of renewable generation on medium voltage feeders-impact and benefits for the utility. In: IEEE conference on innovative technologies for an efficient and reliable electricity supply, pp 86–93
4. Ravindra H, Faruque MO, Schoder K, Steurer M, McLaren P, Meeker R (2012) Dynamic interactions between distribution network voltage regulators for large and distributed pv plants. In: IEEE PES transmission and distribution conference and exposition, pp 1–8
5. Hossain MJ, Saha TK, Mithulananthan N, Pota HR, Lu J (2012) Dynamic interactions among multiple der controllers in distribution systems. In: IEEE PES power system technology. Auckland, New Zealand, pp 1–6
6. Dent CJ, Ochoa LF, Harrison GP (2012) Network distributed generation capacity analysis using OPF with voltage step constraints. IEEE Trans Power Syst 25(1):296–304
7. Katiraei F, Iravani M (2006) Power management strategies for a microgrid with multiple distributed generation units. IEEE Trans Power Syst 21(4):1821–1831
8. Pota HR (2013) Droop control for islanded microgrids. In: IEEE power engineering society general meeting, Vancouver, Canada, pp 1–5
9. Kashem A, Ledwich G (2005) Multiple distributed generators for distribution feeder voltage support. IEEE Trans Energy Convers 20(3):676–684
10. Diaz G, Gonzalez-Moran C, Gomez-Aleixandre J, Diez A (2009) Complex-valued state matrices for simple representation of large autonomous microgrids supplied by PQ and VF generation. IEEE Trans Power Syst 24(4):1720–1730
11. Kwon JM, Kwon BH, Nam KH (2008) Three-phase photovoltaic system with three-level boosting MPPT control. IEEE Trans Power Electron 23(5):2319–2327
12. Hossain MJ, Mithulananthan TKSN, Pota HR (2012) Robust control strategy for PV system integrations in distribution systems. Appl Energy 99:355–362
13. Vaccaro A, Velotto G, Zobaa AF (2011) A decentralized and cooperative architecture for optimal voltage regulation in smart grids. EEE Trans Ind Electron 58(10):4593–4602
14. Etemadi AH, Davison EJ, Iravani R (2012) A decentralized robust control strategy for multi-DER microgrids-part II: performance evaluation. EEE Trans Power Delivery 27(4):1854–1861
15. Mohamed YARI, El-Saadany EF (2008) Adaptive decentralized droop controller to preserve power sharing stability of paralleled inverters in distributed generation microgrids. IEEE Trans Power Electron 23(6):2806–2816
16. Mahmud MA, Pota HR, Hossain MJ (2012) Dynamic stability of three-phase grid-connected photovoltaic system using zero dynamic design approach. IEEE J Photovoltaics 12(4):564–571
17. Mahmud MA, Pota HR, Hossain MJ (2011) Modeling guidelines and a benchmark for power system simulation studies of three-phase single-stage photovoltaic systems. IEEE Trans Power Deliv 26(2):1247–1264
18. Tan YT, Kirschen DS, Jenkins N (2004) A model of PV generation suitable for stability analysis. IEEE Trans Energy Convers 19:748–755
19. Nosrat A, Pearce JM (2011) Dispatch strategy and model for hybrid photovoltaic and tri generation power systems. Appl Energy 88(9):3270–3276
20. Yazdani A, Dash PP (2009) A control methodology and characterization of dynamics for a photovoltaic (pv) system interfaced with a distribution network. IEEE Trans Power Deliv 24(3):1538–1551
21. Li S, Tomsovic K, Hiyama T (2000) Load following functions using distributed energy resources. In: IEEE/PES summer meeting, Seattle, WA, pp 1756–1761
22. Zhu Y, Tomsovic K (2010) Real-time control of distributed energy resources. In: IEEE Power and Energy Society general meeting, Minneapolis, MN, pp 1–6
23. Mahmud MA, Hossain MJ, Pota HR (2013) Voltage variation on distribution networks with distributed generation: worst case scenario. IEEE Syst J. doi:10.1109/JSYST.2013.2265176
24. Cai LJ, Erlich E (2006) Identification of the interactions among the power system dynamic voltage stability controllers using relative gain array. In: IEEE PES power systems conference and exposition, San Antonio, Texas, USA, pp 970–977

25. Milanovic JV, Duque ACS (2004) Identification of electromechanical modes and placement of PSSs using relative gain array. IEEE Trans Power Syst 19(1):410–417
26. Halvarsson B (2008) Comparison of some gramian based interaction measures. In: IEEE international symposium on computer-aided control system, San Antonio, Texas, USA, pp 138–143
27. Gibbard MJ, Vowles DJ, Pourbeik P (2000) Interactions between, and effectiveness of, power system stabilizers and FACTS device stabilizers in multimachine systems. IEEE Trans Power Syst 15(2):748–755
28. Zou ZY, Jiang QY, Cao YJ, Wang HF (2005) Normal form analysis of interactions among multiple SVC controllers in power systems. IEE Proc-Gener Transm Distrib 152(4):469–474
29. Clark K, Fardanesh B, Adapa R (1995) Thyristor controlled series compensation application study-control interaction considerations. IEEE Trans Power Deliv 10(2):1031–1037
30. Kim HJ, Nam T, Hur K, Chang B, Chow JH, Entriken R (2011) Dynamic interactions among multiple FACTS controllers—a survey. In: IEEE power and energy society general meeting, pp 1–8
31. Martins N, Corsi S, Andersson G, Gibbard MK, Silva JJSGA, Taranto GN, Impact of the interaction among power system controls. Technical report, Status report of CIGRE TF 38.02.16
32. Thapar J, Vital V, Kliemann W, Fouad AA (1997) Application of the normal form of vector fields to predict interarea separation in power systems. IEEE Trans Power Syst 12(2):844–850
33. Hossain MJ, Pota HR, Ugrinovski V, Ramos RA (2010) Voltagemode stabilisation in power systems with dynamic loads. Int J Electr Power Energy Syst 32(8):911–920
34. Hossain MJ, Saha TK, Mithulananthan N, Pota HR (2013) Control strategies for augmenting LVRT capability of dfigs in interconnected power systems. IEEE Trans Ind Electr 60(6):2510–2522
35. Petersen IR, Ugrinovskii VA, Savkin AV (2000) Robust control design using H_∞ Methods. Springer, London
36. Lalili D, Mellit A, Lourci N, Medjahed B, Berkouk EM (2011) Decentralized quadratic stabilization of large-scale systems. Renew Energy 36(12):3282–3291
37. Zhai G, Yasuda K, Ikeda M (1994) Decentralized quadratic stabilization of large-scale systems. In: IEEE conference on decision and control, FL, USA, pp 2337–2339

Chapter 10
Conclusions

This chapter summarises the findings from this research study, presents the conclusions drawn and discusses proposed future areas of research. This work will become more relevant as developed countries look to become energy-independent and more conscious of the negative environmental impacts of increasing emissions from conventional power plants.

As the penetration levels of wind farm capacities in power systems increase, more research is required. Also when sufficient data for actual wind farm production becomes available, the developed wind power prediction tools and the wind farm model can be further examined and analysed for the prediction of large-scale wind farm production. As wind power generation is characterised by its variability and uncertainty, the integration of wind facilities into utility grids has several impacts on their optimum power flow, transmission congestion, power quality issues, system stability, load dispatch, protection system, economic analysis and electricity market clearing prices. These impacts present major challenges to power system operators. This book tackles some of these challenges.

This book presents several case studies for capturing the mechanisms of voltage instability caused by the dynamics of the induction motor (IM) load, the on-load tap changer (OLTC) and the fixed-speed induction generator (FSIG). The case studies conducted are on: (i) short-term voltage instability; (ii) long-term voltage instability; (iii) the effects of different wind generator integration on the available (dynamic) transfer capability (ATC); (iv) the compensation required to restore or augment the ATC when it is reduced to allow for fixed-speed wind turbine (FSWT) integration; (v) the effects of large-scale wind generator integration; (vi) the effects of high compensations; and (vii) the effects of wind turbine integration and static synchronous compensator (STATCOM) ratings on the low-voltage ride-through (LVRT) capability of wind generators. The effects of different power system devices on power system dynamic behaviours are also analysed. The devices considered in this book are synchronous generators, induction generators (IGs), exciters, on-load tap changers (OLTCs), induction motors (IMs) and flexible AC transmission system (FACTS) devices. The analyses provide insights into the ways in which short- and long-term

J. Hossain and H. R. Pota, *Robust Control for Grid Voltage Stability: High Penetration of Renewable Energy*, Power Systems,
DOI: 10.1007/978-981-287-116-9_10,

voltage instabilities occur in a power system. The effect of large-scale wind power integration into power system has been investigated by modal analysis as well as by detailed nonlinear simulations.

The central contribution of this book is the design of linear robust controllers which enhance voltage instability while preserving the nonlinearity of power system devices. This is achieved by reformulating the nonlinear power system model using Cauchy's remainder formula with mean-value theorem.

The performances of the proposed controllers are validated through simulations. Different test case power systems are selected and controllers are designed for them. Different of simulation cases are conducted which include load and generation changes, fault conditions and reference voltage changes. The test systems considered here include: (i) a simple but representative single generator and single large IM infinite-bus system; (ii) an 11 bus three generators 2 area system; (iii) a large wind farm infinite-bus system; (iv) an IEEE benchmark 10 machine 39 bus New England system; and (v) a 16 machine 69 bus New England and New York test system. The performances of the proposed controllers are also compared with those of conventional controllers simulated under similar conditions. The simulation results show that the proposed controllers are capable of providing better responses during normal and abnormal power system operating conditions.

As mentioned earlier, (i) the minimax linear quadratic gaussian (LQG), (ii) decentralised minimax output-feedback LQ, and (iii) the output-feedback linear robust control approaches are used to design robust controllers. Before designing the controllers, modal analysis and different case studies are conducted to obtain information about the control problems. The control objectives considered here are: (i) stabilisation of a voltage mode with dynamic loads, (ii) enhancement of the ATC, (iii) stabilisation of both voltage and electromechanical modes, and (iv) augmentation of LVRT capability of wind generators. Within the minimax optimal control design framework, robustness is achieved via optimisation of the worst-case quadratic performance in the underlying uncertain system. This helps to achieve an acceptable trade-off between control performance and system robustness.

From this work, the following conclusions can be drawn:

- Dynamic load modelling is crucial for both short- and long-term dynamic voltage instability analyses.
- The modes related to operating slip and shaft stiffness are critical as they may limit the large-scale integration of wind generation.
- The interactions between dynamic load and generators should be taken into account while designing excitation controller.
- As the penetration level of fixed-speed conventional wind turbines increases, the dynamic transfer capability substantially decreases.
- The STATCOM with reduced capacitors provides better performance in terms of transfer capability than a capacitor bank alone.
- The STATCOM controller designed with interconnection effects and uncertainty, increases transfer capability significantly compared to conventional STATCOM controller.

- The critical speed and voltage, as well as the fault ride-through capability of the wind generator decrease with the integration of fixed-speed wind generators.
- The robust performance of the proposed STATCOM control during a significant voltage sag allows the voltage at the point of common coupling (PCC) to be kept stable. It also avoids a large increase in turbine speed, reducing the likelihood of the turbine exceeding the critical speed and thus increasing the fault-ride through capability.
- The STATCOM control with structured uncertainty representation reduces the calculation burden and produces a less conservative controller.
- The combined strategy of robust STATCOM and pitch angle control is more effective in recovering the system operation and such combined control makes the system ride-through the fault without having to disconnect the generators from the system.
- The designed decentralised STATCOM control with supercapacitor increase both the transient and voltage stability and, thus, low-voltage ride-through capability of fixed-speed wind turbines.
- Distribution systems with PV units in close proximity will have oscillations due to control interactions. The emerging trend of high penetration of PV units needs novel control design for stable operation.
- With the proposed reformulation, it becomes easier to explicitly account for the effect of nonlinearities in the system dynamics, which enables us to more accurately represent the system and also provides guaranteed performance and stability characteristics over a pre-specified region around the equilibrium point.

10.1 Directions for Future Research

Although this research achieved promising results in analysing dynamic voltage instability with wind generators and in applying robust control techniques to the design of control systems for the enhancement of dynamic voltage stability, the work does not end here. The proposed power system design method may be further improved and consolidated by the following processes

(i) Implementing the proposed controllers in a real power system will provide more confidence in the proposed method.
(ii) As the synthesis of controllers for large power systems involving many generators is found to be infeasible because the proposed linear matrix inequality (LMI) optimisation method does not yield solutions for large systems, developing a better numerical solution method could facilitate controller design for large systems.
(iii) Treating loads connected to the power system as proper combinations of both constant and dynamic loads could achieve improvement in the model.
(iv) Computing the power flow solutions corresponding to the gradual load variations and including them in the simulation loop as, in the power system

simulation in this book, loads are varied from one value to another instantaneously, which may not reflect the actual load dynamics happening in a real power system.

(v) Extending this formulation to deal with strong interconnections and overlapping subsystems as the proposed decentralised robust controller is limited to the considerations of nonlinear systems where only weak interconnections arise and, therefore, this passive analysis may be over-conservative in terms of providing a quantitative measure of the input threshold.

(vi) Establishing coordination between excitation and FACTS controllers.

(vii) Applying the proposed reformulation technique to future micro-grids and smart grids as the proposed control algorithms in this study are only successfully applied in a current conventional grid.

(viii) Conducting a detailed study to ascertain the impacts and possible benefits of the inherent variability of renewable sources, particularly wind and solar, and their correlations with load profiles.

Chapter 11
Appendices

11.1 Appendix-I

Power flow and dynamic data of single machine infinite bus system with large induction motors (Table 11.1):

It should be noted that the synchronous machine parameters are on a 500 MVA base, the induction motor parameters on a 800-MVA base, the exciter on a 100 MVA base and the system reactances on a 100-MVA base.

11.2 Appendix-II

Transformation of induction motor equations to common reference axis:

Let

$E'_m = \sqrt{e'^2_{dm} + e'^2_{qm}}$, $\delta_m = \tan^{-1} \frac{-e'_{dm}}{e'_{qm}}$, $(I_{dm} + jI_{qm}) = -(i_{dm} + ji_{qm})e^{-j\delta_m}$, $(V_d + jV_q) = v_d + jv_q)e^{-j\delta_m}$, $T_m = -T_L$, $T_e = e'_{qm} i_{qm} + e'_{qm} i_{qm}$, $i_{dm} = -I_{dm} \cos\delta + I_{qm} \sin\delta$ and $i_{qm} = -I_{qm} \cos\delta - I_{dm} \sin\delta$.

Taking derivative of E'_m, we get

$$\begin{aligned}
\dot{E}'_m &= \frac{1}{2\sqrt{e'^2_{dm} + e'^2_{qm}}}(2e'_{dm}\dot{e}'_{dm} + 2e'_{qm}\dot{e}'_{qm}), \\
&= \frac{1}{T'_{dom}\sqrt{e'^2_{dm} + e'^2_{qm}}}\left[-e'^2_{dm} - e'^2_{qm} - (X - X')(e'_{dm} i_{qm} - e'_{qm} i_{dm}\right],
\end{aligned}$$

J. Hossain and H. R. Pota, *Robust Control for Grid Voltage Stability: High Penetration of Renewable Energy*, Power Systems,
DOI: 10.1007/978-981-287-116-9_11, © Springer Science+Business Media Singapore 2014

Table 11.1 System parameters (in pu)

Machine	Motor	System	AVR	OLTC
300 MW	750 MW	$P_L = 1{,}500$ MW, $Q_L = 150$ MVAR		
$X_d = 2.1$	$X_s = 0.1$	$X_\infty = 0.01$	$K_a = 100$	$r_{\max} = 1.1$
$X_q = 2.1$	$X_r = 0.18$	$X_{34} = 0.0554$	$T_a = 0.01s$	$r_{\min} = 0.8$
$X'_d = 0.4$	$X_m = 3.2$	$X_{14} = 0.016$	$E_{fd}^{\min} = 0$	$\Delta r = 0.01$
$T'_{do} = 8$s	$R_r = 0.018$	$X_{24} = 0.004$	$E_{fd}^{\max} = 5$	$V_{ref} = 0.8$ pu
$\omega_s = 377$	$H_m = 0.5$	C = 150 MVAR		d = 0.01 pu
$H = 3.5$	m = 2			$T_D = 20$
$D = 4$				$T_C = 5$

$$
\begin{aligned}
&= \frac{1}{T'_{dom}}\left[-E'_m - (X - X')\frac{e'_{dm}i_{qm} - e'_{qm}i_{dm}}{\sqrt{e'^2_{dm} + e'^2_{qm}}}\right],\\
&= \frac{1}{T'_{dom}}\left[-E'_m - (X - X')I'_{dm}\right].
\end{aligned} \tag{11.1}
$$

$$
\begin{aligned}
T_e &= e'_{qm}(-I_{dm}\cos\delta + I_{qm}\sin\delta) + e'_{qm}(-I_{qm}\cos\delta - I_{dm}\sin\delta),\\
&= -e'_{qm}I_{dm}\frac{e'_{dm}}{\sqrt{e'^2_{dm} + e'^2_{qm}}} + e'_{qm}I_{qm}\frac{e'_{qm}}{\sqrt{e'^2_{dm} + e'^2_{qm}}} = -E'_m I_{dm}.
\end{aligned} \tag{11.2}
$$

Differentiating δ_m, we get:

$$
\begin{aligned}
\dot{\delta}_m &= \frac{1}{1 + (\frac{e'_{dm}}{e'_{qm}})^2} \times \frac{e'_{qm}\dot{e}'_{dm} - e'_{dm}\dot{e}'_{qm}}{e'^2_{qm}},\\
&= \frac{e^2_{qm}}{e'^2_{dm} + e'^2_{qm}} \times 1/e'^2_{qm}\Big[-e'_{qm}e'_{dm} + e'_{qm}e'_{dm}\\
&\quad -(X - X')(e'_{qm}i_{qm} + e'_{dm}i_{dm}) + T'_{dom}\omega_s(s-1)(e'^2_{dm} + e'^2_{qm})\Big],\\
&= s\omega_s - \omega_s - \frac{X - X'}{T'_{dom}E'_{qm}}I_{qm}.
\end{aligned} \tag{11.3}
$$

11.3 Appendix-III

Expression of $\tilde{\phi}$ for excitation control design:

The following expression for $\tilde{\phi}$ is used to design the voltage controller in Chap. 5. We define $\tilde{\phi} = \left[\tilde{\phi}_1 \ldots \tilde{\phi}_7\right]^T$. The expressions for $\left[\tilde{\phi}_1 \ldots \tilde{\phi}_7\right]^T$ are given below:

$$\tilde{\phi}_1 = \begin{bmatrix} 0 \; 0 \; 0 \; 0 \end{bmatrix}.$$

$$\tilde{\phi}_2 = \begin{bmatrix} x_{11} \; x_{12} \; x_{13} \; x_{14} \end{bmatrix} \begin{bmatrix} y_{11} & 0 & 0 & 0 \\ y_{21} & 0 & 0 & 0 \\ y_{31} & 0 & y_{33} & y_{34} \\ y_{41} & 0 & y_{43} & y_{44} \end{bmatrix},$$

where

$$\begin{aligned}
x_{11} &= \cos\delta - \cos\delta_0, & y_{11} &= -V_\infty G_{13}, & y_{33} &= -y_{31}, \\
x_{12} &= \sin\delta - \sin\delta_0, & y_{21} &= -V_\infty B_{13}, & y_{43} &= -y_{41}, \\
x_{13} &= \cos(\delta_m - \delta) - \cos(\delta_{m0} - \delta_{10}), & y_{31} &= -(E'_m - E'_{m0})G_{12}, & y_{34} &= B_{12}, \\
x_{14} &= -\sin(\delta_m - \delta) + \sin(\delta_{m0} - \delta_{10}), & y_{41} &= -(E'_m - E'_{m0})B_{12}, & y_{44} &= G_{12}.
\end{aligned}$$

$$\tilde{\phi}_3 = \begin{bmatrix} a_{11} \; a_{12} \; a_{13} \; a_{14} \end{bmatrix} \begin{bmatrix} b_{11} & b_{12} & 0 & 0 \\ b_{21} & b_{22} & 0 & 0 \\ b_{31} & b_{32} & b_{33} & b_{34} \\ b_{41} & b_{42} & b_{43} & b_{44} \end{bmatrix} + \frac{1}{2H} \begin{bmatrix} 0 \; -2G_{11}(E'_q - E'_{q0}) \; 0 \; 0 \end{bmatrix},$$

where

$$\begin{aligned}
a_{11} &= -\sin\delta + \sin\delta_0, & b_{11} &= -V_\infty(E'_q - E'_{q0})G_{13}, & b_{33} &= -b_{31}, \\
a_{12} &= \cos\delta - \cos\delta_0, & b_{12} &= V_\infty B_{13}, & b_{34} &= -b_{32}, \\
a_{13} &= \sin(\delta_m - \delta) - \sin(\delta_{m0} - \delta_0), & b_{21} &= -V_\infty(E'_q - E'_{q0})B_{13}, & b_{43} &= -b_{41}, \\
a_{14} &= \cos(\delta_m - \delta) - \cos(\delta_{m0} - \delta_0)), & b_{42} &= -(E'_m - E'_{m0})G_{12}, & b_{44} &= -b_{42}, \\
b_{31} &= -(E'_m E'_q - E'_{m0}E'_{q0})G_{12}, & b_{32} &= (E'_m - E'_{m0})B_{12}, & b_{22} &= -V_\infty G_{13}, \\
b_{41} &= -(E'_m E'_q - E'_{m0}E'_{q0})B_{12}.
\end{aligned}$$

$$\tilde{\phi}_4 = \begin{bmatrix} x_{11} \; x_{12} \; x_{13} \; x_{14} \end{bmatrix} \beta \begin{bmatrix} b_{11} & b_{22} & 0 & 0 \\ b_{21} & b_{22} & 0 & 0 \\ b_{31} & b_{32} & b_{33} & b_{34} \\ b_{41} & b_{42} & b_{43} & b_{44} \end{bmatrix} + \frac{1}{2H} \begin{bmatrix} x_{11} \\ x_{12} \\ x_{13} \\ x_{14} \end{bmatrix}^T \kappa \begin{bmatrix} p_{11} & p_{12} & 0 & 0 \\ p_{21} & p_{22} & 0 & 0 \\ p_{31} & p_{32} & p_{33} & p_{34} \\ p_{41} & p_{42} & p_{43} & p_{44} \end{bmatrix}$$
$$+ \begin{bmatrix} 0 \; q_2(\frac{E'_q}{V_t} - \frac{E'_{q0}}{V_{t0}}) \; 0 \; 0 \end{bmatrix},$$

where

$$\begin{aligned}
p_{11} &= -V_\infty(E'_q - E'_{q0})G_{11}G_\infty + B_{11}B_{13}, & p_{12} &= V_\infty(G_{11}B_{13} - B_{11}G_{13}), \\
p_{31} &= -(E'_q E'_m - E_{q0}E'_{m0})(G_{11}G_{12} + B_{11}B_{12}), & \beta &= X'_d(\tfrac{1}{V_t} - \tfrac{1}{V_{t0}}), \\
p_{21} &= -V_\infty(E'_q - E'_{q0})(G_{11}B_{13} - B_{11}G_{13}), & p_{22} &= -V_\infty G_{11}G_\infty + B_{11}B_{13}), \\
p_{41} &= -(E'_q E'_m - E_{q0}E'_{m0})(G_{11}B_{12} - B_{11}G_{12}), & p_{32} &= (E'_m - E'_{m0})(G_{11}B_{12} - B_{11}G_{12}), \\
p_{42} &= -(E'_m - E'_{m0})(G_{11}G_{12} + B_{11}B_{12}), & q_2 &= 1 + 2X'_d B_{11} + X'^2_d(G^2_{11} + B^2_{11}), \\
\kappa &= X'_d\beta,\; p_{44} = -p_{42},\; p_{43} = -p_{41}, & p_{33} &= -p_{31},\; p_{34} = -p_{32}.
\end{aligned}$$

$$\tilde{\phi}_5 = \begin{bmatrix} k_{11} & k_{12} & k_{13} & k_{14} \end{bmatrix} \begin{bmatrix} 0 & 0 & m_{13} & 0 \\ 0 & 0 & m_{23} & 0 \\ m_{31} & m_{32} & m_{33} & 0 \\ m_{41} & m_{42} & m_{43} & 0 \end{bmatrix},$$

where

$$\begin{aligned} k_{11} &= \cos\delta_m - \cos\delta_{m0}, & m_{13} &= -V_\infty G_{31}, & m_{33} &= -m_{31}, \\ k_{12} &= \sin\delta_m - \sin\delta_{m0}, & m_{23} &= -V_\infty B_{31}, & m_{41} &= (E'_q - E'_{q0})B_{12}, \\ k_{13} &= \cos(\delta - \delta_m) - \cos(\delta_0 - \delta_{m0}), & m_{31} &= (E'_q - E'_{q0})G_{12}, & m_{42} &= -G_{12}, \\ k_{14} &= -\sin(\delta - \delta_m) + \sin(\delta_0 - \delta_{m0}), & m_{32} &= B_{12}, & m_{43} &= -m_{41}. \end{aligned}$$

$$\tilde{\phi}_6 = \begin{bmatrix} c_{11} & c_{12} & c_{13} & c_{14} \end{bmatrix} \gamma \begin{bmatrix} 0 & 0 & d_{13} & d_{14} \\ 0 & 0 & d_{23} & d_{24} \\ d_{31} & d_{32} & d_{33} & d_{34} \\ d_{41} & d_{42} & d_{43} & d_{44} \end{bmatrix},$$

where

$$\begin{aligned} c_{11} &= -\sin\delta_m + \sin\delta_{m0}, & d_{13} &= -V_\infty(E'_m - E'_{m0})G_{31}, & d_{34} &= -d_{32}, \\ c_{12} &= \cos\delta_m - \cos\delta_{m0}, & d_{14} &= -V_\infty B_{31}, & d_{33} &= -d_{31}, \\ c_{13} &= \sin(\delta - \delta_m) - \sin(\delta_0 - \delta_{m0}), & d_{23} &= -V_\infty(E'_m - E'_{m0})B_{31}, & d_{43} &= -d_{41}, \\ c_{14} &= \cos(\delta - \delta_m) - \cos(\delta_0 - \delta_{m0}), & d_{32} &= (E'_m - E'_{m0})B_{12}, & d_{44} &= -d_{42}, \\ \gamma &= \frac{1}{E'^2_m} - \frac{1}{E'^2_{m0}}, & d_{42} &= -(E'_m - E'_{m0})G_{12}, & d_{24} &= V_\infty G_{31}, \\ d_{41} &= -(E'_m E'_q - E'_{m0}E'_{q0})B_{12}, & d_{31} &= (E'_m E'_q - E'_{m0}E'_{q0})G_{12}. \end{aligned}$$

$$\tilde{\phi}_7 = \begin{bmatrix} c_{11} & c_{12} & c_{13} & c_{14} \end{bmatrix} \begin{bmatrix} 0 & 0 & d_{13} & d_{14} \\ 0 & 0 & d_{23} & d_{24} \\ d_{31} & d_{32} & d_{33} & d_{34} \\ d_{41} & d_{42} & d_{43} & d_{44} \end{bmatrix} + \begin{bmatrix} 0 & 0 & 0 & -2(E'_m - E'_{m0})G_{33} \end{bmatrix}.$$

11.4 Appendix-IV

3 machine 2 area test system power-flow data:
Transmission line data- R, X and B in pu on 100 MVA base:

Line	R	X	B
5–6	0.0000	0.0040	0.0000
6–7	0.0015	0.0288	1.173
9–10	0.0010	0.0030	0.0000

Transformer data- R and X in pu on 100 MVA base:

Line	R	X	$Ratio$
T_1	0.0000	0.0020	0.8857
T_2	0.0015	0.0045	0.8857
T_3	0.0010	0.0125	0.9024
T_4	0.0010	0.0030	1.0664
T_5	0.0010	0.0026	1.0800
T_6	0.0010	0.0010	0.9750
T_7	0.0010	0.0030	1.0000
T_8	0.0010	0.0030	1.0000

Loads:

Bus	P (MW)	Q (MVAR)
8	3,271	1,015
11	3,384	971

Generation:

Bus	P (MW)	V (pu)
G_1	3,981	0.98
G_2	1,736	0.98
G_3	1,154	0.98

Shunt capacitors:

Bus	MVAr
7	763
8	600
9	1,710

3-machine 2-area system dynamic data:
Machine 1: Infinite bus, Machine 2: H = 2.09 and Macine 3: H = 2.33.
Machine parameters- machine 2 pu on 2,200 MVA base and machine 3 on 1,400 MVA:

R_a	X_d	X_q	R_a	X_l	X'_d	X''_d	X''_q	T'_{d0}	T'_{q0}	T''_{d0}	T''_{q0}
0.0046	2.07	1.99	0.155	0.28	0.49	0.215	0.215	4.10	0.56	0.033	0.062

Exciters: Both machine 2 and machine 3 have thyristor exciters with a gain of 400 and the sensing circuit-time constant of 0.02 s.

Motor data- pu on 3,600 MVA base:

$X_m = 3.3$	$R_s = 0.01$	$X_s = 0.145$
$R_r = 0.008$	$X_r = 0.145$	$H = 0.6$ s
0.0015	0.0045	0.8857

Load torque component m = 2 (that is $T_L = T_o\omega_r^2$).

Overexcitation limiter for machine 3:

$$I_{fd_{\max 1}} = 3.02 \text{ pu } I_{fd_{\max 1}} = 3.02 \text{ pu } I_{fd_{\max 1}} = 3.02 \text{ pu}$$
$$Time_1 = 3.02 \text{ s} \quad Time_1 = 3.02 \text{ s} \quad Time_1 = 3.02 \text{ s.}$$

OLTC for transformer T_6 between buses 10 and 11:

Time delay for the first tap movement: 30 s
Time delay for the subsequent tap movement: 5 s
Dead band: $\pm 1\%$ pu bus voltage
Tap range: ± 16 steps
Step size: 0.00625 pu.

11.5 Appendix-V

Single wind farm infinite bus test system power flow and dynamic data (Table 11.2).

11.6 Appendix-VI

Expression of $\tilde{\phi}$ for STATCOM controller with unstructured uncertainty representation:

We define $\tilde{\phi} = \left[\tilde{\phi}_1, \ldots, \tilde{\phi}_7\right]^T$, $\tilde{\psi} = \left[\tilde{\psi}_1, \ldots, \tilde{\psi}_7\right]^T$, $z_1 = T_0'/(X - X')$, $z_2 = -\omega_s(E_{qr}' - E_{qr0}')$, $z_3 = -\omega_s(s - s_0)$, $z_4 = \omega_s(E_{dr}' - E_{dr0}')$, $\Lambda = (k - k_0)$, $v_1 = -(E_{dr}' v_{dc} - E_{dr0} v_{dc0})$, $v_2 = -(E_{qr1}' v_{dc} - E_{qr10} v_{dc0})$ and $z = \frac{1}{V_t} - \frac{1}{V_{to}}$ where

$$\tilde{\phi}_1 = \begin{bmatrix} a_{11} \\ a_{12} \\ a_{13} \\ a_{14} \end{bmatrix}^T \begin{bmatrix} 0 & b_{12} & b_{13} & b_{14} \\ 0 & b_{22} & b_{23} & b_{24} \\ 0 & V_\infty G_{13} & V_\infty B_{13} & 0 \\ 0 & V_\infty B_{13} & V_\infty G_{13} & 0 \end{bmatrix} + \begin{bmatrix} 0 \\ c_{12} \\ c_{13} \\ 0 \end{bmatrix}^T,$$

Table 11.2 Machines and grid parameters of test system

Asynchronous machines	
Power: 2 MW	$R_s = 0.0121$ pu
Voltage: 690 V	$X_s = 0.0742$ pu
Frequency, $f = 50$ Hz	$X_m = 2.7626$ pu
Self damping, 0.008 pu	$R_r = 0.008$ pu
Rated Slip: 0.02	$X_r = 0.1761$ pu
Two mass model	STATCOM
$H_m = 2.6$ s, $H_G = 0.22$ s	Capacity: 10 MVA
$D_m = 3$ pu, $K_s = 141$ pu	$R_C = 0.01$ pu
Gearbox ratio: 23.75	$C = 300\ \mu$F
Grid and line	Turbine parameters
$x_g = 0.1126$ pu	Hub height: 30 m
$r_g = 0.01126$ pu	Rotor diameter: 23.2 m
$x_{13} = 0.75$ pu	Rated speed: 42 r.p.m.
$r_{13} = 0.075$ pu	
Load	Shunt compensator
P=50 MW, Q=15 MVAr	Capacitor: 25 MVAr

where

$$\begin{aligned}
a_{11} &= \sin(\alpha-\delta)-\sin(\alpha_0-\delta_0), a_{13} = -\cos\alpha+\cos\alpha_0,\\
a_{12} &= \cos(\alpha-\delta)-\cos(\alpha_0-\delta_0), a_{14} = -\sin\alpha+\sin\alpha_0,\\
b_{12} &= G_{12}\Lambda(v_{dc}-v_{dc0}),\ b_{22} = -B_{12}\Lambda(v_{dc}-v_{dc0}),\\
b_{13} &= -B_{12}\Lambda(v_{dc}-v_{dc0}),\ b_{23} = -G_{12}\Lambda(v_{dc}-v_{dc0}),\\
b_{14} &= -\Lambda B_{12}(E'_{qr}-E'_{qr0})+\Lambda G_{12}(E'_{dr}-E'_{dr0}),\\
b_{24} &= -\Lambda G_{12}(E'_{qr}-E'_{qr0})-\Lambda B_{12}(E'_{dr}-E'_{dr0}),\\
c_{12} &= 2G_{11}(E'_{dr}-E'_{dr0}), c_{13} = -2B_{11}(E'_{qr}-E'_{qr0}).
\end{aligned}$$

$$\tilde{\phi}_2 = \begin{bmatrix} a_{11}\\ a_{12}\\ a_{13}\\ a_{14}\end{bmatrix}^T \begin{bmatrix} 0\,0\,0 & -\Lambda G_{12}\\ 0\,0\,0 & \Lambda B_{12}\\ 0\,0\,0 & 0\\ 0\,0\,0 & 0\end{bmatrix} + z_1 \begin{bmatrix} z_2\\ 0\\ z_3\\ 0\end{bmatrix}^T.$$

$$\tilde{\phi}_3 = \begin{bmatrix} a_{11}\\ a_{12}\\ a_{13}\\ a_{14}\end{bmatrix}^T \begin{bmatrix} 0\,0\,0 & \Lambda B_{12}\\ 0\,0\,0 & \Lambda G_{12}\\ 0\,0\,0 & 0\\ 0\,0\,0 & 0\end{bmatrix} + z_1 \begin{bmatrix} z_4\\ z_3\\ 0\\ 0\end{bmatrix}.$$

$$\tilde{\phi}_4 = \tilde{\phi}_6 = [0,0,0,0],\quad \tilde{\phi}_5 = \tilde{\phi}_1,$$

$$\tilde{\phi}_7 = \begin{bmatrix} a_{11} \\ a_{12} \\ a_{13} \\ a_{14} \end{bmatrix} \begin{bmatrix} 0 & -\Lambda B_{12} & -\Lambda G_{21} & 0 \\ 0 & -\Lambda G_{12} & \Lambda B_{21} & 0 \\ 0 & 0 & 0 & 0 \\ 0 & 0 & 0 & 0 \end{bmatrix}$$

$$\tilde{\psi}_1 = \begin{bmatrix} a_{12} & a_{11} \end{bmatrix} \begin{bmatrix} s_{11} & s_{12} \\ s_{21} & s_{22} \end{bmatrix},$$

where

$$s_{11} = v_1 B_{12} + v_2 G_{12},\ s_{21} = v_1 G_{12} + v_2 B_{12},$$
$$s_{12} = \Lambda v_1 G_{12} - \Lambda v_2 B_{12},\ s_{22} = -\Lambda v_1 B_{12} - \Lambda v_2 G_{12}.$$

$$\tilde{\psi}_2 = \begin{bmatrix} r_{11} & r_{12} \end{bmatrix} \begin{bmatrix} t_{11} & t_{12} \\ t_{21} & t_{22} \end{bmatrix},$$

where

$$t_{11} = -(v_{dc} - v_{dc0}) B_{12},\ t_{21} = -(v_{dc} - v_{dc0}) G_{12},$$
$$t_{12} = -\Lambda (v_{dc} - v_{dc0}) G_{12},\ t_{22} = \Lambda (v_{dc} - v_{dc0}) B_{14}.$$

$$\tilde{\psi}_3 = \begin{bmatrix} r_{11} & r_{12} \end{bmatrix} \begin{bmatrix} u_{11} & u_{12} \\ u_{21} & u_{22} \end{bmatrix},$$

where

$$u_{11} = -(v_{dc} - v_{dc0}) G_{12},\ u_{21} = (v_{dc} - v_{dc0}) B_{12},$$
$$u_{12} = \Lambda (v_{dc} - v_{dc0}) B_{12},\ u_{22} = \Lambda (v_{dc} - v_{dc0}) G_{12}.$$

$$\tilde{\psi}_4 = \tilde{\psi}_6 = [0, 0], \quad \tilde{\psi}_5 = \tilde{\psi}_1,$$
$$\tilde{\psi}_7 = z \begin{bmatrix} a_{12} & -a_{11} \end{bmatrix} \begin{bmatrix} -s_{12} & -s_{11} \\ -s_{22} & -s_{21} \end{bmatrix}.$$

11.7 Appendix-VII

Expression of $\tilde{\phi}$ and $\tilde{\psi}$ for simultaneous STATCOM and pitch angle control design:

The mathematical expressions for uncertain matrices $\tilde{\phi}$ and $\tilde{\psi}$ are useful in determining the scaling parameter β. The expressions for $\tilde{\phi}$ and $\tilde{\psi}$ are given in the rest of this section. We define $\tilde{\phi} = \left[\tilde{\phi}_1, \ldots, \tilde{\phi}_7\right]^T$, $\tilde{\psi} = \left[\tilde{\psi}_1, \ldots, \tilde{\psi}_7\right]^T$, where

$$\tilde{\phi}_1 = \left[r_{11}\ 0\ 0\ 0\right], \tag{11.4}$$

where

$$r_{11} = \frac{\rho}{2\omega_m^2} A_{wt} V_w^3 \{(0.44 - 0.0167\theta) \sin\left[\frac{\pi(\lambda - 3)}{15 - 0.3\theta}\right] - 0.00184(\lambda - 3)\theta\}$$
$$- \frac{\rho}{2\omega_{m0}^2} A_{wt} V_w^3 \{(0.44 - 0.0167\theta_0) \sin\left[\frac{\pi(\lambda - 3)}{15 - 0.3\theta_0}\right] - 0.00184(\lambda - 3)\theta_0\}.$$

$$\tilde{\phi}_2 = \begin{bmatrix} a_{11} \\ a_{12} \\ a_{13} \\ a_{14} \end{bmatrix}^T \begin{bmatrix} 0 & b_{12} & b_{13} & b_{14} \\ 0 & b_{22} & b_{23} & b_{24} \\ 0 & V_\infty G_{13} & V_\infty B_{13} & 0 \\ 0 & V_\infty B_{13} & V_\infty G_{13} & 0 \end{bmatrix} + \begin{bmatrix} 0 \\ c_{12} \\ c_{13} \\ 0 \end{bmatrix}^T, \tag{11.5}$$

where

$$a_{11} = \sin(\alpha - \delta) - \sin(\alpha_0 - \delta_0),\ a_{13} = -\cos\alpha + \cos\alpha_0,$$
$$a_{12} = \cos(\alpha - \delta) - \cos(\alpha_0 - \delta_0),\ a_{14} = -\sin\alpha + \sin\alpha_0,$$
$$b_{12} = G_{12}\kappa(v_{dc} - v_{dco}),\ b_{22} = -B_{12}\kappa(v_{dc} - v_{dco}),$$
$$b_{13} = -B_{12}\kappa(v_{dc} - v_{dco}),\ b_{23} = -G_{12}\kappa(v_{dc} - v_{dco}),$$
$$b_{14} = -\kappa B_{12}(E'_{qr} - E'_{qro}) + \kappa G_{12}(E'_{dr} - E'_{dro}),$$
$$b_{24} = -\kappa G_{12}(E'_{qr} - E'_{qro}) - \kappa B_{12}(E'_{dr} - E'_{dro}),$$
$$c_{12} = -2G_{11}(E'_{dr} - E'_{dro}) - 2B_{11}(E'_{qr} - E'_{qro}),$$
$$c_{13} = -2G_{11}(E'_{qr} - E'_{qro}) - 2B_{11}(E'_{dr} - E'_{dro}),\ \kappa = \sqrt{\frac{3}{8}}(m - m_0).$$

$$\tilde{\phi}_3 = \left[0\ 0\ 0\ 0\right]. \tag{11.6}$$

$$\tilde{\phi}_4 = \tilde{\phi}_2 = \begin{bmatrix} a_{11} \\ a_{12} \\ a_{13} \\ a_{14} \end{bmatrix}^T \begin{bmatrix} 0 & b_{12} & b_{13} & b_{14} \\ 0 & b_{22} & b_{23} & b_{24} \\ 0 & V_\infty G_{13} & V_\infty B_{13} & 0 \\ 0 & V_\infty B_{13} & V_\infty G_{13} & 0 \end{bmatrix} + \begin{bmatrix} 0 \\ c_{12} \\ c_{13} \\ 0 \end{bmatrix}^T. \tag{11.7}$$

$$\tilde{\phi}_5 = \begin{bmatrix} a_{11} \\ a_{12} \\ a_{13} \\ a_{14} \end{bmatrix}^T \begin{bmatrix} 0\,0\,0 & -\kappa G_{12} \\ 0\,0\,0 & \kappa B_{12} \\ 0\,0\,0 & 0 \\ 0\,0\,0 & 0 \end{bmatrix} + \begin{bmatrix} z_1 \\ 0 \\ z_3 \\ 0 \end{bmatrix}^T, \tag{11.8}$$

where

$$z_1 = -T_0'/(X - X')\omega_s(E_{qr}' - E_{qr0}'),\ z_3 = T_0'/(X - X')\omega_s(s - s_0).$$

$$\tilde{\phi}_6 = \begin{bmatrix} a_{11} \\ a_{12} \\ a_{13} \\ a_{14} \end{bmatrix}^T \begin{bmatrix} 0\,0\,0 & \kappa B_{12} \\ 0\,0\,0 & \kappa G_{12} \\ 0\,0\,0 & 0 \\ 0\,0\,0 & 0 \end{bmatrix} + \begin{bmatrix} z_3 \\ z_4 \\ 0 \\ 0 \end{bmatrix}^T, \tag{11.9}$$

where

$$z_4 = -T_0'/(X - X')\omega_s(E_{dr}' - E_{dr0}'),\ z_3 = T_0'/(X - X')\omega_s(s - s_0).$$

$$\tilde{\phi}_7 = \begin{bmatrix} p_{11} & p_{12} & 0 & 0 \end{bmatrix} \begin{bmatrix} 0 & \kappa B_{21} & \kappa G_{21} & 0 \\ 0 & \kappa G_{21} & k B_{21} & 0 \\ 0 & 0 & 0 & 0 \\ 0 & 0 & 0 & 0 \end{bmatrix}, \tag{11.10}$$

where

$$p_{11} = \sin(\delta - \alpha) - \sin(\delta_0 - \alpha_0),\ p_{12} = \cos(\delta - \alpha) - \cos(\delta_0 - \alpha_0).$$

$$\tilde{\phi}_8 = z \begin{bmatrix} a_{11} \\ a_{12} \\ a_{13} \\ a_{14} \end{bmatrix}^T \begin{bmatrix} 0\,0\,0 & \kappa R_s G_{12} \\ 0\,0\,0 & -\kappa R_s B_{12} \\ 0\,0\,0 & 0 \\ 0\,0\,0 & 0 \end{bmatrix} + z \begin{bmatrix} 0 \\ R_s G_{11} \\ R_s B_{11} \\ 0 \end{bmatrix}^T,$$

where

$$z = \frac{1}{v_t} - \frac{1}{v_{t0}}.$$

$$\tilde{\psi}_1 = \begin{bmatrix} m_{12} & m_{11} & 0 & 0 \end{bmatrix} \begin{bmatrix} n_{11} & 0\,0\,0 \\ n_{21} & 0\,0\,0 \\ 0 & 0\,0\,0 \\ 0 & 0\,0\,0 \end{bmatrix}, \tag{11.11}$$

where

$$m_{11} = \sin\left[\frac{\pi(\lambda-3)}{15-0.3\theta}\right] - \sin\left[\frac{\pi(\lambda-3)}{15-0.3\theta_0}\right],\ n_{11} = \frac{0.0835\rho}{2} A_{wt} V_w^3 \left(\frac{1}{\omega_m} - \frac{1}{\omega_{m_0}}\right),$$
$$m_{12} = \cos\left[\frac{\pi(\lambda-3)}{15-0.3\theta}\right] - \sin\left[\frac{\pi(\lambda-3)}{15-0.3\theta_0}\right],\ n_{21} = \frac{0.025\rho(\lambda-3)}{2} A_{wt} V_w^3 \left(\frac{\theta}{\omega_m(15-0.3\theta)} - \frac{\theta_0}{\omega_{m_0}(15-0.3\theta_0)}\right).$$

$$\tilde{\psi}_2 = \begin{bmatrix} a_{12} & a_{11} & 0 & 0 \end{bmatrix} \begin{bmatrix} 0 & s_{12} & s_{13} & 0 \\ 0 & s_{22} & s_{23} & 0 \\ 0 & 0 & 0 & 0 \\ 0 & 0 & 0 & 0 \end{bmatrix}, \tag{11.12}$$

where

$$s_{21} = -\kappa G_{12} p + \kappa B_{12} q,\quad s_{22} = -\kappa B_{12} p - \kappa G_{12} q,$$
$$s_{13} = -\tfrac{3}{8} G_{12} q + \kappa B_{12} p,\quad s_{23} = -\tfrac{3}{8} B_{12} q - \kappa G_{12} p,$$
$$p = (E'_{dr} v_{dc} - E_{dr_0} V_{dc_0}),\ q = (E'_{qr} v_{dc} - E_{qr_0} V_{dc_0}).$$

$$\tilde{\psi}_3 = \begin{bmatrix} 0 & 0 & 0 & 0 \end{bmatrix}. \tag{11.13}$$

$$\tilde{\psi}_4 = \tilde{\psi}_2 = \begin{bmatrix} a_{12} & a_{11} & 0 & 0 \end{bmatrix} \begin{bmatrix} 0 & s_{12} & s_{13} & 0 \\ 0 & s_{22} & s_{23} & 0 \\ 0 & 0 & 0 & 0 \\ 0 & 0 & 0 & 0 \end{bmatrix}. \tag{11.14}$$

$$\tilde{\psi}_5 = \begin{bmatrix} a_{12} & a_{11} & 0 & 0 \end{bmatrix} \begin{bmatrix} 0 & t_{12} & t_{13} & 0 \\ 0 & t_{22} & t_{23} & 0 \\ 0 & 0 & 0 & 0 \\ 0 & 0 & 0 & 0 \end{bmatrix}, \tag{11.15}$$

where

$$t_{12} = -\kappa(v_{dc} - v_{dc_0}) G_{12},\ t_{22} = \kappa(v_{dc} - v_{dc_0}) B_{14},$$
$$t_{13} = -\tfrac{3}{8}(v_{dc} - v_{dc_0}) B_{12},\ t_{23} = -\tfrac{3}{8}(v_{dc} - v_{dc_0}) G_{12}.$$

$$\tilde{\psi}_6 = \begin{bmatrix} a_{12} & a_{11} & 0 & 0 \end{bmatrix} \begin{bmatrix} 0 & u_{12} & u_{12} & 0 \\ 0 & u_{22} & u_{23} & 0 \\ 0 & 0 & 0 & 0 \\ 0 & 0 & 0 & 0 \end{bmatrix}, \tag{11.16}$$

where

$$u_{12} = \kappa(v_{dc} - v_{dc_0})B_{12}, \quad u_{22} = \kappa(v_{dc} - v_{dc_0})G_{12},$$
$$u_{11} = -\tfrac{3}{8}(v_{dc} - v_{dc_0})G_{12}, \; u_{21} = \tfrac{3}{8}(v_{dc} - v_{dc_0})B_{12}.$$

$$\tilde{\psi}_7 = z\left[p_{11}\; p_{12}\; 0\; 0\right]\begin{bmatrix} 0 & \kappa G_{21} & \frac{3}{8}B_{21} & 0 \\ 0 & \kappa B_{21} & \frac{3}{8}G_{21} & 0 \\ 0 & 0 & 0 & 0 \\ 0 & 0 & 0 & 0 \end{bmatrix}. \tag{11.17}$$

$$\tilde{\psi}_8 = \left[a_{11}\; a_{12}\; 0\; 0\right]\begin{bmatrix} 0 & q_{12} & q_{13} & 0 \\ 0 & q_{22} & q_{23} & 0 \\ 0 & 0 & 0 & 0 \\ 0 & 0 & 0 & 0 \end{bmatrix}, \tag{11.18}$$

where

$$q_{12} = \kappa R_s B_{12}(v_{dc} - v_{dc_0}), \; q_{22} = \kappa R_s G_{12}(v_{dc} - v_{dc_0}),$$
$$q_{12} = \tfrac{3}{8}R_s G_{12}(v_{dc} - v_{dc_0}), \; q_{22} = \tfrac{3}{8}R_s B_{12}(v_{dc} - v_{dc_0}).$$

11.8 Appendix-VIII

Expression of $\tilde{\phi}$ and $\tilde{\psi}$ for STATCOM control design with structured uncertainty representation:

The expressions for $\tilde{\phi}_k$ and $\tilde{\psi}_k$ are given below, $k = 0, \ldots, 7$, $\delta_{ab} = \sin(\delta_a - \delta_b)$ and $\Lambda = (K - K_0)$:

$$\tilde{\phi}_0 = \begin{bmatrix} a_{11} \\ a_{12} \\ a_{13} \\ a_{14} \end{bmatrix}^T \begin{bmatrix} 0 & b_{12} & b_{13} & 0 & b_{15} & b_{16} & 0 \\ 0 & b_{22} & b_{23} & 0 & b_{25} & b_{26} & 0 \\ 0 & b_{32} & b_{33} & 0 & 0 & 0 & b_{37} \\ 0 & b_{42} & b_{43} & 0 & 0 & 0 & b_{47} \end{bmatrix} + \left[0\; t_{12}\; t_{13}\; 0\; 0\; 0\; 0\right],$$

where

$$a_{11} = \cos\delta_{21} - \cos\delta_{210}, \; a_{12} = \sin\delta_{21} - \sin\delta_{210},$$
$$a_{13} = \cos(\alpha - \delta_1) - \cos(\alpha_0 - \delta_{10}), \; a_{14} = \sin(\alpha - \delta_1) - \sin(\alpha_0 - \delta_{10}),$$
$$b_{12} = -(E'_{dr2} - E_{dr20})G_{12} - (E'_{qr2} - E_{qr20})B_{12},$$
$$b_{22} = (E'_{dr2} - E_{dr20})B_{12} - (E'_{qr2} - E_{qr20})G_{12},$$
$$b_{32} = -\Lambda(v_{dc} - v_{dc0})B_{14}, \; b_{42} = -\Lambda(v_{dc} - v_{dc0})G_{14},$$

$$
\begin{aligned}
b_{13} &= -(E'_{dr2} - E_{dr20})B_{12} - (E'_{qr2} - E_{qr20})G_{12},\\
b_{23} &= -(E'_{dr2} - E_{dr20})G_{12} + (E'_{qr2} - E_{qr20})B_{12},\\
b_{33} &= -\Lambda(v_{dc} - v_{dc0})G_{14},\ b_{43} = -\Lambda(v_{dc} - v_{dc0})B_{14},\\
b_{15} &= -(E'_{dr1} - E_{dr10})G_{12} - (E'_{qr1} - E_{qr10})B_{12},\\
b_{25} &= (E'_{dr1} - E_{dr10})B_{12} - (E'_{qr1} - E_{qr10})G_{12},\\
b_{16} &= -(E'_{dr1} - E_{dr10})B_{12} - (E'_{qr1} - E_{qr10})G_{12},\\
b_{26} &= -(E'_{dr1} - E_{dr10})G_{12} + (E'_{qr1} - E_{qr10})B_{12},\\
b_{37} &= -\Lambda(E'_{dr1} - E_{dr10})B_{14} - (E'_{qr1} - E_{qr10})G_{14},\\
b_{47} &= -\Lambda(E'_{dr1} - E_{dr10})G_{14} + (E'_{qr1} - E_{qr10})B_{14},\\
t_{12} &= -G_{11}(E'_{dr1} - E'_{dr10}),\ t_{13} = -2B_{11}(E'_{qr1} - E'_{qr10}).
\end{aligned}
$$

$$
\tilde{\phi}_1 = \begin{bmatrix} a_{11} \\ a_{12} \\ a_{13} \\ a_{14} \end{bmatrix}^T \begin{bmatrix} 0 & 0 & 0 & 0 & c_{15} & c_{16} & 0 \\ 0 & 0 & 0 & 0 & c_{25} & c_{26} & 0 \\ 0 & 0 & 0 & 0 & 0 & 0 & c_{37} \\ 0 & 0 & 0 & 0 & 0 & 0 & c_{47} \end{bmatrix} + \left[\beta_{11}\ 0\ \beta_{13}\ 0\ 0\ 0\ 0\right],
$$

where

$$
\begin{aligned}
c_{15} &= -B_{12}, & c_{26} &= B_{12},\\
c_{25} &= -G_{12}, & c_{37} &= -G_{14},\\
c_{16} &= -G_{12}, & c_{47} &= -B_{14},\\
\beta_{11} &= -\omega_s(E'_{qr1} - E_{qr10}), & \beta_{13} &= -\omega_s(s_1 - s_{10}).
\end{aligned}
$$

$$
\tilde{\phi}_2 = \begin{bmatrix} a_{11} \\ a_{12} \\ a_{13} \\ a_{14} \end{bmatrix}^T \begin{bmatrix} 0 & 0 & 0 & 0 & d_{15} & d_{16} & 0 \\ 0 & 0 & 0 & 0 & d_{25} & d_{26} & 0 \\ 0 & 0 & 0 & 0 & 0 & 0 & d_{37} \\ 0 & 0 & 0 & 0 & 0 & 0 & d_{47} \end{bmatrix} + \left[\kappa_{11}\ \kappa_{12}\ 0\ \ 0\ 0\ 0\ 0\right],
$$

where

$$
\begin{aligned}
d_{15} &= -B_{12}, & d_{26} &= B_{12},\\
d_{25} &= G_{12}, & d_{37} &= -B_{14},\\
d_{16} &= G_{12}, & d_{47} &= G_{14},\\
\kappa_{11} &= -\omega_s(E'_{dr1} - E_{dr10}), & \kappa_{12} &= -\omega_s(s_1 - s_{10}).
\end{aligned}
$$

$$
\tilde{\phi}_3 = \begin{bmatrix} m_{11} \\ m_{12} \\ m_{13} \\ m_{14} \end{bmatrix}^T \begin{bmatrix} 0 & n_{12} & n_{13} & 0 & n_{15} & n_{16} & 0 \\ 0 & n_{22} & n_{23} & 0 & n_{25} & n_{26} & 0 \\ 0 & 0 & 0 & 0 & n_{35} & 0 & n_{37} \\ 0 & 0 & 0 & 0 & n_{45} & 0 & n_{47} \end{bmatrix} + \left[0\ 0\ 0\ 0\ g_{15}\ g_{16}\ 0\right],
$$

where

$$\begin{aligned}
&m_{11} = \cos\delta_{12} - \cos\delta_{120},\ m_{12} = \sin\delta_{12} - \sin\delta_{120},\\
&m_{13} = \cos(\alpha - \delta_2) - \cos(\alpha_0 - \delta_{20}),\ m_{14} = \sin(\alpha - \delta_2) - \sin(\alpha_0 - \delta_{20}),\\
&n_{12} = -(E'_{dr2} - E_{dr20})G_{21} - (E'_{qr2} - E_{qr20})B_{21},\\
&n_{22} = (E'_{dr2} - E_{dr20})B_{21} - (E'_{qr2} - E_{qr20})G_{21},\\
&n_{13} = -(E'_{qr2} - E_{qr20})G_{21} - (E'_{dr2} - E_{dr20})B_{21},\\
&n_{23} = -(E'_{qr2} - E_{qr20})B_{21} + (E'_{dr2} - E_{dr20})G_{21},\\
&n_{15} = -(E'_{dr1} - E_{dr10})G_{21} - (E'_{qr1} - E_{qr10})B_{21},\\
&n_{25} = (E'_{dr1} - E_{dr10})B_{21} - (E'_{qr1} - E_{qr10})G_{21},\\
&n_{35} = \Lambda(v_{dc} - E_{dc0})B_{24},\ n_{45} = (v_{dc} - E_{dc0})G_{24},\\
&n_{16} = -(E'_{qr1} - E_{qr10})G_{21} - (E'_{dr1} - E_{dr10})B_{21},\\
&n_{26} = (E'_{qr1} - E_{qr10})B_{21} - (E'_{dr1} - E_{dr10})G_{21},\\
&n_{37} = -\Lambda(E'_{dr2} - E_{dr20})B_{24} - (E'_{qr2} - E_{qr20})G_{24},\\
&n_{47} = -\Lambda(E'_{dr2} - E_{dr20})G_{24} + (E'_{qr2} - E_{qr20})B_{24},\\
&g_{15} = -G_{22}(E'_{dr2} - E_{dr20}),\ g_{16} = -B_{22}(E'_{qr2} - E_{qr20}).
\end{aligned}$$

$$\tilde{\phi}_4 = \begin{bmatrix} m_{11}\\ m_{12}\\ m_{13}\\ m_{14}\end{bmatrix}^T \begin{bmatrix} 0 & p_{12} & p_{13} & 0 & 0 & 0 & 0\\ 0 & p_{22} & p_{23} & 0 & 0 & 0 & 0\\ 0 & 0 & 0 & 0 & 0 & 0 & p_{37}\\ 0 & 0 & 0 & 0 & 0 & 0 & p_{47}\end{bmatrix} + \begin{bmatrix}0 & 0 & 0 & -\omega_s(E'_{qr2} - E_{qr20}) & 0 & -\omega_s(s_2 - s_{20}) & 0\end{bmatrix},$$

where

$$p_{12} = -B_{21},\quad p_{22} = -G_{21},\ p_{13} = -G_{21},\quad p_{23} = B_{21},\ p_{37} = -\Lambda B_{24},\quad p_{47} = -\Lambda G_{24}.$$

$$\tilde{\phi}_5 = \begin{bmatrix} m_{11}\\ m_{12}\\ m_{13}\\ m_{14}\end{bmatrix}^T \begin{bmatrix} 0 & v_{12} & v_{13} & 0 & 0 & 0 & 0\\ 0 & v_{22} & v_{23} & 0 & 0 & 0 & 0\\ 0 & 0 & 0 & 0 & 0 & 0 & v_{37}\\ 0 & 0 & 0 & 0 & 0 & 0 & v_{47}\end{bmatrix} + \begin{bmatrix}0 & 0 & 0 & 0 & -\omega_s(E'_{dr2} - E_{dr20}) & -\omega_s(s_2 - s_{20}) & 0\end{bmatrix},$$

where

$$\begin{aligned}
&v_{12} = -G_{21},\quad v_{22} = B_{21},\\
&v_{13} = -B_{21},\quad v_{23} = G_{21},\\
&v_{37} = -\Lambda G_{24},\ v_{47} = \Lambda B_{24}.
\end{aligned}$$

$$\tilde{\phi}_6 = \begin{bmatrix} x_{11} \\ x_{12} \\ x_{13} \\ x_{14} \end{bmatrix}^T \begin{bmatrix} 0 & y_{12} & y_{13} & 0 & 0 & 0 & 0 \\ 0 & y_{22} & y_{23} & 0 & 0 & 0 & 0 \\ 0 & 0 & 0 & 0 & y_{35} & y_{36} & 0 \\ 0 & 0 & 0 & 0 & y_{45} & y_{46} & 0 \end{bmatrix},$$

where

$$\begin{aligned} x_{11} &= \cos(\delta_1 - \alpha) - \cos(\delta_{10} - \alpha_0), & y_{12} &= -\Lambda G_{41}, \\ x_{12} &= \sin(\delta_1 - \alpha) - \sin(\delta_{10} - \alpha_0), & y_{22} &= -\Lambda B_{41}, \\ x_{13} &= \cos(\delta_2 - \alpha) - \cos(\delta_{20} - \alpha_0), & y_{13} &= -\Lambda B_{41}, \\ x_{14} &= \sin(\delta_2 - \alpha) - \sin(\delta_{20} - \alpha_0), & y_{23} &= \Lambda G_{41}, \\ y_{35} &= -\Lambda B_{42}, & y_{45} &= -\Lambda G_{42}, \\ y_{36} &= -\Lambda G_{42}, & y_{46} &= \Lambda B_{42}. \end{aligned}$$

$$\tilde{\phi}_7 = z \begin{bmatrix} a_{11} \\ a_{12} \\ a_{13} \\ a_{14} \end{bmatrix}^T \begin{bmatrix} 0 & -b_{13} & -b_{12} & 0 & -b_{16} & -b_{15} & 0 \\ 0 & -b_{23} & -b_{22} & 0 & -b_{26} & -b_{25} & 0 \\ 0 & -b_{33} & b_{23} & 0 & 0 & 0 & -b_{47} \\ 0 & -b_{43} & -b_{33} & 0 & 0 & 0 & -b_{37} \end{bmatrix} + \begin{bmatrix} 0 & z_{12} & z_{13} & 0 & z_{15} & z_{16} & 0 \end{bmatrix},$$

where

$$\begin{aligned} z &= \frac{1}{V_t} - \frac{1}{V_{t0}}, \\ z_{12} &= 2X_1'(E_{dr1} - E_{dr10}) + 4G_{12}B_{12}(E_{qr1} - E_{qr10}) \\ &\quad + 2(E_{dr1} - E_{dr10}) + 2B_{11}(E_{dr1} - E_{dr10}) - 2t_{12} - b_{37}, \\ z_{13} &= 2X_1'(E_{qr1} - E_{qr10}) + 4G_{12}B_{12}(E_{dr1} - E_{dr10}) \\ &\quad + 2(E_{qr1} - E_{qr10}) + 2B_{11}(E_{qr1} - E_{qr10}) - t_{13} - b_{47}, \\ z_{15} &= 2X_1'(E_{dr2} - E_{dr20}) + 4G_{12}B_{12}(E_{qr2} - E_{qr20}) \\ &\quad + 2(E_{dr2} - E_{dr20}) + 2B_{11}(E_{dr2} - E_{dr20}) - 2g_{15} - n_{37}, \\ z_{16} &= 2X_1'(E_{qr2} - E_{qr20}) + 4G_{12}B_{12}(E_{dr2} - E_{dr20}) \\ &\quad + 2(E_{qr2} - E_{qr20}) + 2B_{11}(E_{qr2} - E_{qr20}) - 2g_{16} - n_{47}. \end{aligned}$$

$$\tilde{\psi}_0 = \begin{bmatrix} r_{11} & r_{12} \end{bmatrix} \begin{bmatrix} s_{11} & s_{12} \\ s_{21} & s_{22} \end{bmatrix},$$

where

$$r_{11} = \cos(\alpha - \delta_1) - \cos(\alpha_0 - \delta_{10}),$$
$$r_{12} = \sin(\alpha - \delta_1) - \sin(\alpha_0 - \delta_{10}),$$
$$s_{11} = -(E'_{dr1}v_{dc} - E_{dr10}v_{dc0})B_{14} - (E'_{qr1}v_{dc} - E_{qr10}V_{dc0})G_{14},$$
$$s_{21} = -(E'_{dr1}v_{dc} - E_{dr10}v_{dc0})G_{14} - (E'_{qr1}v_{dc} - E_{qr10}v_{dc0})B_{14},$$
$$s_{12} = -\Lambda(E'_{dr1}v_{dc} - E_{dr10}v_{dc0})G_{14} + \Lambda(E'_{qr1}v_{dc} - E_{qr10}v_{dc0})B_{14},$$
$$s_{22} = \Lambda(E'_{dr1}v_{dc} - E_{dr10}v_{dc0})B_{14} + \Lambda(E'_{qr1}v_{dc} - E_{qr10}v_{dc0})G_{14}.$$

$$\tilde{\psi}_1 = \begin{bmatrix} r_{11} & r_{12} \end{bmatrix} \begin{bmatrix} t_{11} & t_{12} \\ t_{21} & t_{22} \end{bmatrix},$$

where

$$t_{11} = -(v_{dc} - v_{dc0})B_{14},\ t_{21} = -(v_{dc} - v_{dc0})G_{14},$$
$$t_{12} = -\Lambda(v_{dc} - v_{dc0})G_{14},\ t_{22} = \Lambda(v_{dc} - v_{dc0})B_{14}.$$

$$\tilde{\psi}_2 = \begin{bmatrix} r_{11} & r_{12} \end{bmatrix} \begin{bmatrix} u_{11} & u_{12} \\ u_{21} & u_{22} \end{bmatrix},$$

where

$$u_{11} = -(v_{dc} - v_{dc0})G_{14},\ u_{21} = (v_{dc} - v_{dc0})B_{14},$$
$$u_{12} = \Lambda(v_{dc} - v_{dc0})B_{14},\ u_{22} = \Lambda(v_{dc} - v_{dc0})G_{14}.$$

$$\tilde{\psi}_3 = \begin{bmatrix} h_{11} & h_{12} \end{bmatrix} \begin{bmatrix} k_{11} & k_{12} \\ k_{21} & k_{22} \end{bmatrix},$$

where

$$h_{11} = \cos(\alpha - \delta_2) - \cos(\alpha_0 - \delta_{20}),$$
$$h_{12} = \sin(\alpha - \delta_2) - \sin(\alpha_0 - \delta_{20}),$$
$$k_{11} = -(E'_{dr2}v_{dc} - E_{dr20}v_{dc0})B_{24} - (E'_{qr2}v_{dc} - E_{qr20}v_{dc0})G_{24},$$
$$k_{21} = -(E'_{dr2}v_{dc} - E_{dr20}v_{dc0})G_{24} - (E'_{qr2}v_{dc} - E_{qr20}v_{dc0})B_{24},$$
$$k_{12} = -\Lambda(E'_{dr2}v_{dc} - E_{dr20}v_{dc0})G_{24} + \Lambda(E'_{qr2}v_{dc} - E_{qr20}v_{dc0})B_{24},$$
$$k_{22} = \Lambda(E'_{dr2}v_{dc} - E_{dr20}v_{dc0})B_{24} + \Lambda(E'_{qr2}v_{dc} - E_{qr20}v_{dc0})G_{24}.$$

$$\tilde{\psi}_4 = \begin{bmatrix} h_{11} & h_{12} \end{bmatrix} \begin{bmatrix} g_{11} & g_{12} \\ g_{21} & g_{22} \end{bmatrix},$$

where

$$g_{11} = -(v_{dc} - v_{dc0})B_{24},\ g_{21} = -(v_{dc} - v_{dc0})G_{24},$$
$$g_{12} = -\Lambda(v_{dc} - v_{dc0})G_{24},\ g_{22} = \Lambda(v_{dc} - v_{dc0})B_{24}.$$

$$\tilde{\psi}_5 = \begin{bmatrix} h_{11} \ h_{12} \end{bmatrix} \begin{bmatrix} e_{11} \ e_{12} \\ e_{21} \ e_{22} \end{bmatrix},$$

where

$$e_{11} = -(v_{dc} - v_{dc0})G_{24},\ e_{21} = (v_{dc} - v_{dc0})B_{24},$$
$$e_{12} = \Lambda(v_{dc} - v_{dc0})B_{24},\ e_{22} = \Lambda(v_{dc} - v_{dc0})G_{24}.$$

$$\tilde{\psi}_6 = \begin{bmatrix} x_{11} \ x_{12} \ x_{13} \ x_{14} \end{bmatrix} \begin{bmatrix} k_{11} \ k_{12} \\ k_{21} \ k_{22} \\ k_{31} \ k_{32} \\ k_{41} \ k_{42} \end{bmatrix},$$

where

$$\begin{aligned}
k_{11} &= -(E'_{dr1} - E_{dr10})B_{41} - (E'_{qr1} - E_{qr10})G_{41}, \\
k_{21} &= -(E'_{dr1} - E_{dr10})G_{41} + (E'_{qr1} - E_{qr10})B_{41}, \\
k_{31} &= -(E'_{dr2} - E_{dr20})B_{42} - (E'_{qr2} - E_{qr20})G_{42}, \\
k_{41} &= -(E'_{dr2} - E_{dr20})G_{42} + (E'_{qr2} - E_{qr20})B_{42}, \\
k_{12} &= \Lambda(E'_{dr1} - E_{dr10})G_{41} - \Lambda(E'_{qr1} - E_{qr10})B_{41}, \\
k_{22} &= -\Lambda(E'_{dr1} - E_{dr10})B_{41} - \Lambda(E'_{qr1} - E_{qr10})G_{41}, \\
k_{32} &= -\Lambda(E'_{dr2} - E_{dr20})G_{42} - \Lambda(E'_{qr2} - E_{qr20})B_{42}, \\
k_{42} &= -\Lambda(E'_{dr2} - E_{dr20})B_{42} - \Lambda(E'_{qr2} - E_{qr20})G_{42}.
\end{aligned}$$

$$\tilde{\psi}_7 = z \begin{bmatrix} r_{11} \ r_{12} \end{bmatrix} \begin{bmatrix} -s_{12} & -s_{11} \\ -s_{22} & -s_{21} \end{bmatrix}.$$

11.9 Appendix-IX

Expression of $\tilde{\phi}$ and $\tilde{\psi}$ for decentralised STATCOM/ESS control design:

We define $\tilde{\phi}_i = \left[\tilde{\phi}_{i1}, \ldots, \tilde{\phi}_{i9}\right]^T$, $\tilde{\psi} = \left[\tilde{\psi}_{i1}, \ldots, \tilde{\psi}_{i9}\right]^T$, where

$$\tilde{\phi}_{i1} = \begin{bmatrix} 0 & 0 & 0 & 0 \end{bmatrix}.$$

$$\tilde{\phi}_{i2} = \begin{bmatrix} x_{11} \\ x_{12} \\ x_{13} \\ x_{14} \\ x_{15} \\ x_{16} \end{bmatrix}^T \begin{bmatrix} 0 & y_{12} & y_{13} & 0 \\ 0 & y_{22} & y_{23} & 0 \\ 0 & y_{32} & y_{33} & 0 \\ 0 & y_{42} & y_{43} & 0 \\ 0 & y_{52} & y_{53} & y_{54} \\ 0 & y_{62} & y_{63} & y_{64} \end{bmatrix},$$

where

$$x_{11} = \cos\delta_{ji} - \cos\delta_{ji0},\ x_{13} = \cos(\delta_{m_k} - \delta_i) - \cos(\delta_{m_{k0}} - \delta_{i0}),$$

$$x_{12} = \sin\delta_{ji} - \sin\delta_{ji0},\ x_{14} = \sin(\delta_{m_k} - \delta_i) + \sin(\delta_{m_{k0}} - \delta_{i0}),$$

$$y_{12} = \sum_{j=1}^{n} (E'_{dr_j} - E'_{dr_{j0}})G_{ij} + \sum_{j=1}^{n} (E'_{qr_j} - E'_{qr_{j0}})B_{ij},$$

$$y_{22} = -\sum_{j=1}^{n} (E'_{dr_j} - E'_{dr_{j0}})B_{ij} + \sum_{j=1}^{n} (E'_{qr_j} - E'_{qr_{j0}})G_{ij},$$

$$y_{32} = \sum_{k=1}^{n} (E'_{q_k} - E'_{q_{k0}})B_{ik},$$

$$x_{15} = \cos(\alpha_l - \delta_i) - \cos(\alpha_{l0} - \delta_{i0}),$$

$$y_{42} = \sum_{k=1}^{n} (E'_{q_k} - E'_{q_{k0}})G_{ik},$$

$$x_{16} = \sin(\alpha_l - \delta_i) - \sin(\delta_{l0} - \delta_{i0}),$$

$$y_{52} = -\sum_{l=1}^{n} k_l (v'_{dc_l} - v_{dc_{l0}})B_{il},$$

$$y_{62} = \sum_{l=1}^{n} k_l (v'_{dc_l} - v_{dc_{l0}})G_{il},$$

$$y_{13} = \sum_{j=1}^{n} (E'_{dr_j} - E'_{dr_{j0}})B_{ij} + \sum_{j=1}^{n} (E'_{qr_j} - E'_{qr_{j0}})G_{ij},$$

$$y_{23} = -\sum_{j=1}^{n} (E'_{dr_j} - E'_{dr_{j0}})G_{ij} - \sum_{j=1}^{n} (E'_{qr_j} - E'_{qr_{j0}})B_{ij},$$

$$y_{33} = \sum_{k=1}^{n} (E'_{q_k} - E'_{q_{k0}})G_{ik},\ y_{43} = -\sum_{k=1}^{n} (E'_{q_k} - E'_{q_{k0}})B_{ik},$$

$$y_{53} = \sum_{l=1}^{n} \kappa_l (v'_{dc_l} - v_{dc_{l0}})G_{il},\ y_{63} = \sum_{l=1}^{n} \kappa_l (v'_{dc_l} - v_{dc_{l0}})B_{il},$$

$$y_{54} = \sum_{l=1}^{n} \kappa_l (E'_{dr_i} - E'_{dr_{i0}}) B_{il} + \sum_{l=1}^{n} (E'_{qr_i} - E'_{qr_{i0}}) G_{il},$$

$$y_{64} = \sum_{l=1}^{n} \kappa_l (E'_{dr_i} - E'_{dr_{i0}}) G_{il} + \sum_{l=1}^{n} (E'_{qr_i} - E'_{qr_{i0}}) G_{il}.$$

$$\tilde{\phi}_{i3} = \begin{bmatrix} 0\,0\,0\,0 \end{bmatrix}, \ \tilde{\phi}_{i4} = -\tilde{\phi}_{i2}. \tag{11.19}$$

$$\tilde{\phi}_{i5} = \begin{bmatrix} x_{11} \\ x_{12} \\ x_{13} \\ x_{14} \\ x_{15} \\ x_{16} \end{bmatrix}^T \begin{bmatrix} 0 & G_{ij} & B_{ij} & 0 \\ -B_{ij} & B_{ij} & G_{ij} & 0 \\ 0 & 0 & 0 & 0 \\ 0 & 0 & 0 & 0 \\ 0 & 0 & 0 & -\kappa_l B_{il} \\ 0 & 0 & 0 & \kappa_l G_{il} \end{bmatrix} + \frac{1}{2H} \begin{bmatrix} C_{11} \\ C_{12} \\ 0 \\ 0 \end{bmatrix}^T,$$

where

$$C_{11} = -\omega_s (E'_{dr_i} - E'_{dr_{i0}}), \ C_{12} = -\omega_s (s_i - s_{i0}).$$

$$\tilde{\phi}_{i6} = \begin{bmatrix} x_{11} \\ x_{12} \\ x_{13} \\ x_{14} \\ x_{15} \\ x_{16} \end{bmatrix}^T \begin{bmatrix} 0 & -B_{ij} & -G_{ij} & 0 \\ -G_{ij} & G_{ij} & B_{ij} & 0 \\ 0 & 0 & 0 & 0 \\ 0 & 0 & 0 & 0 \\ 0 & 0 & 0 & \kappa_l B_{il} \\ 0 & 0 & 0 & -\kappa_l G_{il} \end{bmatrix} + \frac{1}{2H} \begin{bmatrix} D_{11} \\ 0 \\ D_{13} \\ 0 \end{bmatrix}^T,$$

where

$$D_{11} = \omega_s (E'_{qr_i} - E'_{qr_{i0}}), \ D_{13} = \omega_s (s_i - s_{i0}).$$

$$\tilde{\phi}_{i7} = \begin{bmatrix} p_{11} \\ p_{12} \\ p_{13} \\ p_{14} \\ p_{15} \\ p_{16} \end{bmatrix}^T \begin{bmatrix} 0 & d_{12} & d_{13} & d_{14} \\ 0 & d_{22} & d_{23} & d_{24} \\ 0 & 0 & 0 & d_{34} \\ 0 & 0 & 0 & d_{44} \\ 0 & 0 & 0 & d_{54} \\ 0 & 0 & 0 & d_{64} \end{bmatrix} + \frac{1}{2H} \begin{bmatrix} 0 \\ 0 \\ 0 \\ C_{14} \end{bmatrix}^T,$$

where

$$p_{11} = \cos\alpha_{lj} - \cos\delta_{lj0},\ p_{13} = \cos(\delta_i - \alpha_l) - \cos(\delta_{i0} - \alpha_{l0}),$$
$$p_{12} = \sin\alpha_{lj} - \sin\delta_{lj0},\ p_{14} = \sin(\delta_i - \alpha_l) + \sin(\delta_{l0} - \alpha_{l0}),$$
$$d_{12} = \sum_{l=1}^{n} \kappa_l(v_{dc_l} - v_{dc_{l0}})G_{il},\ d_{22} = \sum_{l=1}^{n} \kappa_l(v_{dc_l} - v_{dc_{l0}})B_{il},$$
$$d_{13} = \sum_{l=1}^{n} \kappa_l(v_{dc_l} - v_{dc_{l0}})G_{il},$$
$$p_{15} = \cos(\delta_{m_k} - \alpha_l) - \cos(\delta_{m_{k0}} - \alpha_{l0}),$$
$$d_{23} = \sum_{l=1}^{n} \kappa_l(v_{dc_l} - v_{dc_{l0}})G_{il},$$
$$p_{16} = \sin(\delta_{m_k} - \alpha_l) - \sin(\delta_{m_{k0}} - \alpha_{l0}),$$
$$d_{14} = \sum_{j=1}^{n} \kappa_l(E'_{dr_j} - E'_{dr_{j0}})B_{il} + \sum_{j=1}^{n} \kappa_l(E'_{qr_j} - E'_{qr_{j0}})G_{il},$$
$$d_{24} = -\sum_{j=1}^{n} \kappa_l(E'_{dr_j} - E'_{dr_{j0}})G_{ij} - \kappa_l \sum_{j=1}^{n} (E'_{qr_j} - E'_{qr_{j0}})B_{ij},$$
$$d_{34} = -\sum_{l=1}^{n} \kappa_l(E'_{q_k} - E'_{q_{k0}})G_{il},\ d_{44} = \sum_{l=1}^{n} \kappa_l(E'_{q_k} - E'_{q_{k0}})B_{il},$$
$$d_{54} = \sum_{k=1}^{n} \kappa_l(v_{dc_j} - v_{dc_{j0}})G_{lj},\ d_{64} = \sum_{l=1}^{n} \kappa_l(v_{dc_j} - v_{dc_{j0}})B_{lj},$$
$$C_{14} = \sum_{l=1}^{n} 2\kappa_l(v_{dc_l} - v_{dc_{l0}})G_{ll}.$$

$$\tilde{\phi}_{i8} = \beta\psi_{i2} + \frac{1}{2H}\begin{bmatrix} x_{11} \\ x_{12} \\ x_{13} \\ x_{14} \\ x_{15} \\ x_{16} \end{bmatrix}^T z \begin{bmatrix} 0 & P_{12} & p_{13} & 0 \\ 0 & P_{22} & p_{23} & 0 \\ 0 & P_{32} & p_{33} & 0 \\ 0 & P_{42} & p_{43} & 0 \\ 0 & P_{52} & p_{53} & p_{54} \\ 0 & P_{62} & p_{63} & p_{64} \end{bmatrix} + \begin{bmatrix} 0 \\ q_2 \\ 0 \\ 0 \end{bmatrix}^T,$$

where

$$p_{12} = \sum_{\substack{s=1 \\ s\neq j}}^{n} (E'_{qr_s} - E'_{qr_{s0}})(E'_{dr_i} - E'_{dr_{i0}})(G_{is}G_{ij} + B_{is}B_{ij}),$$
$$p_{22} = -\sum_{\substack{s=1 \\ s\neq j}}^{n} (E'_{qr_s} - E'_{qr_{s0}})(E'_{dr_i} - E'_{dr_{i0}})(G_{is}B_{ij} - B_{is}G_{ij}),$$
$$p_{32} = \sum_{\substack{k=1 \\ k\neq j}}^{n} (E'_{q_k} - E'_{q_{k0}})(E'_{qr_j} - E'_{qr_{j0}})(G_{ik}G_{ij} + B_{ik}B_{ij}),$$

$$p_{42} = -\sum_{\substack{k=1\\k\neq j}}^{n}(E'_{q_k} - E'_{qk0})(E'_{qr_j} - E'_{qr_{j0}})(G_{ik}B_{ij} - B_{ik}G_{ij}),$$

$$p_{52} = \sum_{\substack{l=1\\l\neq j}}^{n}\kappa_l(v_{dc_l} - v_{dc_{l0}})(E'_{qr_j} - E'_{qr_{j0}})(G_{il}G_{ij} + B_{il}B_{ij}),$$

$$p_{62} = -\sum_{\substack{l=1\\l\neq j}}^{n}\kappa_l(v_{dc_l} - v_{dc_{l0}})(E'_{qr_j} - E'_{qr_{j0}})(G_{il}B_{ij} - B_{il}G_{ij}),$$

$$p_{13} = \sum_{\substack{s=1\\s\neq j}}^{n}(E'_{dr_s} - E'_{dr_{s0}})(E'_{qr_i} - E'_{qr_{i0}})(G_{is}B_{ij} - B_{is}G_{ij}),$$

$$p_{23} = \sum_{\substack{s=1\\s\neq j}}^{n}(E'_{dr_s} - E'_{dr_{s0}})(E'_{qr_i} - E'_{qr_{i0}})(G_{is}G_{ij} + B_{is}B_{ij}),$$

$$p_{33} = \sum_{\substack{k=1\\k\neq j}}^{n}(E'_{q_k} - E'_{qk0})(E'_{dr_j} - E'_{dr_{j0}})(G_{ik}B_{ij} - B_{ik}G_{ij}),$$

$$p_{43} = \sum_{\substack{k=1\\k\neq j}}^{n}(E'_{q_k} - E'_{qk0})(E'_{dr_j} - E'_{dr_{j0}})(G_{ik}G_{ij} + B_{ik}B_{ij}),$$

$$p_{53} = \sum_{\substack{l=1\\l\neq j}}^{n}\kappa_l(v_{dc_l} - v_{dc_{l0}})(E'_{dr_j} - E'_{dr_{j0}})(G_{il}B_{ij} - B_{il}G_{ij}),$$

$$p_{63} = \sum_{\substack{l=1\\l\neq j}}^{n}\kappa_l(v_{dc_l} - v_{dc_{l0}})(E'_{dr_j} - E'_{dr_{j0}})(G_{il}G_{ij} + B_{il}B_{ij}),$$

$$p_{54} = \sum_{\substack{l=1\\l\neq j}}^{n}\kappa_l(E'_{dr_j} - E'_{dr_{j0}})(E'_{qr_j} - E'_{qr_{j0}})(G_{li}B_{ij} - B_{li}G_{ij}),$$

$$p_{64} = \sum_{\substack{l=1\\l\neq j}}^{n}\kappa_l(E'_{dr_j} - E'_{dr_{j0}})(E'_{qr_j} - E'_{qr_{j0}})(G_{li}G_{ij} + B_{li}B_{ij}),$$

$$z = X'_{di}\beta_i,\ \beta_i = X'_{di}(\frac{1}{V_{ti}} - \frac{1}{V_{ti0}}),$$

$$q_2 = (\frac{E'_{qr_i}}{V_{t_i}} + \frac{E'_{dr_i}}{V_{t_i}} - \frac{E'_{dr_{i0}}}{V_{t0}} - \frac{E'_{qr_{i0}}}{V_{t0}}).$$

$$\tilde{\phi}_{i9} = \begin{bmatrix} 0 \; 0 \; 0 \; 0 \end{bmatrix}. \tag{11.20}$$

$$\tilde{\psi}_{i1} = \begin{bmatrix} 0 \; 0 \; 0 \; 0 \end{bmatrix}. \tag{11.21}$$

$$\tilde{\psi}_{i2} = \begin{bmatrix} x_{11} \\ x_{12} \\ x_{13} \\ x_{14} \\ x_{15} \\ x_{16} \end{bmatrix}^T \begin{bmatrix} 0 & 0 & 0 \; 0 \\ 0 & 0 & 0 \; 0 \\ 0 & 0 & 0 \; 0 \\ 0 & 0 & 0 \; 0 \\ \frac{y_{62}}{\kappa_i} & \sum_{l=1}^{n} \sqrt{\frac{3}{8}} G_{lj} & 0 \; 0 \\ \frac{y_{52}}{\kappa_i} & \sum_{l=1}^{n} \sqrt{\frac{3}{8}} B_{lj} & 0 \; 0 \end{bmatrix}. \tag{11.22}$$

$$\tilde{\psi}_{i3} = \begin{bmatrix} 0 \; 0 \; 0 \; 0 \end{bmatrix}, \; \tilde{\psi}_{i4} = -\tilde{\psi}_{i2}. \tag{11.23}$$

$$\tilde{\psi}_{i5} = \begin{bmatrix} x_{11} \\ x_{12} \\ x_{13} \\ x_{14} \\ x_{15} \\ x_{16} \end{bmatrix}^T \begin{bmatrix} 0 & 0 & 0 \; 0 \\ 0 & 0 & 0 \; 0 \\ 0 & 0 & 0 \; 0 \\ 0 & 0 & 0 \; 0 \\ 0 & \sum_{l=1}^{n} \sqrt{\frac{3}{8}} G_{il} & 0 \; 0 \\ 0 & \sum_{l=1}^{n} \sqrt{\frac{3}{8}} G_{il} & 0 \; 0 \end{bmatrix}. \tag{11.24}$$

$$\tilde{\psi}_{i6} = \begin{bmatrix} x_{11} \\ x_{12} \\ x_{13} \\ x_{14} \\ x_{15} \\ x_{16} \end{bmatrix}^T \begin{bmatrix} 0 & 0 & 0 \; 0 \\ 0 & 0 & 0 \; 0 \\ 0 & 0 & 0 \; 0 \\ 0 & 0 & 0 \; 0 \\ 0 & \sum_{l=1}^{n} \sqrt{\frac{3}{8}} G_{il} & 0 \; 0 \\ 0 & -\sum_{l=1}^{n} \sqrt{\frac{3}{8}} G_{il} & 0 \; 0 \end{bmatrix}. \tag{11.25}$$

$$\tilde{\psi}_{i7} = \begin{bmatrix} p_{11} \\ p_{12} \\ p_{13} \\ p_{14} \\ p_{15} \\ p_{16} \end{bmatrix}^T \begin{bmatrix} \sum_{l=1}^{n} k_l v_{dc_l} & \sum_{l=1}^{n} \sqrt{\frac{3}{8}} v_{dc_l} & 0 & 0 \\ \sum_{l=1}^{n} k_l v_{dc_l} & \sum_{l=1}^{n} \sqrt{\frac{3}{8}} v_{dc_l} & 0 & 0 \\ \sum_{l=1}^{n} k_l v_{dc_l} & \sum_{l=1}^{n} \sqrt{\frac{3}{8}} v_{dc_l} & 0 & 0 \\ \sum_{l=1}^{n} k_l v_{dc_l} & \sum_{l=1}^{n} \sqrt{\frac{3}{8}} v_{dc_l} & 0 & 0 \\ \sum_{l=1}^{n} k_l v_{dc_l} & \sum_{l=1}^{n} \sqrt{\frac{3}{8}} v_{dc_l} & 0 & 0 \\ \sum_{l=1}^{n} k_l v_{dc_l} & \sum_{l=1}^{n} \sqrt{\frac{3}{8}} v_{dc_l} & 0 & 0 \end{bmatrix}. \tag{11.26}$$

$$\tilde{\psi}_{i8} = \beta \psi_{i2} + \frac{1}{2H} \begin{bmatrix} x_{11} \\ x_{12} \\ x_{13} \\ x_{14} \\ x_{15} \\ x_{16} \end{bmatrix}^T z \begin{bmatrix} 0 & 0 & o & 0 \\ 0 & 0 & 0 & 0 \\ 0 & 0 & 0 & 0 \\ 0 & 0 & 0 & 0 \\ r_{51} & r_{52} & 0 & 0 \\ r_{61} & r_{62} & 0 & 0 \end{bmatrix} + \begin{bmatrix} 0 \\ q_2 \\ 0 \\ 0 \end{bmatrix}^T,$$

where

$$r_{51} = \sum_{\substack{l=1 \\ l \neq j}}^{n} w_l (E'_{dr_j} - E'_{dr_{j0}})(E'_{qr_j} - E'_{qr_{j0}})(G_{li} B_{ij} - B_{li} G_{ij}),$$

$$r_{52} = \sum_{\substack{l=1 \\ l \neq j}}^{n} w_l (E'_{dr_j} - E'_{dr_{j0}})(E'_{qr_j} - E'_{qr_{j0}})(G_{li} G_{ij} + B_{li} B_{ij}),$$

$$r_{61} = \sum_{\substack{l=1 \\ l \neq j}}^{n} \kappa_l w_l (E'_{dr_j} - E'_{dr_{j0}})(E'_{qr_j} - E'_{qr_{j0}})(G_{li} G_{ij} + B_{li} B_{ij}),$$

$$r_{62} = \sum_{\substack{l=1 \\ l \neq j}}^{n} \kappa_l w_l (E'_{dr_j} - E'_{dr_{j0}})(E'_{qr_j} - E'_{qr_{j0}}) G_{li} B_{ij} - B_{li} G_{ij}),$$

$$w_l = (v_{dc_l} - v_{dc_{l0}}).$$

$$\tilde{\phi}_{i9} = \begin{bmatrix} 0 & 0 & 0 & 0 \end{bmatrix}. \tag{11.27}$$

The expression for $\tilde{\Xi}$ is not presented here, however, it can be derived in a similar way as $\tilde{\phi}$.

11.10 Appendix-X

16 machine 5 area study system power-flow data

A1. Machine bus data

Bus number	Voltage (pu)	Power generation (pu)
1	1.0450	2.50
2	0.9800	5.45
3	0.9830	6.50
4	0.9870	6.32
5	1.0110	5.05
6	1.0500	7.00
7	1.0630	5.60
8	1.0300	5.40
9	1.0250	8.00
10	1.0100	5.00
11	1.0000	10.00
12	1.0156	13.50
13	1.0110	35.91
14	1.0000	17.85
15	1.0000	10.00
16	1.0000	40.00

A2. Load bus data

Bus number	Real load (pu)	Reactive load (pu)
17	60.00	3.0000
18	24.70	1.2300
19	0	0
20	6.80	1.0300
21	2.74	1.1500
22	0	0
23	2.48	0.8500
24	3.09	−0.9200
25	2.24	0.4700
26	1.39	0.1700
27	2.81	0.7600
28	2.06	0.280
29	2.84	0.2700
30	0	0
31	0	0
32	0	0
33	1.12	0
34	0	0

(continued)

A2. (continued)

Bus number	Real load (pu)	Reactive load (pu)
35	0	0
36	1.02	−0.1946
37	0	0
38	0	0
39	2.67	0.1260
40	0.6563	0.2353
41	10.00	2.5000
42	11.50	2.5000
43	0	0
44	2.6755	0.0484
45	2.08	0.2100
46	1.507	0.2850
47	2.0312	0.3259
48	2.412	0.0220
49	1.64	0.2900
50	1.00	−1.4700
51	3.37	−1.2200
52	1.58	0.300
53	2.527	1.1856
54	0	0
55	3.22	0.0200
56	2.00	0.7360
57	0	0
58	0	0
59	2.34	0.8400
60	2.088	0.7080
61	1.04	1.2500
62	0	0
63	0	0
64	0.09	0.8800
65	0	0
66	0	0
67	3.20	1.5300
68	3.29	0.3200
69	0	0

A3. Line data:

From bus	To bus	R (pu)	X (pu)	Line charging (pu)	Tap ratio
54	1	0	0.0181	0	1.0250
58	2	0	0.0250	0	1.0700
62	3	0	0.0200	0	1.0700
19	4	0.0007	0.0142	0	1.0700
20	5	0.0009	0.0180	0	1.0090

(continued)

A3. (continued)

From bus	To bus	R (pu)	X (pu)	Line charging (pu)	Tap ratio
22	6	0	0.0143	0	0250
23	7	0.0005	0.0272	0	0
25	8	0.0006	0.0232	0	1.0250
29	9	0.008	0.0156	0	1.0250
31	10	0	0.0260	0	1.0400
32	11	0	0.0130	0	1.0400
36	12	0	0.0075	0	1.0400
17	13	0	0.0033	0	1.0400
41	14	0	0.0015	0	1.0000
42	15	0	0.0015	0	1.0000
18	16	0	0.0030	0	1.0000
36	17	0.0005	0.0045	0.3200	0
49	18	0.0076	0.1141	1.1600	0
68	19	0.0016	0.0195	0.3040	0
19	20	0.0007	0.0138	0	1.0600
68	21	0.0008	0.0135	0.2548	0
21	22	0.0008	0.0140	0.2565	0
22	23	0.0006	0.0096	0.1846	0
23	24	0.0022	0.0350	0.3610	0
68	24	0.0003	0.0059	0.0680	0
54	25	0.0070	0.0086	0.1460	0
25	26	0.0032	0.0323	0.5310	0
37	27	0.0013	0.0173	0.3216	0
26	27	0.0014	0.0147	0.2396	0
26	28	0.0043	0.0474	0.7802	0
26	29	0.0057	0.0625	1.0290	0
28	29	0.0014	0.0151	0.2490	0
53	30	0.0008	0.0074	0.4800	0
61	30	0.0019	0.0183	0.2900	0
61	30	0.0019	0.0183	0.2900	0
30	31	0.0013	0.0187	0.3330	0
53	31	0.0016	0.0163	0.2500	0
30	32	0.0024	0.0288	0.4880	0
32	33	0.0008	0.0099	0.1680	0
33	34	0.0011	0.0157	0.2020	0
35	34	0.0001	0.0074	0	0.9460
34	36	0.0033	0.0111	1.4500	0
61	36	0.0022	0.0196	0.3400	0
61	36	0.0022	0.0196	0.3400	0
68	37	0.0007	0.0089	0.1342	0
31	38	0.0011	0.0147	0.2470	0
33	38	0.0036	0.0444	0.6930	0
41	40	0.0060	0.0840	3.1500	0
48	40	0.0020	0.0220	1.2800	0
42	41	0.0040	0.0600	2.2500	0
18	42	0.0040	0.0600	2.2500	0
17	43	0.0005	0.0276	0	0
39	44	0	0.0411	0	0
43	44	0.0001	0.0011	0	0

(continued)

A3. (continued)

From bus	To bus	R (pu)	X (pu)	Line charging (pu)	Tap ratio
39	45	0	0.0839	0	0
44	45	0.0025	0.0730	0	0
38	46	0.0022	0.0284	0.4300	0
53	47	0.0013	0.0188	1.3100	0
47	48	0.0025	0.0268	0.4000	0
47	48	0.0025	0.0268	0.4000	0
46	49	0.0018	0.0274	0.2700	0
45	51	0.0004	0.0105	0.7200	0
50	51	0.0009	0.0221	1.6200	0
37	52	0.0007	0.0082	0.1319	0
55	52	0.0011	0.0133	0.2138	0
53	54	0.0035	0.0411	0.6987	0
54	55	0.0013	0.0151	0.2572	0
55	56	0.0013	0.0213	0.2214	0
56	57	0.0008	0.0128	0.1342	0
57	58	0.0002	0.0026	0.0434	0
58	59	0.0006	0.0092	0.1130	0
57	60	0.0008	0.0112	0.1476	0
59	60	0.0004	0.0046	0.078	0
60	61	0.0023	0.0363	0.3804	0
58	63	0.0007	0.0082	0.1389	0
62	63	0.0004	0.0043	0.0729	0
64	63	0.0016	0.0435	0	1.0600
62	65	0.0004	0.0043	0.0729	0
64	65	0.0016	0.0435	0	1.0600
56	66	0.0008	0.0129	0.1382	0
65	66	0.0009	0.0101	0.1723	0
66	67	0.0018	0.0217	0.3660	0
67	68	0.0009	0.0094	0.1710	0
53	27	0.0320	0.3200	0.4100	1.0000
69	18	0.0006	0.0144	1.0300	0
50	69	0.0006	0.0144	1.0300	0

16 machine 5 area test system dynamic data

B1. Machine data in pu:

Machine	Bus	Base MVA	X_{ls}	R_s	X_d	X'_d	X''_d	T'_{do} (s)	T''_{do} (s)
1	1	100	0.0125	0.0	0.1	0.031	0.025	10.2	0.05
2	2	100	0.035	0.0	0.295	0.0697	0.05	6.56	0.05
3	3	100	0.0304	0.0	0.2495	0.0531	0.045	5.7	0.05
4	4	100	0.0295	0.0	0.262	0.0436	0.035	5.69	0.05
5	5	100	0.027	0.0	0.33	0.066	0.05	5.4	0.05
6	6	100	0.0224	0.0	0.254	0.05	0.04	7.3	0.05
7	7	100	0.0322	0.0	0.295	0.049	0.04	5.66	0.05
8	8	100	0.0028	0.0	0.29	0.057	0.045	6.7	0.05

B1. (continued)

Machine	Bus	Base MVA	X_{ls}	R_s	X_d	X'_d	X''_d	T'_{do} (s)	T''_{do} (s)
9	9	100	0.00298	0.0	0.2106	0.057	0.045	4.79	0.05
10	10	100	0.0199	0.0	0.169	0.0457	0.04	9.37	0.05
11	11	100	0.0103	0.0	0.128	0.018	0.012	4.1	0.05
12	12	100	0.022	0.0	0.101	0.031	0.025	7.4	0.05
13	13	200	0.0030	0.0	0.0296	0.0055	0.004	5.9	0.05
14	14	100	0.0017	0.0	0.018	0.00285	0.0023	4.1	0.05
15	15	100	0.0017	0.0	0.018	0.00285	0.0023	4.1	0.05
16	16	200	0.0041	0.0	0.0356	0.0071	0.0055	7.8	0.05

B1. Machine data :(continued)

Machine	X_q (pu)	X'_q (pu)	X''_q (pu)	T'_{qo} (s)	T''_{qo} (s)	H (s)	D
1	0.069	0.028	0.025	1.5	0.035	42.0	4.0
2	0.282	0.060	0.05	1.5	0.035	30.2	9.75
3	0.237	0.050	0.045	1.5	0.035	35.8	10
4	0.258	0.040	0.035	1.5	0.035	28.6	10
5	0.31	0.060	0.05	0.44	0.035	26.0	3
6	0.241	0.045	0.04	0.4	0.035	34.8	10
7	0.292	0.045	0.04	1.5	0.035	26.4	8
8	0.280	0.050	0.045	0.41	0.035	24.3	9
9	0.205	0.050	0.045	1.96	0.035	34.5	14
10	0.115	0.045	0.04	1.5	0.035	31.0	5.56
11	0.0123	0.015	0.012	1.5	0.035	28.2	13.6
12	0.095	0.028	0.025	1.5	0.035	92.3	13.5
13	0.0286	0.005	0.004	1.5	0.035	248.0	33
14	0.0173	0.0025	0.0023	1.5	0.035	300.0	100
15	0.0173	0.0025	0.0023	1.5	0.035	300.0	100
16	0.0334	0.006	0.0055	1.5	0.035	225.0	50

C1. DC excitation system data:

Machine	T_r (s)	K_A	T_A (s)	$V_{r\max}$ (pu)	$V_{r\min}$ (pu)	K_E (s)	T_E	A_{ex}	B_{ex}
1	0.01	40	0.02	10	−10	1	0.785	0.07	0.91
2	0.01	40	0.02	10	−10	1	0.785	0.07	0.91
3	0.01	40	0.02	10	−10	1	0.785	0.07	0.91
4	0.01	40	0.02	10	−10	1	0.785	0.07	0.91
5	0.01	40	0.02	10	−10	1	0.785	0.07	0.91
6	0.01	40	0.02	10	−10	1	0.785	0.07	0.91
7	0.01	40	0.02	10	−10	1	0.785	0.07	0.91
8	0.01	40	0.02	10	−10	1	0.785	0.07	0.91

C1. DC excitation system data:

Machine	T_r (s)	K_a	$V_{r\max}$ (pu)	$V_{r\min}$ (pu)	K_{pss} (s)	T_1 (s)	T_2 (s)	T_3 (s)	T_4 (s)
9	0.01	200	5	−5	$\frac{12}{377}$	0.1	0.2	0.1	0.2

11.11 Appendix-XI

10-machine, New England System Power Flow Data

A1. Machine bus data

Bus number	Voltage (pu)	Power generation (MW)
1	1.04550	250.00
2	1.04360	572.93
3	1.02038	650.00
4	0.99780	632.00
5	0.99355	508.00
6	0.99651	650.00
7	0.94083	560.00
8	0.95460	540.00
9	1.01114	830.00
10	1.00861	1005.729

A2. Load bus data

Bus number	Real load (pu)	Reactive load (pu)
1	0.00	0.00
2	0.00	0.00
3	322.00	2.40
4	500.00	1.0300
5	0.00	0.00
6	0.00	0.00
7	233.8	840.00
8	522	176
9	0.00	0.00
10	0.00	0.00
11	0.00	0.00
12	8.50	88.00
13	0.00	0.00
14	0.00	0.00
15	320.00	153.00
16	329.40	323.00
17	0.00	0.000
18	158.0	30.00
19	0.00	0.00
20	680.00	103.00
21	680.00	103.00
22	0.00	0.00

(continued)

A2. (continued)

Bus number	Real load (pu)	Reactive load (pu)
23	247.00	84.00
24	308.60	−92.200
25	224.00	47.20
26	139.00	17.00
27	281.00	75.50
28	206.00	27.60
29	283.50	126.90
31	9.20	4.60
39	1104.00	250.00

A3. Line data:

From bus	To bus	R (pu)	X (pu)	Line charging (pu)
1	2	0	0.0181	0
1	39	0	0.0250	0
2	3	0	0.0200	0
2	25	0.0007	0.0142	0
3	4	0.0009	0.0180	0
3	18	0	0.0143	0
4	5	0.0005	0.0272	0
4	14	0.0006	0.0232	0
5	6	0.008	0.0156	0
5	8	0	0.0260	0
6	7	0	0.0130	0
6	11	0	0.0075	0
7	8	0	0.0033	0
8	9	0	0.0015	0
9	39	0	0.0015	0
10	11	0	0.0030	0
10	13	0.0005	0.0045	0.3200
13	14	0.0076	0.1141	1.1600
14	15	0.0016	0.0195	0.3040
15	16	0.0007	0.0138	0
16	17	0.0008	0.0135	0.2548
16	19	0.0016	0.0195	0.3040
16	21	0.0008	0.0135	0.2548
16	24	0.0003	0.0059	0.0680
17	18	0.0007	0.0082	0.1319
17	27	0.0013	0.0173	0.3216
21	22	0.0008	0.0140	0.2565
22	23	0.0006	0.0096	0.1846
23	24	0.0022	0.0350	0.3610
25	26	0.0032	0.0323	0.5130
26	27	0.0014	0.0147	0.2396
26	28	0.0043	0.0474	0.7802
26	29	0.0057	0.0625	1.02900
28	29	0.0014	0.0151	0.2490

10-machine New England System Dynamic Data

B1. Machine data in pu:

Machine	Bus	Base MVA	X_{ls}	R_s	X_d	X'_d	X''_d	T'_{do} (s)	T''_{do} (s)
1	30	100	0.0125	0.0	0.1	0.031	0.025	10.2	0.05
2	31	100	0.035	0.0	0.295	0.0697	0.05	6.56	0.05
3	32	100	0.0304	0.0	0.2495	0.0531	0.045	5.7	0.05
4	33	100	0.0295	0.0	0.262	0.0436	0.035	5.69	0.05
5	34	100	0.027	0.0	0.33	0.066	0.05	5.4	0.05
6	35	100	0.0224	0.0	0.254	0.05	0.04	7.3	0.05
7	36	100	0.0322	0.0	0.295	0.049	0.04	5.66	0.05
8	37	100	0.0028	0.0	0.29	0.057	0.045	6.7	0.05
9	38	100	0.00298	0.0	0.2106	0.057	0.045	4.79	0.05
10	39	100	0.0199	0.0	0.169	0.0457	0.04	9.37	0.05

B1. Machine data in pu:

Machine	X_q (pu)	X'_q (pu)	X''_q (pu)	T'_{qo} (s)	T''_{qo} (s)	H (s)	D
1	0.069	0.028	0.025	1.5	0.035	42.0	4.0
2	0.282	0.060	0.05	1.5	0.035	30.2	9.75
3	0.237	0.050	0.045	1.5	0.035	35.8	10
4	0.258	0.040	0.035	1.5	0.035	28.6	10
5	0.31	0.060	0.05	0.44	0.035	26.0	3
6	0.241	0.045	0.04	0.4	0.035	34.8	10
7	0.292	0.045	0.04	1.5	0.035	26.4	8
8	0.280	0.050	0.045	0.41	0.035	24.3	9
9	0.205	0.050	0.045	1.96	0.035	34.5	14
10	0.115	0.045	0.04	1.5	0.035	31.0	5.56

C1. DC excitation system data:

Machine	T_r (s)	K_A	T_A (s)	$V_{r\max}$ (pu)	$V_{r\min}$ (pu)	K_E (s)	T_E	A_{ex}	B_{ex}
1	0.01	40	0.02	10	−10	1	0.785	0.07	0.91
2	0.01	40	0.02	10	−10	1	0.785	0.07	0.91
3	0.01	40	0.02	10	−10	1	0.785	0.07	0.91
4	0.01	40	0.02	10	−10	1	0.785	0.07	0.91
5	0.01	40	0.02	10	−10	1	0.785	0.07	0.91
6	0.01	40	0.02	10	−10	1	0.785	0.07	0.91
7	0.01	40	0.02	10	−10	1	0.785	0.07	0.91
8	0.01	40	0.02	10	−10	1	0.785	0.07	0.91
9	0.01	40	0.02	10	−10	1	0.785	0.07	0.91
10	0.01	40	0.02	10	−10	1	0.785	0.07	0.91

Index

J. Hossain and H. R. Pota, *Robust Control for Grid Voltage Stability: High Penetration of Renewable Energy*, Power Systems,
DOI: 10.1007/978-981-287-116-9, © Springer Science+Business Media Singapore 2014

MIX
Papier aus verantwortungsvollen Quellen
Paper from responsible sources
FSC® C105338

If you have any concerns about our products,
you can contact us on
ProductSafety@springernature.com

In case Publisher is established outside the EU,
the EU authorized representative is:
Springer Nature Customer Service Center GmbH
Europaplatz 3, 69115 Heidelberg, Germany

Printed by Libri Plureos GmbH
in Hamburg, Germany